GENETICS IN THE MADHOUSE

GENETICS *in the* MADHOUSE

The UNKNOWN HISTORY *of* HUMAN HEREDITY

THEODORE M. PORTER

PRINCETON UNIVERSITY PRESS
Princeton and Oxford

Published by Princeton University Press
41 William Street, Princeton, New Jersey 08540

In the United Kingdom: Princeton University Press
6 Oxford Street, Woodstock, Oxfordshire OX20 1TR

press.princeton.edu

Jacket image: From a report of the 1880 US census (1888), an early example of graphical representation of asylum statistics, here comparing the prevalence of hereditary taint of women to that of men and indicating the importance of different relatives as sources of insanity. From Frederick Howard Wines, US Census Office, *Report on the Defective, Dependent, and Delinquent Classes of the Population of the United States as Returned at the Tenth Census* (June 1, 1880) [vol. 21 of 1880 census], Washington: Government Printing Office, 1888.

ISBN 978-0-691-16454-0

Library of Congress Control Number: 2018935173

British Library Cataloging-in-Publication Data is available

This book has been composed in Miller and Century Modern

Printed on acid-free paper. ∞

Printed in the United States of America

1 3 5 7 9 10 8 6 4 2

CONTENTS

ILLUSTRATIONS

SOME WORDS OF INTEREST

I use a few old-fashioned, odd, foreign, and, to modern ears, rude words where translation or anachronism demands too much sacrifice of meaning. Here a few that recur.

Alienist. From a French word for "madness," this was the name for doctors specializing in madness in English and French (*aliéniste*) until the early twentieth century. The German word was ***Irrenarzt***, literally "physician to the errant (or mad)."

Anlage. A German word (equivalent to the Danish/Norwegian *Anlæg*) that often appears where English used "predisposition." But it suggests a factor underlying the predisposition, and it persisted into the 1940s along with ***Erbanlage*** (hereditary factor), as the German word for "gene." See also my brief etymological discussion in chapter 3.

Madness. The oldest and least medicalized word in a cluster of terms. **Insanity** implied a lack of legal responsibility and usually extended to "idiots and imbeciles," while **lunacy** referred only to the mad. **Mental illness** is a more medicalized form. These terms were all in use by 1800. **Idiot**, a very old word with Greek roots, referred to the most extreme intellectual disability. **Imbecile** was introduced for a less severe form, while **feebleminded** (often written as **feeble-minded**) was increasingly used in the late nineteenth century. In Britain, **feebleminded** referred only to those with modest disabilities, while **mental deficiency** included the full spectrum. **Mental defect** usually extended to insanity as well as **mental weakness**.

Moral treatment refers to a relatively gentle, psychological form of care that emphasized cultivating and manipulating the basic rationality that most patients were said to retain. It inspired great optimism in the early nineteenth century and helped to stimulate the first wave of asylum expansion.

Abbreviations in Text

ABA American Breeders Association

AJI *American Journal of Insanity*

AMP *Annales médico-psychologiques*
(Medical-Psychological Annals, France)

AZP *Allgemeine Zeitschrift für Psychiatrie*
(General Journal of Psychiatry, Germany)

ARGB *Archiv für Rassen- und Gesellschaftsbiologie*
(Archive for Racial and Social Biology, Germany)

ERO Eugenics Record Office

JMS *Journal of Mental Science*
(began as the *Asylum Journal of Mental Science*)

ZgNP *Zeitschrift für die gesamte Neurologie und Psychiatrie*
(Journal for All of Psychiatry and Neurology, Germany)

GENETICS IN THE MADHOUSE

INTRODUCTION

Data-Heredity-Madness

A Medical-Social Dream

> The plan of the institution, the budget, the rules for its administration were not calculated merely to pursue cures for the mentally ill; science itself was also to be advanced.
>
> —Report of a Rhine Asylum Committee (1830)

> Heredity has an undeniably great importance for mental illness and psychical deficiencies. So it is no accident that attention was focused earlier and more intensely on the inheritance question in psychiatry than in any other area of medicine.
>
> —Wilhelm Schallmayer (1918)

Genetics has been supported by compelling images. We think first of DNA, whose helical structure, announced in 1953, is still often exalted as the secret of life. For half a century before that, the science of heredity was identified with neat diagrams of green and yellow or smooth and wrinkled peas bred by Gregor Mendel in the garden of an Augustinian monastery.[1] The mutant eyes or wings of the fruit fly also assumed an iconic form. Images like these distract us from a science of mass reproduction. The agricultural breeding factories that already had sprung up before anyone cared about Mendel, and the industrialized laboratories of recombinant DNA, have never been appealing in the way of a ladder swirling heavenward. Graceful curves can only be part of the story. Let the reader cast an eye over the great filing cabinets of data from armies, prisons, immigration offices, census bureaus, and insurance offices that have been brought to bear on the topic of human heredity. Already by 1830, the investigation of heredity was saturated with numbers. A century later, the data of human heredity still were produced principally in two related institutions: insane asylums and special schools for children who were called feebleminded. DNA does not flow gracefully in unbounded space but is bent and twisted to fit onto stubby chromosomes. The science of human heredity arose first amid the moans, stench, and unruly despair of mostly hidden places where data were recorded, combined, and grouped into tables and graphs.

In practice, human genetics has always depended on mundane tools to classify and record bodily traits. Phenotypic heredity, which deals in quantities such as egg or milk production, IQ scores, and medical conditions, persists alongside the analysis of genetic factors that may be supposed to code for such traits. Its importance for breeding and other practical endeavors was and remains much greater than is commonly realized. Statistical techniques, from ordered lists and correlation tables to regressions and cluster analysis, have been fundamental to both sorts of hereditary research, genotypic and phenotypic. The public knows little of this. A bitter debate in the early twentieth century between "biometricians" and "Mendelians" about how best to study biological inheritance seemed to end in a victory for genetics, defined by a focus on discrete nuggets of hereditary causation for which Wilhelm Johannsen in 1909 coined the term "gene." The new genetics emphasized microscopy, agricultural breeding, and model organisms. Despite geneticists' intense engagement with eugenics and medicine, *Homo sapiens* was not their preferred organism. It was too resistant to laboratory manipulation and had too long a generation time in comparison to fruit flies, nematodes, and viruses. Historians of genetics, until recently, almost always echoed laboratory scientists and breeders in their focus on genes and then DNA.

This book brings historical focus to that other science of heredity, the tradition of amassing, ordering, and depicting data of biological inheritance, especially in humans. The deployment of hereditary data in medical and social institutions preceded academic genetics by about a century and continued thereafter as a set of tools and approaches loosely interwoven with classical genetic methods and understandings. In the dance of influence and appropriation, data work was never a passive partner, and in recent decades it has reclaimed the limelight, supported by our present enthusiasm for Big Data. The Human Genome Initiative, sold with a promise to find the genes for talents, diseases, and every kind of personal characteristic, has returned to its roots as a data science. Historical writing on genetics, now tapping into the wider conditions of hereditary knowledge, has begun to pay more heed to data practices.[2]

These provided my inspiration for taking up this work. Karl Pearson, the subject of my previous book, combined extraordinarily wide-ranging intellectual ambitions with an unwavering commitment to statistics, eugenics, and "scientific method." He also took data to be highly diverse and even personal. An invitation to contribute to an edited volume on the history of heredity prompted me to suppose that an inquiry into the sources of Pearson's data might open up broader cultural dimensions. I found that experts on the treatment of the insane and feebleminded in 1910 were not sleepily awaiting the magic touch of a geneticist or statistician to give meaning and purpose to their data. For decades already, asylum doctors had regarded themselves as medical scientists, and they took a vital interest in the role of heredity in reproducing the conditions they treated. My discovery of these efforts recapitulated Pearson's own, as the institutions he looked to for data turned into sites of collaboration. Right from the start, his journal *Biometrika* published studies initiated by institutional doctors and psychologists and prison administrators. Although they engaged sometimes in fierce disputes, many were eager to adopt his tools to raise the statistical quality of research in which they were already engaged. He, in turn, readily acknowledged his dependence on them for access to human subjects and for diagnostic expertise as well as family data.

Pearson's experience was echoed a few years later by Charles B. Davenport at the Eugenics Record Office in Cold Spring Harbor, New York. He held, in opposition to Pearson, that Mendel's experiments on heredity of plant hybrids had changed everything. Beginning about 1908, he built up a vast data enterprise to identify Mendelian factors for the most disabling and costly human defects. He very quickly realized that on almost every topic of practical importance, and especially for insanity and feeble-mindedness, hereditary investigation was already proceeding on a massive scale in special schools and asylums. Doctors and psychologists were as active as he was in developing the basic tools of hereditary data work.[3]

Davenport and his collaborators organized their pedigree data to reveal characteristic Mendelian ratios for the most worrisome mental conditions. Many in Britain and Germany as well as the United States were persuaded by his conclusions, to the point that hereditary research in psychiatry and psychology often required validation by these "Mendel numbers." Early critiques, mostly from Pearson and his associates, gradually developed into a broad scientific rejection of Davenport's work during the 1930s. Yet geneticists continued to teach basic Mendelism as the prototype for every sort of hereditary transmission, and the gene has sustained its supremacy in ordinary discourse. Molecular genetics, like other high-tech enterprises, has been fond of histories based on transformative discoveries. In the era of recombinant DNA and genomics, these continued to promise the discovery of the gene or genes that code for great abilities and dread diseases, with mental illness as a particular focus of interest. Grand historical narratives about learning the secret of life or an eighth day of creation have been written in support of scientific entrepreneurs and venture capitalists, that is, as present-day interventions.[4] Historians, too, write for the present, but (we hope) by challenging easy present-minded assumptions and by immersing themselves in primary sources and careful scholarship. The problem of gaining independence from the stories that scientists tell remains a pressing one for histories of genetics. History can provide the basis for a deeper understanding of the work of science, even in the present.

When I began this work, most of its characters were completely unknown to me. Commencing, innocently enough, with Pearson's allies and collaborators, I began following my sources backward in time. While there can be no definitive point of origin, a few months of digging brought me to events in 1789 that could anchor my narrative. For a European historian, no starting point could be more obvious. This one, however, was not the French Revolution, but the furor unleashed during a bout of madness suffered by King George III of England, the occasion for strenuous debates about insanity and recordkeeping. The decade that followed is well known to historians of psychiatry for the beginnings in France and England of a gentler and more hopeful "moral treatment," which in turn provided a rationale for a vast expansion of Western asylum systems.

At first we detect no more than a shadowy premonition of those sprawling, amorphous, yet insistent institutions of medicine that now absorb limitless resources and intrude into every dimension of our lives. Yet by 1850, mental hospitals were becoming, in parts of Europe and North America, the costliest of social programs. We might think of them as a trial run for the welfare state, working to relieve the suffering of the mentally ill and to lighten the terrible burden on their families and communities. As sites of medical treatment, they soon appeared to be failing. Yet it was scarcely possible to set free or even to hold back the intake of so many thousands of deranged, unruly, inconvenient, and (at times) dangerous persons. The eugenics movement, which has so often been characterized as an (illegitimate) outgrowth of Darwinian biology, is better understood as a reaction to the failure of asylum care to check the hyper-Malthusian increase of the institutionalized insane. Alienists (as doctors for the mad were called) set out on the basis of statistics to ascertain the principles by which insanity was reproduced. About 1880, with the spread of mandatory schooling, a parallel crisis of "feeblemindedness" began raising alarms. Institutions founded to treat these conditions supplied not only the incentive for a science of human heredity, but also its experts and its abundant reserves of data.

The numbers were issued first in asylum reports and census tallies. They could then be deployed to assess the performance of institutions and to plan for the future. Bragging rights at the asylum depended first of all on statistics of cures, which provided legitimacy to the asylums. Their status as curative institutions, in fact, was never altogether secure. While legislatures and ministries shared the medical hope for treatment, they also were concerned to maintain order in local communities. In practice, the asylums complemented and competed with prisons and poorhouses, promising better outcomes, but at an elevated cost. Many patients, especially in the early years, were brought to an asylum from such allied institutions, and some at least were discharged to them. A cure was highly advantageous from a budgetary standpoint, yet the abundance of reported cures did not suffice to reduce or even to stabilize patient numbers. Asylum doctors began to argue that the reduction of insanity was not within the power of medicine but must depend on public-health efforts. These, in turn, were to be guided by statistics of causes, which supported alienist preaching against alcohol, overwork, and masturbation. Right from the start, heredity sometimes appeared as the most fundamental source of insanity, and by the 1840s this view was widespread. Happily, heredity was not destiny. It was, on the contrary, singularly amenable to intervention, if only those contemplating marriage would pay heed to the mental health of the families of their intended partners.

Social medicine achieved an unwonted importance in the nineteenth century, especially for battles against epidemic disease. Insanity, too, was understood this way, and if, as the new alienists insisted, the appropriate treatment was "moral," it made sense to emphasize moral causes. It was not even self-evident that treatment of the insane belonged in the hands of doctors. While they insisted on the curative power of their potions, they also wielded impressive behavioral technologies to maintain order: opiates to calm, painful "remedies" to punish unruliness, and better rooms or lightened restraints with the prospect of eventual release to reward good behavior. Although the asylum was a closed space, the battle against insanity reached out into the larger society. Data, especially on heredity, demanded a web of information to connect the prison-like interiors of these institutions with the towns and countryside stretching out beyond its walls.

Mental or psychological medicine was always only partly about cures. It was a statistical human science, increasingly focused on insanity as a social problem. Almost from the beginning, it addressed the hereditary characteristics of the healthy almost as much as those of the sick. By the twentieth century, especially under the Nazis, the disregard for individuals appears ugly, but there had never been a clear boundary between collective and individual health. The investigation of inheritance of mental illness ("lunacy") and intellectual disability ("feeblemindedness") was at once medical and social, reflecting the political orders of Europe and North America as they evolved from the 1790s to the 1940s and beyond.

As we enter onto this history, it is pertinent to note that, against all expectations, the history of data has become fashionable. Data manipulation now generates great fortunes, not least in medicine and science. Apart from social media and algorithmic marketing, genomics is among the most celebrated of big-data projects. University statistics departments unexpectedly find themselves no longer stereotyped as tedious data crunchers. Money and opportunity have transformed them into brilliant data crunchers, and "data science" into one of the most exalted and absorbing of vocations. Google and Amazon were not the first to imagine that data mining was the answer to every problem of knowledge. The data visionaries of the 1850s and 1860s, like so many in our own time, saw no fundamental distinction between scientific statistics and commercial or bureaucratic numbers. Human understanding, they proclaimed, can be relegated to a secondary role, and it is often more comforting not to dig too deeply.

On the ground, the accumulation and management of statistics has always been a humdrum pursuit, though teeming ambiguities lie hidden in data, and techniques of design and analysis are often highly ingenious. Since the rise of the state mental hospital in the early nineteenth century, ordered, standardized statistics have had a paradoxical relationship to the disorderly scenes of madness and of suffering that they are supposed to sum up. Behind the classical or Gothic asylum façade lurked misery and filth, and beneath the ordered statistical surface, perhaps, a chaos of tabulated unreason. The phenomena might be made to conform to the accounts, but rarely without a struggle. Data projects, often conceived in a utopian spirit, run up against quiet or even organized resistance. For history, the fascination of statistics arises not alone from its technical power, but still more from its irrepressible human characteristics.

The three parts of this book correspond to three basic technologies of data and information. The first involves the introduction within asylums of systematic recordkeeping and the amassing of numbers into list-like tables. Data on heredity first materialized on the pages of case books and admission forms as answers to innocent medical-administrative questions about causes of illness. The doctors had no need to explain why they asked, since the assignment of a cause was already routine in medical case histories. Almost by instinct, nineteenth-century asylum doctors converted the marks in their case books into statistics, which in most cases required little more than totaling up the entries in each column of a registration book, or perhaps dividing them into numerical intervals, for example, of age. The resulting numbers were also calculated to inform and win over the public. The "supposed causes," including heredity, had news value from the beginning, second only to the statistics of cures. The force of these numbers reflected the trajectory of asylum medicine, whose initial optimism proved fleeting. Cure rates began high and then declined, putting ever more pressure on the investigation of causes, hereditary ones in particular. Alienists wanted to cure patients, but they also had a key role in shaping infrastructures of public health. They looked to knowledge of causes as a guide in blocking the production of madness at its source.

These numbers stimulated efforts to improve and standardize tables. John Thurnam at the York Retreat in England gained an international reputation for his excellent tables and for some basic calculations. He also took a lead role in the drive to improve data on heredity by tracking down sick relatives of asylum patients. Another special focus of early statistical inquiries was the question of whether, as patient numbers seemed to imply, madness really was increasing. The French alienist Étienne Esquirol launched a debate by speaking of insanity as a disease of civilization. It was a discouraging finding, challenging hopes for progress by its suggestion that madness was bound to increase. The census of insanity was partly a scheme to settle basic questions of causation, and partly an administrative tool for planning public asylum systems. By 1840, several states had regularized such a census, and others thought their dignity required one.

Jules Baillarger's template for uniform data entry, published in 1844 in the new French journal of medical psychology, may be seen as a harbinger of a second wave of paper technologies for hereditary data, one that took off in the late 1850s. He aimed to make the table into a tool of research and reason. One clear sign of a more systematic approach is the push for standardized statistics, first in France, England, and Scandinavia. A more encompassing standardization appeared on the agenda of international statistical congresses as early as 1855 and was taken up more systematically in 1867. By 1871, this French-led effort had failed internationally, but it provided the basis for an impressive German initiative to integrate asylum statistics with census results on insanity throughout the empire. Still more consequentially, the Germans introduced in the early 1870s a flexible technology of census cards, one card for each individual, to be sorted and counted with simple hand movements.

In the end, however, the most important data work for the study of heredity was carried out on a smaller scale, at the level of individual asylums. Most asylum directors preferred the flexibility and the fine judgments made possible with long-term statistics from a single institution over simultaneous mass statistics from many institutions. The most promising technology of the new era, taking off about 1860, was the correlation table, which placed a variable that mattered on each axis, such as hereditary relationship and form of illness, in an effort to clarify causal relationships. Alienists also prepared intricate tables to test claims for hereditary degeneration and to measure the risk involved when the insane were allowed to reproduce.

Much of this work required elaborate systems to collect and process information from outside as well as within the institutions. Among asylum directors who took these investigations seriously, it was not enough merely to add up numbers accumulated to meet bureaucratic requirements. Now, data gathering was to be adapted to specific aims. Among the most impressive innovations in hereditary research from this era were the family pedigrees of mental illness published in Norway in 1859, more than three decades before pedigree tables emerged as the principal template for eugenic data. The author of this work, Ludvig Dahl, relied on intense local medical-social research, made possible by detailed census records.

The third phase of hereditary research, extending from the 1890s to the 1930s, began with the pedigree table, which then gave way to the full surveys of local populations. In this era, at last, we encounter famous researchers on human heredity, including Francis Galton, Karl Pearson, Charles Davenport, Wilhelm Weinberg, Ernst Rüdin, and Lionel Penrose. The eugenics movement took off about 1900, with Galton and Pearson as its most effective early advocates in Britain, and Schallmayer and Rüdin in Germany. Galton had been working toward a science of human heredity since 1865. If I surmise correctly, his ambitions were linked to asylum studies right from the start, and certainly by 1875. The biometric approach, emphasizing statistical tools for understanding the transmission of human traits, fit well with data work on inheritance of insanity and feeblemindedness, and Pearson built up a considerable network of connections with doctors and alienists, most of them entirely friendly.

Mendelian genetics, which appeared suddenly on the scene in 1900, began to be integrated with asylum data on insanity and feeblemindedness about 1908. In practice, this meant tracking down discontinuous variables, like Mendel's tall/short or smooth/wrinkled peas, for traits like mental ability that appeared to be continuous. Such traits should be distributed among siblings according to familiar Mendelian ratios, typically 3:1 or 1:1. By this time, many agricultural breeders as well as experimental biologists were insisting on the indispensability of Mendelian genetics. Its extension by Davenport to eugenic issues involved close collaboration with psychiatrists and psychologists. Few if any human geneticists gave up on Mendelism, but in psychiatric heredity, it was moving to the back burner by 1920. Rüdin's group in Munich, working statistically at "empirical hereditary prognosis," was at the top of the prestige hierarchy in the 1920s and early 1930s. Their research depended on teeming files of data cards on family traits of asylum patients, students in special schools, and prisoners. These data files expanded by another order of magnitude under the Nazis. It is impossible not to see them as ominous. Yet Penrose, whose politics were diametrically opposed to Rüdin's, worked in the same scientific tradition, and for decades afterward, psychiatric geneticists continued to cite and to praise this German research until its alliance with Nazi policies came to seem too disreputable.

An appropriately critical reader will ask how this tradition of asylum statistics, if it was really as central to human genetics as I claim, could have remained in the shadows for so long. In the first place, this is not the tradition that postwar human and medical geneticists wanted to understand as their own. It was too tightly allied with eugenic interventions and with social and medical inquiries rather than distinctively scientific investigation. The new Mendelism, after all, coincided in time with a new discipline of genetics and soon became inseparable from it. Historians of biology at first took their lead from the historical verdicts of scientists, most of whom structured their histories in terms of theoretical novelties or experimental innovations without recognizing a role for such mundane recordkeeping. Although some of the twentieth-century characters in this book are well known to historians of genetics or eugenics, their data work has mostly remained in the shadows. Much is changed when we examine these figures from the standpoint of institutional and statistical practices. Once we take the numbers seriously, we must notice that mental hospitals and institutions for the "feebleminded" retained their status as key sites of hereditary investigation right through the 1930s and beyond. A focus on data lends specificity and concreteness to arguments about the relations of human genetics to eugenics, and more generally to ideologies of racial and social inequality. These are now often downplayed. Here, they appear as fundamental.

This book also reveals a much deeper history of human heredity, linking the twentieth-century story of statistics and genetics to a set of nineteenth-century developments that have rarely even been mentioned in histories of genetics. How could these vast storehouses of data and statistics on heredity have remained so long unnoticed? This question is, for me, not at all rhetorical. More than three decades ago I wrote a dissertation and book on the history of statistical thinking from 1820 to 1900 without ever noticing the contemporaneous flourishing of asylum statistics. Part of the explanation is that these statistical discourses were somewhat isolated in their own time. It may be significant that medicine, as a profession, was in certain respects a closed world, and that asylum work was largely distinct from ordinary medicine. Although alienists did reach out to the public in hopes of altering behaviors that contributed to insanity, most of their work appeared in their own journals or in official reports of various kinds. Finally around 1900, when statistics of insanity, feeblemindedness, and degeneration emerged as evidence of a crisis of modern life, alienists began to make common cause with biologists, statisticians, and social scientists for the sake of a future now shrouded in fear.

The research in this book proceeds on the supposition, for which it gives arguments, that the investigation of human genetics was and remains a human science. The human sciences treat schools, militaries, factories, offices, and hospitals as objects of investigation. To speak of asylums as the context for an emerging science of human heredity is not enough. These institutions took the lead in defining research problems, tracking down the human subjects, and selecting and training the researchers who gathered, analyzed, and circulated the data that gave shape to this science. Although much of it happened within walled spaces, the care—often the confinement—of the insane was closely watched by state ministries, health officials, political leaders, and directors of intersecting institutions such as poorhouses, prisons, and special schools. The objects studied, mostly human behaviors and bodily traits, were sensible to lay observers in a way that genes, molecules, and viruses are not. In a later phase of development, scientists and doctors could focus more intensely on more technical, less accessible objects of investigation. But hopeful or suffering humans continue often to intrude into the story, even when researchers preferred to write them out. There are drawbacks to treating medicine as technical.

Unless we reckon with the past of science, including its wide diversity of relevant actors, we cannot comprehend well the choices that defined the trajectory of human and medical genetics even in more recent times. Scientists have never yet built an impermeable laboratory space, not even for manipulating DNA. Neither can scholars and other citizens embrace a definition of science that sets it apart from history. The broader perspective I favor is evident in the outline of this book and intrudes everywhere in points of detail. This study, while advancing a large argument, seeks also to recover the texture of the past, which often is polyphonic or contrapuntal. The narrative includes extended discussions of developments in at least six countries (depending on how we count the German states) and scores of institutions. Since the research activity was highly decentralized, I can never assume without evidence that a medical paper, book, or institutional report, however brilliant, weird, or ingenious it appears to me, was readily made known to others engaged in the study of inherited mental defect. Yet the alienists of many lands saw themselves as allies in a shared endeavor. They sent around reports and journals, traveled repeatedly to other institutions both domestic and foreign, and engaged in extensive reviewing of meetings, reports, researches, and statistics.

This is the history not of a clearly demarcated discipline or scientific specialty, but of the circulation and reshaping of knowledge within a loosely structured yet self-consciously international field. I have framed it not as a comparative study of autonomous nations, but transnationally, in terms of systematic interchanges, from local to international. I do indulge some comparisons, but not always at the same scale. Individual asylums, as the story reveals clearly, had distinct cultures and traditions, of which they were keenly conscious. In many countries, including Germany and the United States, institutions were funded and regulated by states and provinces rather than by nations or empires. Comparison is thus possible at multiple levels. I have chosen not to privilege a single unit of analysis but to recognize stable entities (of whatever sort) where I find them, while emphasizing shared problem situations and ubiquitous exchanges. This is a history of professional knowledge in the world.

It has been important for me not only to investigate places and genres of knowledge but also to identify the circumstances of their contact. Again and again, during the protracted process of writing, I was drawn back into sources to try to identify a connection, untangle an odd detail, or just dig up basic information about an event or a person in my story. Often, what I discovered went far beyond what I had sought. It was necessary to sacrifice many intriguing but wayward tidbits to maintain forward motion. This is not the story of a coherent enterprise, a group of researchers working at a single problem that could be solved or even formulated in a unified way. The characters here were loosely joined by a shared ambition to measure and comprehend outcomes of hereditary processes associated with mental disability. Their specific methods, techniques, and tools were diverse. Since this is not a novel, I cannot simply bring dispersed enterprises and characters together to tie up the loose ends. I have tried to be true to my topic, depicting as clearly as I can the scenes and forms of activity through which a science of human heredity took shape in asylums, clinics, schools, and the occasional laboratory. A stream of narrative explanation flowing straight to the sea without tributaries, pools, eddies, log jams, and storms would forsake its twisted splendor to be made more simplistic and less truthful.

PART I

Recording Heredity

> Looking over the annual reports of American Asylums, we need not say how almost entirely they are filled by numerical tables.
>
> —John P. Gray (1861)

Asylum management was a quantitative business. Nineteenth-century states, having been seduced by the promise of abundant cures, demanded proper accounts of patients and of money. Was the investment in patient care worthwhile? The extraordinary increase in numbers of the insane, after so many institutions had been set up to cure them, led to grave doubts and then to demands for evidence. Registers of cures and deaths were to be weighed against costs on an implicit institutional balance sheet. These bureaucratic numbers were also public ones, laid out as if their meaning was transparent. The reports circulated among institutions, internationally as well as domestically, typifying the scientific ambitions and the cosmopolitan spirit of the quantified lunatic asylum. From their first beginnings in the 1840s, alienist journals of every nation reported regularly on each other. Asylum statisticians labored to make their numbers comparable, often juxtaposing figures from different institutions.

Purely administrative accounts were denominated in money terms as amounts spent on staff, food, fuel, and paper. Alienists presented patient numbers, including admissions, outcomes, and causes, as medical. They expected at first that printed tables would provide a public demonstration of the fruits of state investment in mental health. Indeed, they registered many cures, yet the patients continued to multiply, suggesting a radical failure of asylum medicine. It came to seem necessary to supplement institutional records with scientific counts. Doctors and administrators relied more and more on censuses to provide valid numbers as well as to indicate causes on a population level. They soon began investigating families to clarify the increasingly urgent question of hereditary causation.

Recordkeeping is at the heart of the chapters in this section, which extends from about 1789 to the 1850s. The perceived need to maintain and to publicize institutional records grew up more or less contemporaneously with the asylum as a social and medical institution. In the debates set off by George III's bout of madness and by a series of scandals involving English madhouses, poor or missing records became an embarrassment. For a condition like madness, which did not readily yield to a rational accounting, the proper standard of accuracy was far from clear. Asylum doctors, who could as yet have no special training in the diagnosis or treatment of insanity, pronounced on the success of their own remedies while relying on a family member to fix the cause of illness and the moment of first onset. Passive recording seems to have the advantage over expert judgment in tabular records of mental illness. This is the usual way of data, so often idealized for thin factuality and innocence of interpretive distortion.

Medical concepts of heredity, at least as applied to madness, are not readily distinguished here from folk understandings. A case was called hereditary when an ancestor or (perhaps) a collateral relative had been similarly afflicted. Although doctors often complained about the unreliability of data provided by lay informants, especially in regard to the first manifestation of disease and its cause, they seem to have been willing to record as cause almost anything a family member told them. This was data-driven science avant la lettre. Yet patient heredity, or at least the observable characteristics of ancestors and siblings, could be tracked down. Heredity was among the few entries that almost all doctors regarded as a legitimate cause. Beginning about 1838 with John Thurnam's inquiries from the York Retreat, family investigation quickly emerged as the basic tool of research on inheritance of insanity. Thurnam was also a great champion of systematization and standardization, and his reports provided an international model of asylum statistics.

Along with the effort to improve the quality of hereditary information on insanity, the other great push to get beyond asylum routines involved census counts. These took up the question of causation on a collective level rather than individually. Did the incessant increase of asylum populations point to a genuine increase of insanity, or was it merely an artifact of having acceptable institutions to record, treat, and care for odd or disorderly family members? The hottest statistical debate was provoked by Étienne Esquirol's argument that insanity was at bottom a disease of civilization. In this period, censuses were rarely used to gauge the inheritance of insanity, but for those who were already convinced of the key role of heredity, census figures showed the urgency of controlling its reproduction.

CHAPTER 1

Bold Claims to Cure a Raving King Let Loose a Cry for Data, 1789–1816

> It would be a most fortunate circumstance for medicine and mankind . . . were the parliament to examine physicians on every disease, as they have lately done on the unfortunate malady of a Great Personage. The utility of the arithmetical system would then be as universally conspicuous throughout every disease, in any great emergency, as it was in insanity, or as it is in politicks and commerce.
>
> —William Black (1789)

The King Is Mad; Long Live Statistics

The collection and manipulation of data, so central to the study of human heredity, knows no unique point of origin. The assembly of a committee of the House of Commons in January 1789 to inquire into George III's apparent madness, potentially a crisis situation, brought to the fore some vital questions about proper recordkeeping. A highly politicized disagreement about the king's medical prognosis had roused a furor among the royal physicians extending to Parliament and the nation. How could they anticipate whether he would recover in time to preclude any need for a regency? What confidence was to be placed in the Reverend Dr. Francis Willis, who had been brought in from the provinces to care for the king? The committee questioned Dr. Richard Warren, one of the king's physicians:

> Whether if Nine Persons out of Ten, placed under the Care of a Person who had made this Branch of Medicine his particular Study, had recovered, if they were placed under his Care within Three Months after they had begun to be afflicted with the Disorder, Doctor Warren would not deem such Person, either very skilful or very successful?

Warren answered yes, if so many really did recover.

> Whether, in order to induce Doctor Warren to believe, that, for Twenty-seven years, Nine persons out of Ten had been cured, he would not require some other Evidence than the Assertions of the Man pretending to have performed such Cures?
>
> I certainly should.[1]

Willis, having made very specific claims, could not back them up with written records. Dr. Warren's sharp skepticism points to an emerging ethic of data and accountability as a basis for public knowledge. The administration of madness participated fully in an explosion of statistical activity across a range of scientific, bureaucratic, and professional projects that already was beginning to take shape. Public numbers were not only for legislators but also for common people, who were urged to calculate in order to judge their rulers as well as to better their lives. The debates sparked by Thomas Malthus's *Essay on Population* in 1798 were unprecedentedly lively, if far from new. Medicine, too, experienced sharp controversies, most famously regarding the advantages of inoculation campaigns against smallpox. Many physicians, however, were not eager to be judged by their numbers. If patients and diseases were unique, undiscerning data could not reliably guide a treatment decision. A public campaign for "numerical method" took off finally in the 1820s, launching an era of statistical hope coupled with doubts about the old remedies.[2] The treatment of the insane was more profoundly reshaped by numbers than was ordinary medicine. Asylum medicine, as an area of public health, had only a weak ethic of individualism. In an age when physicians were paid directly by patients they visited in their homes, hospitals were more like poorhouses. Much confinement of the mad was for the sake of public safety, and many patients, lacking the means to pay the costs of their treatment, were designated paupers. As public or charitable institutions, asylums were particularly vulnerable to accountability standards, which meant providing numbers for patient admissions and outcomes as well as for revenues and expenditures.[3]

Asylum statistics originated as a form of bookkeeping. The prototype of the patient reckoning was a balance sheet. John Strype's 1720 edition of Stow's classic survey of London includes tables for Bethlem Hospital (Bedlam) beginning in 1704. The "Disturbed Men and Women then brought in" for the year numbered 64, while 50 were "Cured of their Lunacy and Discharged," and 20 more were "Buried," leaving 130 patients under care. The table for 1705 shows 72 admissions, 34 cures, 29 deaths, and 137 patients remaining. (This arithmetic leaves 2 patients unaccounted for.) The next year, "as it was Published," is "1705 to 1706," with 72 patients admitted, 52 cured, and 13 buried. Strype gave another table extending from Easter 1706 to Easter 1707, followed by three years with no information, a table for 1711, five more unrecorded years, and tables for 1717 and 1718.[4]

If the accounting seems fragmentary, the therapeutic evidence was still more so. The reports from 1680 to 1705 claimed cure rates between 57% and 82%. Yet the patients discharged as cured were not distinguished in the records from those sent away still mad. Some, certainly, did not recover, for Bedlam had an announced policy to limit the residence of patients to about a year. (The evidence of the numbers, with annual admissions about half as numerous as patients remaining under care, implies an average stay of about two years.) Strype did not explain the calculation of cure rates, merely proclaiming: "So that by God's Blessing for Twenty Years past, ending 1703, there have been above two Patients in three cured, as the Physician hath told me." Yet Bethlem was in some manner held responsible for patient outcomes, and the higher powers were not so lax as to leave the numbers to happenstance. Hence the need for techniques of deception. A Bethlem historian remarks that they maintained a low death rate by proactively discharging weak or debilitated patients.[5]

Although many asylum officials continued to treat their records as proprietary into the nineteenth century, others were working to expand public access to data on the mad. Dr. William Black, who was trained up in medical arithmetic by the campaign for smallpox inoculation, had complained in 1781 that at Bethlem, the "relieved, cured and discharged, are jumbled into one list" so that none but its "eminent physician," Dr. John Monro, could resolve what proportion is cured. Was this a backhanded compliment? In 1788, Black described the Bethlem medical books as an "untrodden wilderness." The eminent Dr. Monro, however, introduced him to his estimable son, Thomas, who in turn led Black to yet another praiseworthy individual, the resident apothecary, John Gozna, "whose learning and curiosity induced him to keep a *private* register of all the patients, upon which, as incontrovertible data, I have founded and collected all the following tables and propositions."[6]

At this point, the king's madness turned the barren wilderness of Bethlem records into terrain of extraordinary value. Black, quite unexpectedly, found himself holding the key to unlock a great medical-political mystery. He set to work in great haste on a new edition of his just-published work on human mortality, withdrawing the 1788 version from circulation. In the crucial year of George III's madness, he offered a solution to the urgent problem of prognostication. In a dedication of his book to the younger George, Prince of Wales, who was more than willing to assume the regency that Black's numbers were likely to advance, Black declared: "I trust it will not be arrogant in me to say, there will be found considerable original, useful, and authentick information." Such was the power of the medical numbers he prepared from Gozna's record book. "I may with safety assert, that mine are the only numerical and certain data that ever have been published in any age or country, by which to calculate the probabilities of recovery, of death, and of relapse in every species and stage of insanity, and in every age." The tools of arithmetic, so valuable "in politicks and commerce," would now prove themselves in medicine. Insanity, long written off as "the most difficult and conjectural" topic in all of medicine, would at last yield to "medical arithmetick."[7]

These were heady times for Black. "And it is not a little flattering, that the interrogatories, in this national dilemma, to some of the medical superintendants of Bedlam, were answered by a reference to my calculations." His was the method of "authentick information," the same method by which he had shown how to eliminate a million deaths annually in Europe. Arithmetic was the best medicine.[8]

In the wake of these investigations, the officers at Bethlem began to perceive recordkeeping as a political necessity and as a basis for medical legitimacy in the face of unsubstantiated claims by provincial practitioners. Black, who had spoken of his own data as *certain* and *incontrovertible*, put no faith in theirs. While Gozna's private numbers, collected with the assistance of the Doctors Monro, were exemplary, the official ones had taken great liberties. Instead of accepting responsibility for patients sent away uncured, the authorities claimed to be making room for others with better prospects and, on this pretext, omitted them from the outcome statistics. Black continued: "Of those who are said to be annually cured, it is difficult to say in how many this may be only a lucid interval of reason: a transitory calm of this mental insurrection."[9]

Philosophical writers have liked to suppose that death, being unequivocally real, supplies a basis of unquestioned fact. Black, who knew better, was scathing on manipulations of mortality figures. In his revised text of 1789, he condemned Bethlem's reliance on "the ambiguous term *discharged*" for hundreds of patients "reported as sick and weak, as afflicted with epileptick fits, or with paralytick strokes, and none of them liberated from insanity when discharged." They were entered as incurable, but some could just as well have been "added to the dead list." The hospital sent them away early to evade responsibility for their impending deaths. The statistics, then, were formally correct but misleading. "This is truth, but not the whole truth." These manipulations of Bethlem data tended to raise false hopes of recovery. And Willis was no more credible.[10]

John Haslam, Gozna's successor as Bethlem apothecary, was still more severe. The king did in fact get better for a time, which seemed to lend credence to Willis's bold claims. His supposed cures, indeed, far surpassed the experience at Bethlem, where, according to Haslam's tables, almost two-thirds of the patients were sent away uncured between 1784 and 1794. "Medicine," wrote Haslam, "has generally been esteemed a progressive science, in which its professors have confessed themselves indebted to great preparatory study, and long subsequent experience, for the knowledge they have acquired." Willis, in defiance of human experience, claimed unparalleled success right from the start. The Bethlem data also undercut Willis's pretensions by showing much poorer results for older patients than for persons in their teens or early twenties. The odds for a man of age fifty, such as King George, were 4:1 against recovery. And among patients admitted (against stated policy) after more than a year of illness, Haslam found not a single lasting recovery.

> When the reader contrasts the preceding statement with the account recorded in the report of the Committee who have attended His Majesty, &c. he will either be inclined to deplore the unskilfulness or mismanagement which has prevailed among those medical persons who have directed the treatment of mania in the largest public institution in this kingdom, of its kind, compared with the success which has attended the private practice of an individual; *or to require some other evidence, than the bare assertion of the man pretending to have performed such cures.**

The footnote referred to Willis's testimony in 1789, where he claimed, without evidence, cures at a rate far surpassing Bethlem's most modern results.[11]

There is no agreement among scholars or psychiatrists on the cause of the king's illness or on the significance of Willis's treatment for his temporary recovery. He continued to be troubled by intermittent mental illness and during the 1810s was so disabled that his son, the future George IV, was appointed to govern as regent.

A Statistical Specialty

The infrastructure of public and private medical numbers on which Black relied was still emerging. In Britain, financial and insurance calculations, which relied on extensive tabular data, had become quite sophisticated by early in the eighteenth century. Medical institutions were not exempt from the demands of financial reporting.[12] Patient records were another matter. Medical records of insane asylums were gradually regularized in Europe and North America during the first half of the nineteenth century. In 1815, a parliamentary committee reported scathingly on the efforts of English asylum officials to conceal institutional records. Along with eyewitness accounts of female patients kept naked in bare, squalid cells behind a gate whose keeper claimed no access to a key, the committee called attention to the concealment and falsification of patient books. A magistrate, Godfrey Higgins, describing his efforts to inspect the Yorkshire Asylum, explained when asked about patient registration: "There was a set of books regularly kept by the apothecary, and also another set by the steward, both of which purported to be a correct account of admissions of patients, and how they were disposed of, but I have reason to believe that those accounts were false, and that they were kept falsely on purpose." He pointed to discrepancies between closely held books and the official ones, providing a basis for muckraking newspaper accounts. A steward refused to deliver up the books he kept, claiming they were his own property, and later testified that he had destroyed them. Just after the court of governors ordered an investigation of the Yorkshire Asylum, its buildings caught fire, immolating several patients along with the record books.[13]

These 1815 hearings soon acquired a place in historical accounts as the moment when old abuses in the public asylums were washed away, when the institutions at last became responsible and humane. They presaged a monumental expansion of asylum systems, first in England and France and soon afterward in other parts of Europe and North America, that was to go on for 150 years. While public asylums typically fell to the charge of regional governments such as the American states or English counties, they have a notable place in the genesis of the welfare state, first as curative institutions showcasing public investment in a healthy citizenry, and then as custodial ones to protect the population from degeneracy. In both guises, asylums and hospitals helped give shape to new standards of public accountability. Those developments provided essential background to the development of asylum statistics of heredity.[14]

Patient accounts as well as financial ones appear often to have been driven by external, bureaucratic demands, but they had also a medical logic. The English physician Thomas Percival, in his 1803 treatise *Medical Ethics*, spoke of recordkeeping as integral to sound medical practice. Keeping statistics was for him an ethical imperative, reflecting medical obligations to patients. A physician or surgeon, he wrote, should draw up an account of every case that is "rare, curious, or instructive." Hospital registers ought to include three tables: one for admissions and outcome, a second for illnesses treated, and a third breaking down patients by age, sex, and occupation. These would advance knowledge of healthy and unfavorable "situations, climates, and seasons," the effects of particular trades and manufactures, and the outbreak or cessation of epidemics. Finally, "physicians and surgeons would obtain a clearer insight into the comparative success of their hospital and private practice; and be incited to a diligent investigation of the causes of such difference."[15]

Percival's sense of the hospital as a place for advancing public health applied well to mental hospitals, which already were becoming exemplary sites of statistics. The most fundamental, and the most ubiquitous, object of their tabulations was the flow of persons through the institutions and the outcomes of their treatment, providing, in effect, a medical balance sheet. Tables could demonstrate the value of these institutions, measured as the number of persons cured of this terrible disease who could return to work and family. The register of causes seemed at first less fundamental, but causes were routinely included in case reports as potentially relevant to treatment. Alienists advised the public to avoid behaviors, such as overwork and excessive drink, shown by statistics to induce mental illness.

Asylums also were the primal site for statistical knowledge of human heredity.

The Causes of Insanity

The earliest known statistical table of causes of insanity appeared in Black's new (1789) edition of *An Arithmetical and Medical*

Analysis of the Diseases and Mortality of the Human Species. In the 1788 edition, he had presented causes and numbers in paragraph form. Medical tables go back at least to John Graunt's well-known 1662 report on deaths and their causes, based on the London bills of mortality. Until about 1840, most asylum tables were little more than simple tallies, for example, of assigned causes. Black, however, seized the opportunity to build in a second dimension and in this way to relate causation to the urgent question of curability.[16] Already in 1788, he had noted with surprise that those with "hereditary insanity . . . seem to recover nearly as well as from the less inherent causes; there are several instances of recovery when hereditary from the parents on both sides." In 1789, he repeated the thought while omitting the suggestive word "inherent," which points to internal sources of illness. Black sought out but could not find an anatomical explanation for the "latent predisposition or frailty in the recesses of the brain, which render some more than others liable to this mutiny of reason." Although the mechanism was hidden, its incidence could be tallied. "Family and hereditary" was, apart from a grab-bag list of miscellaneous troubles and disappointments, his most important cause. Black's medical histories listed "predisposing and occasional causes" not just for insanity but for most of the diseases he described. As source for the causes of insanity, his table relied on the Bethlem registers, that is, on Gozna's unofficial book. He elaborated that Gozna acquired this information through inquiries to family members.[17]

The historian of madness Roy Porter once remarked that the causes on such lists were already familiar to writers on insanity two centuries earlier. Black doubted their validity. "Most of the proximate causes assigned in authors for madness, are mere hypotheses; and of no active use to the community, or to medicine."[18] Haslam, though refraining from the compilation of tables of causation, undertook to check and to improve his data. He worried less about ignorance than mendacity:

TABLE VI.

The principal, occaſional, and remote cauſes of inſanity, together with the comparative proportion of cured, incurable, and dead, as influenced by each of theſe different cauſes; and alſo the proportion of relapſes: founded on the Bedlam regiſters.

The principal occaſional, and remote cauſes of inſanity; and proportion of relapſes.	Total number of inſane from each of the different cauſes.	Cured	Incurable.	Dead.
Misfortunes, troubles, diſappointments, grief, vexation, loſſes, croſſes, jealouſy, ill-uſage, anxiety, deſpair, diſtreſs	383	109	235	39
Religion and methodiſm —	166	54	90	22
Frights — —	96	38	51	7
Love — —	136	50	80	6
Study — —	40	10	27	3
Pride — —	23	2	20	1
Drink — —	111	31	68	12
Parturition — —	145	66	69	10
Fever — —	212	100	86	26
Family and hereditary —	213	90	103	20
Venereal — —	24	10	11	3
Contuſion, fracture, and fall	13	8	5	0
Obſtruction — —	18	8	5	
Ulcer and ſcab dried up —	7			
Relapſes from all the preceding cauſes, and alſo from preceding leſion by inſanity —	1205	508	623	74
Total —	2829			

FIGURE 1.1. Table of causes of insanity by William Black. This table may be the first ever to list and give numbers for causes of insanity. Black, one of the leading medical statisticians of his day, went beyond adding up causes to show the bearing of each cause on the prospect of a cure. From Black, *Arithmetical and Medical Analysis*, 133.

> When patients are admitted into Bethlem Hospital, an enquiry is always made of friends who accompany them, respecting the cause supposed to have occasioned their insanity.
>
> It will readily be conceived that there must be great uncertainty attending the information we are able to procure upon this head: and even from the most accurate accounts, it would be difficult to effect. The friends and relatives of patients are, upon many occasions, very delicate upon this point, and cautious of exposing their frailties or immoral habits: and when the disease is a family one, they are oftentimes still more reserved in disclosing the truth.
>
> Fully aware of the incorrect statement frequently made concerning these causes, I have been at no inconsiderable pains to correct or confirm the first information by subsequent enquiries.[19]

The assignment of hereditary and other causes depended principally on observations of family members or close acquaintances and sometimes on opinions given by the patients themselves. The inheritance of insanity, in the basic sense of tending to run in families, was no discovery of medicine or of statistics but an accepted fact of everyday life. Even as professional knowledge, it belonged as much to law as to medicine. Insanity was highly relevant to another kind of inheritance: the right to control property and to pass it on to the next generation. Already in Black's day, medical jurisprudence was keenly alert to the problem of inherited insanity. John Johnstone, who published a treatise on the subject in 1800, referred to madness as "the most constant and persevering" of hereditary diseases, one whose taint can persist right though a generation that has escaped it. He even called the "hereditary disposition to madness" a "fair ground of evidence in cases of imputed derangement of mind," meaning that a court could argue backward from evidence of family insanity to the mental state of the criminal.[20] Decades later, the life of the scoundrel and delusionary Charles Guiteau was minutely documented in preparation for his 1881 trial for the assassination of the American president James Garfield, yet the legal judgment as to his mental competence seemed to hinge on a demonstration that his condition was inherited.[21]

Haslam also wrote on medical jurisprudence, and in the second edition of his book on madness he complained of critics who balked at anything less than complete knowledge of the laws of heredity. There is no "infallible transmission" of mental disease, he conceded. It is not possible to specify why the hereditary tendency skips generations and how the sex of the affected parent affects its transmission to sons versus daughters. Despite these mysteries, he continued, hereditary transmission was well known from human experience and from the evidence of cattle breeding.

> In illustration of the fact, that the offsprings of insane persons are, *ceteris paribus*, more liable to be affected with madness than those whose parents have been of sound minds; it was my intention to have constructed a table, whereon might be seen the probably direct course of this disease, and also its collateral bearings: but difficulties have arisen. It appeared, on consideration, improper to attempt precision with that which was variable.

And yet it was not too soon to face the implications of inheritance of insanity and even to take action to block its reproduction. "The investigation of the hereditary tendency of madness is an object of the utmost importance, both in a legal and moral point of view. Parents and guardians, in the disposal, or direction of the choice of their children in marriage, should be informed that an alliance with a family, where insanity has prevailed, ought to be prohibited."[22]

French Tables of Hereditary Insanity

Black's statistics of mental illness were unmatched for decades. His table of causes, however, made less of an impression on alienists than one compiled in 1816 by Esquirol for his entry on insanity (*folie*) in a massive medical encyclopedia. Perhaps Black was too early; even Esquirol's table achieved fame mainly as reprinted two decades later. Paris medicine led the world in those years, and

CAUSES PHYSIQUES. N° 7.

Salpêtrière.		
Hérédité	105	150
Convulsions de la mère pendant la gestation	11	4
Épilepsie	11	2
Désordre menstruel	55	19
Suite de couches	52	21
Temps critique	27	11
Progrès de l'âge	60	4
Insolation	12	4
Coups ou chutes sur la tête	14	4
Fièvre	13	12
Syphilis	8	1
Mercure	14	18
Vers intestinaux	24	4
Apoplexie	60	10
Total	351	Total 107

FIGURE 1.2. Étienne Esquirol's 1816 table of physical causes of insanity. His neglect of arithmetic accuracy cannot explain why the number of hereditary causes in the right-hand column exceeds the total of all causes. Other listed causes include maternal convolutions, reproductive and menstrual irregularities, fevers, blows to the head, old age, intestinal worms, and apoplexy. He provided moral causes in a separate table. From Esquirol, "Folie," 178.

Esquirol had succeeded Pinel as France's most respected alienist, whereas Black never worked as an asylum doctor. Like Pinel, Esquirol taught that insanity could be cured and that a calm, ordered life was an important part of the therapy. He was especially alert to "moral" causes of insanity, such as domestic sorrows, political events, and wounded vanity. The disproportionate number of cases he assigned to heredity, nevertheless, tipped the balance in favor of physical causes.[23]

Esquirol's table of causes appears weirdly incomprehensible. The confusion begins with the inaccurate sums. These discrepancies are much diminished, though they do not disappear, if we ignore the first and largest entry on each list: heredity. Flawed arithmetic cannot explain how, in the right-hand column, the figure for heredity (150) exceeds the total for all causes (107). Silently, he must have

SALPÊTRIÈRE.		MON ÉTABLISSEMENT.
Hérédité	105	150
Convulsions de la mère pendant la gestation	11	4
Épilepsie	11	2
Désordre menstruel	55	19
Suites de couches	52	21
Temps critique	27	11
Progrès de l'âge	60	4
Insolation	12	4
Coups ou chutes sur la tête.	14	4
Fièvre	13	12
Syphilis.	8	1
Mercure	14	18
Vers intestinaux	24	4
Apoplexie	60	10
TOTAL	466	TOTAL 264

FIGURE 1.3. Étienne Esquirol's table of physical causes, reworked for his chef d'oeuvre, *Des maladies mentales* (64). In correcting the arithmetic, he simply added the hereditary cases with other physical causes and thereby effaced his basic distinction between precipitating causes and hereditary ones.

treated heredity as a different sort of cause from menstrual disorder, advancing age, or apoplexy. Alienists of this era often classified heredity as a "predisposing" cause, which acted only in combination with an "effective" or "triggering" cause. Physicians used the Greek term "diathesis" for susceptibilities linked to bodily constitution, and Esquirol appears to have been thinking in these terms. Although he declined to mark off predisposing causes in his table, he clearly viewed hereditary causation as not quite commensurable with fevers and falls.[24]

Esquirol's magnum opus, published in 1838, brought together his many papers and reports. The chapter on insanity added not one unit to his original table. He now made La Salpêtrière into the heading for the left column and inserted *mon établissement* (my establishment) on the right. In correcting the arithmetic, he set aside issues of incommensurability and simply added in the numbers for heredity, raising the total for his private mental institution from 107 to 264. The revised table thus privileged the logic of addition over the logic of causation, provoking a later German critic to complain of the incoherence of tables whose categories are not mutually exclusive.[25]

It is unlikely, Esquirol wrote, that the figure for heredity at La Salpêtrière could diverge so widely from that for his private patients. He attributed the discrepancy to indigent women at the hospital who, he said, were unable to supply hereditary information and sometimes did not even know the names of their parents. His paying patients were of quite a different class. It seems that he relied on the patients themselves, even the impoverished women, for much of his causal information. For neither group is there any indication that he followed up with probing questions or investigations to improve his tables. The numbers would always be imperfect. The true importance of hereditary causation, he declared, must exceed what the tables show. Yet it was already at the top of his list. He concluded his 1835 report on the Charenton Hospital: "Of all illnesses, mental alienation is the most eminently hereditary."[26]

Nineteenth-century asylum figures on heredity were usually put forth as if anchored in raw data concerning diagnosable conditions of patients and their families. Both Esquirol's original tables and the revised ones reveal tensions between his understanding of hereditary causation and the requirement to work arithmetically with data as he received it. The discourse of heredity, unalterably numerical, was structured by tabular forms and numerical reasoning. Although this emphasis was sometimes interwoven with questions about objects and processes of heredity, the prime mover here was data practices.

CHAPTER 2

Narratives of Mad Despair Accumulate as Information, 1818–1845

> The opinions, therefore, of medical individuals, as to the consequence or duration of the malady, are as far inferior to medical arithmetick in ascertaining truth or probability, as the oracles of old were to the demonstrations of Euclid.
>
> —William Black (1789)

> Facts, facts, facts, reports should represent facts and figures convey facts.
>
> —William A. Awl to Samuel Woodward, 18 April 1842

Gozna, the Bethlem apothecary, recorded patient data that appeared relevant to medical outcomes or to good order in the asylum. The information he provided for each patient included age, duration and causes of illness, whether the patient was "mischievous," and how these factors bore on prospects for recovery. Esquirol's tables indicated disease forms as well as causes. By 1835, asylum reports were becoming formulaic, subsuming medical data into bureaucratic tables. Almost anything put down in the admission book was likely to be summed up as statistics. Even variables like occupation, religion, and place of residence, though recorded for administrative reasons, could be construed as causally relevant. Almost all the basic data of asylum medicine, including cures, causes, and inheritance of mental illness, appeared first in institutional tables.

Asylum numbers circulated widely, sometimes as full reports and often in summaries and snippets. In the United States and Britain, public asylums enveloped the tables in commentaries for an annual report, which was then circulated to town officials and state legislators, the local press, families of patients, and medical colleagues. The reports provided evidence of sound institutional management, offered advice on healthy living, and advertised the benefits of prompt treatment. On the European continent, annual tables and accompanying reports were copied for distribution to local officials and state ministries but usually were not printed. Most alienists, especially German ones, held that multiyear reports were most valuable for science. These appeared in alienist journals, local medical or statistical periodicals, and administrative documents, or occasionally in book form. Heinrich Laehr, the professional leader of German alienists, declared in 1875 that his colleagues were missing an important opportunity to reach out to the public and to counter the prevalent misinformation on insanity. It would be easy to print up annual reports, since the necessary information was already required by state authorities. And their value was not limited to public relations. Laehr subsequently initiated a voluntary consortium of domestic and foreign asylums to exchange reports as valuable data.[1]

Flows of Information

The provision of asylum care, sporadic through the 1830s, soon became systematic. A French law of 1838 requiring facilities for the insane in every *département* was a model for other nations of Western Europe and North America. The systematic circulation of statistics reflected a sense of state responsibility that arose in alliance with these medical-social institutions. State regulation of recordkeeping was light at first, and the presentation of statistics in annual reports from the 1830s and 1840s often surpassed requirements. By 1840, it could be foreseen that good recordkeeping might turn the asylums into data banks for the investigation of causes. Alienist ambitions extended beyond the walls of the asylums to the amelioration of mental disturbance in the whole population. Statistics were intrinsic to that function, not only statistics of patients they treated but also of persons at large who showed signs of mental disorders. Alienists took the lead in efforts to census the insane. Despite their abiding devotion to numbers, they often doubted their own data, especially when it depended on lay observers to make medical assessments of illness and its causes. Massed into collectives and compressed into mean values however, the numbers were hard to dismiss. These "little tools of knowledge" counted for something and, just because they appear so mundane, require careful attention to understand how they functioned as the not-quite-raw material of an inchoate science.[2]

Since the data forms, admission books, and asylum reports are mute on the circumstances of data recording, it has been necessary to look to other sources, most of them archival, to get a sense of the acquisition of patient data. Without the benefit of more discursive sources, we cannot hope to understand the roles and meaning of data in the investigation of heredity.

A Not-Quite-Total Institution

The moral treatment construed the asylum as an alternative family in which the ill were to be treated as if capable of some degree of rationality. The new institutions were dense with the subtle pressures described by Michel Foucault as producing disciplined subjects, persons who need not be tyrannized by an external authority because they have already internalized its values.[3] An array of wards and rewards enabled the superintendent to offer better conditions to those who held their own unruly impulses in check. Alienists were keenly conscious of this soft power. Samuel Hanbury Smith of the Ohio Lunatic Asylum told a story of debt collection in the "Celestial Empire" by means of relentless watching—and nothing more—to illustrate the unbearable effectiveness of moral pressure. While he did this as a warning to families to allow their discharged relatives some space, he also praised the English alienist John Conolly for advancing moral treatment to the point that the huge London asylum at Hanwell could maintain order with no need for shackles, muffs, or straitjackets.[4]

The moral system insulated patients from contact with the outside world, even from family members, who after all had shaped the environment that brought on mental illness in the first place. This ideal of reshaping patients within a little world cut off from their familiar one inspired Erving Goffman's sociological account of the asylum as a total institution, an analysis that helped stimulate the anti-psychiatry movement of the 1960s and 1970s.[5] Goffman was by no means the first critic of these institutional powers. Already in 1840, the pioneering medical statistician William Farr had condemned the policy of sequestering patients as rooted in a self-interested delusion. Alienists, he charged, had invented reasons to shield themselves from observation by potential critics. They liked to recall the bad old days when visitors came into Bethlem for a small contribution to amuse themselves by gawking at the lunatics. Farr, for all his faith in statistics, insisted on the need for direct observation to prevent abuses. It is absurd, he said, to imagine that routine printed reports combined with occasional, preannounced visits by inspectors or commissioners can protect the mad from their attendants. If asylum personnel control the recordkeeping, only the watchful eye of the public can assure that rules will be followed and records honestly kept.[6]

While alienists idealized their institutions as refuges from a turbulent world, the data they most craved could come only from the outside. For most of the century, internal records provided only the sketchiest documentation of developing illness or the results of treatment. Episodes of acute bodily illness interrupted the silence of the books. Otherwise, the case record included just two bursts of intense recording: the moments of admission and of discharge or death. Data production thus coincided with an opening of institutional portals. Records made at the time of discharge, which fixed the result of treatment as cure, improvement, or failure, were based on evidence of the patient's words, behaviors, and bodily condition as experienced by the medical staff. At admission, by contrast, the information flowed inward. Patients arrived with a letter from one or more physicians, perhaps a family doctor, a public health officer, or a consulting physician, to authorize entrance. A "friend" bearing legal responsibility, typically a family member, was the most accessible source for information about personal and family background and the prior history of the illness. Patients transferred from another institution might arrive with no informed attendants, and thus with disconcertingly sparse documentation of their lives outside its walls.

At St. Luke's Hospital for Lunatics in London, the book of case records for 1839–1840 was little more than an admission register, a bound stack of small folded sheets with some printed questions on the left side of each page, leaving space to enter data on the right: name, residence, occupation, religion, and marital state, plus a few entries on the history and characteristics of the disease and its causes, including "relatives affected." The only item that might extend to more than a few words was "Evidence of Patient's Insanity," which met a specific legal requirement. The doctors recorded what they learned about the patient as unadorned data, without mentioning sources or giving explanations. Perhaps an eyebrow was raised when William Shakespeare walked in as the first patient in this new book. But the inscription was handled routinely: Paddington schoolmaster, thirty-five years old, "occasionally violent and sometimes sullen," dangerous "to himself & others," and untainted by mad relatives, his illness owing to "overapplication."[7]

The tendency to compress asylum knowledge into nuggets of data was encouraged by the recording technologies. To the end of the eighteenth century, patient records in madhouses often included no medical information whatsoever. The admission books at Bethlem were like this until 1815. The initial stimulus to expand recordkeeping was more legal than medical. Private institutions were still slower to take up copious data entry. The Bloomingdale Asylum in New York, for example, which focused after 1840 on wealthy patients, seems to have kept no case books until 1872, relying instead on a comparatively large space in the admission book to record descriptive information:

> Suspicious of brother and others—Louis Philippe's sons have spies in U.S.—Applied to Victoria for Generalship (1847);
>
> Thinks people want to poison him. Yesterday attempted to stab one of his friends with scissors, was put in prison, tried to

strike the Keeper with poker and afterwards to strangle himself with his handkerchief (1849);

Has of late been conferring with an 'angelic brother' long dead & has given $13,000 to the 'medium' by his order. Is to be Post M. Gen'l or will commit suicide. boisterous (1853).[8]

Data at the Retreat

The York Retreat, established in 1796 and long associated with the family of William Tuke, adopted a form of moral therapy aligned with Quaker principles. This rural institution for middle-class Quakers had little else in common with those other classic sites of moral treatment, Pinel's great Paris hospitals, which teemed with pauper patients. The York institution was made famous by Samuel Tuke's *Description of the Retreat*, published in 1813. In the early 1840s, thanks to the physician John Thurnam, it became a model of statistical recording, including, as chapter 3 shows, data on patient heredity. The prior directors at York had compiled considerable information on individuals but did not systematically convert it to statistics. Thurnam's great data project exploited these earlier efforts, demonstrating what was possible in the asylums of his era and providing a model for the next generation.

Tuke's little book consigned the data on patients to an appendix, listing them in order of admission with terse entries for age, sex, marital state, whether the case was old or new, disease form, and result of treatment. There was also a column for remarks on causes and heredity. Tuke dutifully recorded the causes assigned by relatives, even though he put no faith in them. "The human mind does not like uncertainty; and the relatives of the insane, are generally anxious to fix on some particular circumstance as the cause of disease." His book includes just two tables, which give outcomes for recent cases of mania and then for melancholia, each allowing three possibilities: perfectly recovered, much improved, and dead. The unimproved, who would soon cease to be recent cases, did not cloud his statistics. Already in 1803, a printed booklet of rules of the institution included numbers for patient outcomes.[9]

The bulkiest records in those years were drawn up not for current patients but for prospective ones. The admission papers began as handwritten letters from a physician certifying mental illness. On 18 May 1818, the Retreat issued its first preprinted certificates, explaining in a preamble that many patients ill-suited to the institution were showing up, inconveniently, at its door. The form allowed for prior approval. At first, it included sixteen questions, front and back, to identify the prospective patient and to sketch out background, medical condition, and behavior, including the "supposed cause" of the illness. The sheets indicate a mix of physical causes, such as fever or illness, and "moral" ones, including stress, overactivity, intemperance, masturbation, and religious enthusiasm. Heredity was ordinarily given as predisposing, for example: "Uterine derangement has probably been the exciting cause, but it appears that there is some family predisposition to the malady."[10]

The Retreat pared back its admission sheet to twelve items in 1832 then expanded it in 1842 to twenty-nine, some with multiple sections, reflecting Thurnam's data gluttony. His form began with a medical rationale: "Successful treatment of the insane frequently depends upon a full knowledge of their respective cases. . . . Information relative to even the more remote history of the case, would frequently be important in directing the right course of moral and medical treatment."[11] It is difficult to imagine that they ever used evidence of family predisposition in this way. That kind of data was for science. By 1840, signatures of two physicians were required to commit a patient. From the handwriting, it appears that these physicians gave signatures only and that someone else, a relative or custodian, supplied descriptions of the habits and behavior of the patient and the history of the disease. The recurring phrase, "supposed cause," on forms like these suggests that asylum doctors were unconvinced.

Although skeptical of such surmises, doctors believed that mental illness required both a triggering cause and a predisposing one. According to the American alienist Isaac Ray, the Bard of Avon already had recognized how these two types of causes acted in combination. "In the tragedy of King Lear, Shakespeare has represented the principal character as driven to madness by the unexpected ingratitude of his daughters; or more scientifically speaking, he has represented a strong predisposition to the disease as being rapidly developed under the application of an adequate exciting cause."[12] Alienist doctrine held that the insane were predisposed by constitution, temperament, or heredity and pushed over the edge sometimes by disease, injury, or abuse, sometimes by alcohol and masturbation, and sometimes by the psychic trauma of business failure, disappointed love, death of a close relation, false or fanatical religion, or pressures of work or study.[13]

Although alienists complained of their reticence regarding hereditary insanity, the relatives frequently volunteered information on peculiarities of family members.

> There is reason to think that the disease is hereditary—at least, one or two near relatives have labored under similar delusions.
>
> There appears to be some doubt respecting this particular. It is certain there was a good deal of eccentricity in the character of his mother & grandmother but whether arising from this source or not opinions vary.
>
> Her sister R.E. died here [at the Retreat].
>
> Her Mother's Father committed suicide under the depressing effect of large pecuniary losses and some first Cousins of her Father are insane.
>
> Not known to be hereditary—his Father during some function of his life acted strangely. What made his friends suspect that his mind was not in a healthy state.[14]

Such reports typify the interactions between family members with personal knowledge and doctors who had charge of the records. Before Thurnam, most alienists did not try to research heredity on their own but accepted what they were told. Hereditary mental illness was a fact of observation, a folk category, shared by the physicians. Although doctors, too, were almost never explicit about its meaning, they discussed and debated its relation to moral and other causes, its apparent tendency to skip generations, and the nature of hereditary transmission. While assuming in most cases that families could be divided between those with hereditary predisposition and those without, they sometimes tried to identify particular relatives and conditions. Nobody argued that heredity by itself brought on insanity, but asylum data seemed to confirm that it heightened vulnerability.

Minding the Numbers

Thurnam modestly remarked that the high cure rate of his institution should be compared not with numbers for the patients in general but with those of the middle classes. Members of the Society of Friends were especially long-lived, he said, and the Retreat encouraged the admission of more readily curable cases by offering a reduced charge to impecunious persons arriving within six weeks of their first attack. Abundant statistical tables from almost every institution seemed to demonstrate that prior duration of illness could be decisive for the success of treatment. Since cure rates were the preferred basis for self-promotion, those who lacked the authority to limit admissions to new and hopeful patients were often driven to create a special category of hopeless ones. The public asylum in Yorkshire, shunning such indirect means, had falsified 144 case results for the sake of better statistics.[15]

American asylum superintendents were particularly energetic in statistical self-promotion. On the authority of comparative tables compiled in 1829 by a New York specialist in forensic medicine, Theodric R. Beck, the Connecticut Retreat boasted of cure rates comparable to those of the most celebrated Old World asylums. For recent cases, their results were second only to a private English asylum. The following year, the Connecticut institution was able to bring its cure rate to over 90%, allowing the board of visitors to claim "unparalleled success."[16] Pliny Earle, soon to become a

	Admitted.	Cured.	Per cent.
Connecticut Asylum.			
Recent cases, - -	97	86	88.66
Old cases, - -	99	14	14.14
These may be compared with the result at the			
Retreat near York, (from 1796 to 1819.)			
Recent cases, - -	92	65	70.65
Old cases, - -	161	47	29.19
Dr. Burrows' Private Asylum.			
Recent cases, - -	242	221	91.32
Old cases, - -	54	19	35.18
Glasgow Lunatic Asylum.			
Recent cases, - -			50.00
Old cases, - -			13.00*

FIGURE 2.1. Comparative table of cures, showing US institutions at least on a par with the most famous Europeans ones. American alienists were particularly given to statistical boasting. From Theodric Romeyn Beck's published lecture, "Statistical Notices," 80.

leading authority on insanity in America, took up the comparative statistics of American and European asylums in an essay in 1838 for the *American Journal of the Medical Sciences*. He presented the Connecticut Retreat as a bright star in the firmament of mental medicine. Carrying forward its statistics of cures, he found cure rates in the early 1830s whose stability looks like the false accounting of an investment fraud: 91.6%, 91.6%, 91.6%, and 91.66%. The raw numbers, 21 cures of 23 new patients in the first year, and 22 of 24 in each of the three following, differ strangely, if trivially, from the calculated percentages, but the odd arithmetic scarcely affected his bold claims for the therapeutic effectiveness of moral therapy in a well-managed asylum. "Remarkable improvements have been made within the last half century in the treatment of insanity." Later in the 1830s, the model institution in Worcester, Massachusetts, achieved similar results. Numbers so excellent, in an age that put faith in progress, were bound to create trouble later. For the moment, the asylums treasured this demonstration of their effectiveness in numerical terms that every citizen and legislator could understand. They should understand, too, the need to hustle their relatives into an asylum at the first sign of odd behavior.[17]

Recognizing the budgetary obsessions of legislatures, alienists did not hesitate to reduce the value of mental health to money terms. The reformer Horace Mann, writing for a commission charged with planning the asylum in Worcester, explained how a small expenditure in the first weeks of illness may accomplish what no sum of money can achieve later. The false economy of inadequate asylum facilities became a favorite topic of calculation.[18] An 1860 report of the US census warned that delay could be debilitating for the patient, leading to ruinous expenses to the state. A decade earlier, Samuel Smith had used figures from the 1850 census to calculate these costs. Even supposing, conservatively, that the true rate of insanity in Ohio was just 1 per 1,000, he argued, there must be more than 2,000 lunatics there. Now compare the $100 needed for a cure with $250 annually to maintain each chronic patient, then multiplying by 200 patients waiting for a space, and you can see the terrible waste. From the "well-known" fact that an average incurable patient cost the community more than $2,000, it follows that if only 500 of them were made incurable by delay of treatment, the avoidable expense climbs to "*one million of dollars.*" This "monstrous sum" made an unanswerable argument. With so much money, you could build five new institutions.[19] Cool calculation, and not just sentimental feelings, demanded adequate public provision of asylum care. This kind of calculation, though it sometimes appeared elsewhere, was an American specialty. Dorothea Dix wielded it passionately in her international campaign to replace prisons with asylums. It was a strikingly social calculation, merging state budgets with costs to families.[20]

Thurnam Calls for Standards

Thurnam, though savvy about how figures could be manipulated or misunderstood, put his faith in numbers, no matter whether he was discussing mental medicine, pauperism, or natural science. He introduced a compilation of asylum numbers with an epigraph from Alexander von Humboldt's *Cosmos* to the effect that mean values reveal "the permanent in the change and in the flight of phenomena." He followed it with an ostensibly barbed epigraph by Thomas Carlyle, who called statistics "a science which ought to be honourable" if "not conducted by steam" (that is, mindlessly and mechanically), but with "a head that already understands and knows."[21] Thurnam aspired to discernment of this kind, declaring in answer to Carlyle's innuendo that the obstacles to good statistics could be overcome. "Even when the facts to which numbers are applied are in themselves of a more doubtful character,—under which head, the causes, forms, and duration of the disorder, may be mentioned,—and when the resulting statements are rather to be received as the approximative expression of the results obtained by different observers, they seem to me to be far from destitute of interest and value." Indeed, the "only means of forming a correct judgment as to the comparative success of any class of institutions . . . is to be found in numerical comparison of the results they have afforded." The preacher and political economist Thomas Chalmers thought otherwise, arguing that a focus on a single parish or household might provide "deeper insight" than could be had from observations "which must be superficial in proportion to their extension." Thurnam, while not disagreeing, countered that proper standardization and basic mathematics could redeem the method of large numbers.[22]

In the first version of his *Statistics of the Retreat*, published three years after his appointment as medical superintendent, in 1841, Thurnam laid out seventeen tables as the potential basis for a uniform system. The list began with admissions, discharges, and outcomes, followed by characteristics of the patients: rank and profession, marital state, age, duration of disease upon admission, and whether Quaker or non-Quaker, urban or rural. Other tables summarized medical experience, providing a breakdown of the patient population and results of treatment.[23] This mixing of institutional and medical-scientific purposes was typical of the nineteenth-century human sciences.[24] Yet even the administrative data was crafted to serve scientific ends. Thurnam invoked the statistician Adolphe Quetelet, for example, as a prophet of mathematical uniformity. "We cannot, indeed, but augur favourably to the interests both of science and humanity," when asylum superintendents join together to advance knowledge. With superabundant data, reliably and coherently presented, the statistical alienist can correct for discrepancies arising from institutional and territorial peculiarities.[25]

In June 1842, the (English) Association of Medical Officers of Hospitals for the Insane endorsed Thurnam's appeal for standardized numbers. A few months later, he opened a big admission book with spaces for a cornucopia of medical facts: age at first attack; number of previous attacks and duration of present attack (providing a choice among four duration classes); "apparent or alleged causes," including predisposing, hereditary, and exciting; form of mental disorder; "particular propensities and hallucination"; accompanying bodily disorder; and alterations of disease form prior to discharge. Standardization cleared the way for printed registration sheets, "available at cost price" from Dr. Samuel Hitch of the Gloucester Asylum. Thurnam bound up a thick sheaf of them but was able to fill only fifteen pages before he moved on to a new position at the Wiltshire County Asylum. His successor had other ideas about recordkeeping, showing that asylum statistics could not be standardized in a day.[26] Asylum doctors in Britain and elsewhere kept at it for the rest of the century.

Massachusetts Case Books and Sites of Information

In 1833, on the opening of the Worcester Lunatic Asylum, the physician began recording patient information in a big admission book. The size of the pages, about 16 by 12 inches, is in the normal range, although the horizontal lines were unusually narrow, more than forty per page, leaving almost no space for inserted remarks. This first volume includes about 6,400 patients, each written all the way across facing pages, verso and recto. Until 1882, the pages were lined by hand, after which they appear with the imprint of

Stanford & Company, a Worcester stationer. In England, such volumes had become available off the shelf from medical suppliers by about 1850. The persistence of disunited states as the loci of asylum law and regulation may have held up standardization in the vast American market.

The opening of the Worcester hospital was a landmark for American asylum statistics. The trustees and the superintendent, Samuel B. Woodward, understood their mission clearly in terms of conditions of health in the state and beyond. They believed in science as the correct frame for public information. Proper tabular forms were central to their drive to cultivate knowledge of causes and treatment, prevention as well as cure, of insanity. For decades, the annual reports included a numbered patient list that provided basic data on each, updated (usually) from the admission register. These data were processed each year into abundant tables, and sometimes into cumulative numbers. On the basis of this inherently statistical admission book, Woodward set about building a data reserve. He and his successors even included every year a complete daily record of weather conditions, which might correlate somehow with disease experience. A later superintendent, Merrick Bemis, expressed grave doubts about the assignment of causes of insanity yet retained these entries in the register out of respect for three decades of accumulated records. Who could say what data might someday prove useful?[27]

The Worcester books had printed on each left page "Admission" and on the right, symmetrically if strangely, "Dismission." The columns defined spaces to record the usual personal information; basic data on the disease; "Supposed Cause," whether hereditary, periodical, suicidal, or homicidal; and entries to provide legal authority for admitting and discharging the patient. Relations of medicine to law were never merely routine. The Worcester directors complained that judges and other officials made admission decisions without regard for the capacity of patients to benefit from treatment. The physician, however, as keeper of the books, had the authority to specify "in what state" the patient was discharged. He also determined, in effect, whether the institution had been given a fair chance with these patients by silently excluding from the statistics those who had been too sick upon entry to be treated or who were removed by a judge before the hospital had time to achieve a cure. This power of inscription was vital in the campaign for good cure rates.[28] "Cause" and "heredity," however, depended on observations made outside the hospital. The physicians might, in principle, pose critical or probing questions to their informants, but they seem to have written down what they were told. Why else were they so skeptical of their own data on causes?

At Worcester, as at most institutions, the annual tables consisted mainly of vertical sequences of numbers corresponding to categories in the admission book. The table giving the distribution of ages at outbreak in five-year intervals, for example, was put together mechanically by grouping recorded figures from a single column in the admission book. Woodward, an energetic statistician, expanded some of his tables into a second dimension, primarily in search of a relationship between cause, disease form, or delayed admission, and recovery rate. Black had already done this in 1789, but it remained uncommon until the 1860s.

Woodward's great expectations for the public asylum of Massachusetts extended also to the recordkeeping. The case books are about 8 by 14 inches and 400 pages in length and begin with an index listing patients by the first letter of their last name. Yet they cannot have functioned well as repositories of usable information. How many pages should be set aside for each arriving patient? Woodward at first allotted ten, allowing space for an entry every day. But usually there was nothing new, and he soon settled into a different pattern, with a cluster of entries during the patient's first weeks, gradually subsiding to a single entry at the end of each month, with a burst of activity as the patient approached release, was injured, or began to succumb to an infection. Even one brief entry per month reflects an unusually high level of recordkeeping in the asylum of his day. Since paper was expensive and not to be wasted, cases that exceeded their assigned space were extended backward to empty pages from the months of profligacy and then forward into new or unfilled pages in a later volume. Once the standard allotment per patient had dropped to two pages, there was less need to fill empty spaces, but the record of a long-term patient was still unmanageable, sometimes spilling over into five or more volumes. Although these leaps were cross-referenced, it would have been immensely cumbersome to inspect the full record of a patient of long duration. Later, as legal requirements for patient records grew stricter, it became common for staff to make these entries in concentrated bursts, perhaps weeks after the physician's visit.[29]

Worcester's first patients, most of them transferred from jails and poorhouses, often arrived with no record of the background to their illness. Soon, however, the data began pouring in. Patient 131 was brought in by the Society of Friends, which knew of "two sisters and one brother insane beside cousins." Woodward entered the case as hereditary. Patient 205 is labeled as hereditary insanity, but with no indication of who said so or on what evidence. Medical descriptions can be chilling. The first patient in the second case book was admitted "after an acute attack of febrile disease affecting the side and heart, for which he was thrice bled by his physician, purged freely, and blistered in addition to this he was put upon low diet and small but nauseating doses of antimony." Somehow the treatment failed to ward off his incipient insanity. In another case marked as hereditary, we find that alcohol was consistently involved in the episodes of derangement: "His father once in a similar situation has been a temperate drinker of Ardent spirits." But no sources are mentioned.[30]

In many institutions, data recording was treated as a meaningless routine. But alienists who did not suppress their curiosity, who labored to establish the causes and treatment of insanity as a medical science, had reason to extend the initial moment of openness to information from outside. They might want to identify sources of insanity in the population at large or to determine if their cures were lasting. Woodward, like Thurnam, was such a man and understood his institution as a node in a larger network of public health. Detailed case histories, which are rare, supply resources for a closer examination of data practices.

Ezra's Tale

A hundred pages into Worcester Case Book 7, the familiar flow of repetitious tidbits is interrupted by a detailed narrative, the case history of Ezra, taken in 1837. "Gave me a History of his insanity to day which as far as I can recollect is as follows," it begins. Woodward had had to write the whole story from memory because Ezra had become alarmed when he began to record some notes, declaring: "You are not going to write it, you must not write it, and the table would not contain it." The narrative, which runs on for nineteen breathless pages, begins by inventorying relatives. "His ancestors and family connections were insane. His grandfather his father one uncle one aunt and two sisters were insane." This line of inquiry must have been on the initiative of Woodward, who was keenly interested in patient heredity.[31]

Ezra was born in 1784. "His parents were poor but by industry and frugality obtained a comfortable living, when he was nine years old in 1793 he had a short turn of insanity it occurred immediately after a season of religious excitement in the town of Houghton Massachusetts, almost every attack followed a close, particular, and distressing attention to the subject of Religion." The case was making sense. Woodward shared the common view that insanity required both some kind of susceptibility and a triggering cause. Immoderate religion assumed here the latter role, and heredity the former. "This first attack was of short duration. He had slight attacks afterwards up to the year 1810 when he had a paroxysm while at work for Col. Dudley in Roxbury."[32]

Dudley and his men were builders. On this fateful day, during a break, Ezra was "attacked with an unaccountable feeling of hopelessness and distressing anxiety, he had attended religious meetings for some time and he had felt deeply interested in the subject. This was a season of revival in the town." After a spell at home with his father, he recovered sufficiently to come back to work and to keep at it for eight more years. In 1819, disordered religion again intruded in the form of a revival. Ezra left his work, "spending nights in the woods and was very furious and highly excited and was often confused, bound, and treated harshly." The tabular record silently picked out this episode as the beginning of his insanity. By 1822 he was working as an independent craftsman, despite episodes of disabling lunacy. His story proceeds with a striking account of precise craftsmanship in the face of acute madness, culminating when the workmen gather to raise his house. Everything fit perfectly. In Woodward's words: "Thus this man while in a paroxysm of insanity which rendered his confinement necessary repeatedly, which induced him in the periods of its impulses to run away into the wood and hide himself almost daily and often more frequently, commenced to saw and hew timber in its natural state framed and erected a house and furnished it completely except to raise it all the carpentry and even the lathing to a great extent."

Day 2 of the interview skips quickly to 1829, by which time Ezra is married with children. His wife recognizes his need sometimes to run away and seclude himself, which seems to reflect a partly rational attempt to protect the family during moments of acute mental disturbance. In one of these episodes, he sees angels and a star, followed by a beautiful golden wheel that stayed with him for three months then disappeared. Woodward no doubt sensed the dangers of exaggerated religion, but for Ezra it was "the most splendid vision that was ever presented to human view he looks at it with amazement and delight a few moments and it vanished away." Now Ezra could return home to his loving wife, so admirably patient and forgiving. According to a contemporary news account, his neighbors thought he should be committed to the house of correction as a lunatic, but she objected "and thus fell victim to her well-meant but mistaken views of humanity." Upon his return home, Ezra remarked, she had received him sweetly. He cannot explain, he adds, why he picked up a heavy object, or why he brought it down on her head from behind, killing her. He had never contemplated such an act, but merely gave in to an impulse "which compelled him to the deed."[33]

The murderer was sent to prison for a time. But his madness was evident, and he was among the first moved to the Worcester hospital upon its completion. Unfortunately for the statistics, the

43	"	14	27	"	Married	Loss of Property . .	1 month
44	"	16	45	"	Single	Disappointed Affection .	16 years
45	"	16	49	"	Widower	Unknown	14 years
46	"	19	46	"	Single	Intemperance . . .	15 years
47	"	27	42	"	"	"	8 years

34	March	6	38	do	Single	do	4 years
44	do	16	49	do	do	do . . .	16 years
45	do	16	53	do	Widower	Religious Anxiety . .	14 years
56	do	28	35	do	Single	Hard Study . . .	6 years
64	April	3	46	Female	do	Indulgence of Temper	13 years

FIGURE 2.2. Selection from the printed register of current patients printed each year in the annual report for Worcester Asylum. The column labels are patient number (Ezra is patient 45), age (49), marital state (widower), cause, how long insane (14 years), and so on. Ezra was admitted on the same day as patient 44, also a murderer. Woodward edited these entries on the basis of new information or changes in the patient's condition. Ezra's cause was given as unknown in the first report then revised to religious anxiety. His illness was entered as hereditary from the start. From "First Annual Report of the Trustees of the State Lunatic Hospital, 1833," in *Reports and other Documents*, 46; and *Fifth Annual Report of the Trustees of the State Lunatic Hospital at Worcester, 1837* (Boston: Dutton and Wentworth, 1838), 22.

deranged murderer could not be discharged as cured, yet he appeared to Woodward as a notable success. He was one of twelve patients featured in the 1835 report.

> No. 12.—A patient six years confined for homicide in close jail, and would probably have been confined for life. After six months here he commenced labor, and has not only continued it daily, but takes excellent care of every thing connected with the farming and gardening establishments. He is pleasant, very mild in his feelings, and ready to perform whatever is required of him. He is trustworthy, and can perform labor without superintendence. We have frequently noticed the novel spectacle of two men ploughing in the field alone, *both insane, both having committed homicide*, and both having been confined in jail for a very long time.[34]

The tabular records show that Ezra was admitted on the same day as another murderer during the asylum's third month of operation,

"	8 months	"	"	"	
"	8 1-2 mths	Remains	Stationary	Incurable	"
"	8 1-2 mths	"	Much impr.		Her. & Period.
"	5 months	Died	Stationary	"	" "
"	8 1-2 mths	Remains	"	"	

do	56 months	do	do	Labors some.
do	56 months	do	do	Hereditary. Homicidal.
do	56 months	do	do	do do
do	56 months	do	Not Improved	Labors well.
do	55 months	do	do	Labors some.

on 16 March 1833. In the first annual report, Ezra was listed as a widower, age forty-nine, insane for fourteen years prior to admission, his disease "hereditary and periodical," and his condition "much improved" after eight and a half months there. In 1835, his condition was assigned a cause, "religious anxiety," and he was for the first time identified as homicidal. The records vacillate on whether he was improving.[35]

In 1837, in the wake of his extended interview, Woodward again featured Ezra in the annual report, discussing him in what amounted to a medical paper on legal responsibility. His homicide exemplified what Woodward proposed to call the "insane impulse . . . , an uncontrollable propensity, as transitory as it is sudden, by which an act is committed without one moment's reflection or premeditation." Ezra and nine fellow inmates had succumbed to it, and none should be legally culpable. The elaboration of this concept, involving public issues of law and mental illness, had inspired Woodward to interview Ezra and fill out the details of his biography. A few years later, in 1844, the Massachusetts courts introduced a doctrine of "irresistible impulse" as grounds for an insanity defense. Woodward, with Ezra's assistance, thus played a part in demonstrating how research on asylum patients could matter for the law. Ezra's highly distinctive case also shows how patient data was filled out or revised in response to new information. Woodward's work on heredity, too, went well beyond passive recording.[36]

London Insanity in Data and Narrative

Although the Bethlem physicians, like so many others, presented numbers as if they had fallen from heaven, their sources often survive in archives. The 1815 parliamentary hearings on madhouses, humiliating for this hospital, coincided with the move to a new facility and with it the introduction of more detailed case books. By 1822, Bethlem had introduced printed data forms with questions about disease history, "lucid intervals," "causes and previous appearances," and "whether hereditary," as well as the patient's habits and education, "diseased ideas," and prior episodes of confinement. Preserved, now and again, between the pages of case books, these forms enable the tracing of data sources.

In June 1822, a man arrived with a doctor's letter addressed "To the Physician, Bethlem Hospital" explaining some problems of bodily health and describing an attempt to cut his own throat with a razor. His "friends," it continues, "are inclined to attribute the attack to a disappointment in a love affair" as well as to business anxiety, but perhaps "habits of drunkenness in which he indulged will account better for it." Edward Thomas Monro, the fourth in that dynasty of physicians at Bethlem, copied some of this onto the printed form and added what was not in the letter: "Sister died insane." Evidently he was supplementing the medical reports with information from patient records. From time to time, there are references also to interviews with a relative of a patient. The institution had a subcommittee for admitting patients, to which "friends" such as parents, siblings, or spouses presented petitions accompanied by medical certificates. This would have provided an occasion for passing along information about causes and patient history. By 1844, there was a new, more detailed form, which later began to be printed on pages of the case book, combining the required legal documentation with information on whether the patient was dangerous or disorderly as well as on bodily health, symptoms, history of the disease, causes, and heredity.[37]

The Hanwell Asylum, established in 1831 on the outskirts of London, was built for pauper patients in Middlesex County and soon became one of the largest mental hospitals in the world. It is famous especially for John Conolly's introduction in 1839 of the principle of non-restraint. Beginning in 1844, his successor turned the case book for males into a treasure house of patient histories. Here, again, the histories provide a sense of the resources at hand for gathering data as well as constructing narratives. These case notes always begin with admission. A few days later, according to the usual formula, the appearance of a relative is noted. The case book typically identifies this person by relationship, often mother, daughter, or wife, and reports what she said, beginning with basic data such as age and profession and followed by information on family and life experiences.

> Relations sound and healthy [says his sister]—the foregoing statement is corroborated by the patient's wife.
>
> . . . his father died insane, one brother also died insane and another brother is becoming insane.
>
> Narrator says that the patient's grandfather was insane and died in that state.[38]

An especially affecting case record involving a boy of thirteen was told by his mother. Her husband had abandoned her with five children and no resources to care for them. At the age of five, Philip was taught by neighborhood boys to stand on his head. Soon this oft-inverted child was performing for the diversion of the public, receiving, she explained, a few pence for his trouble. Perhaps it was no fitting use for a young boy's head, for by age seven he was suffering fits. He began knocking on doors after his performances to solicit money and throwing stones at windows when he was refused. For such infractions, he was brought to prisons and workhouses, and he became superbly adept at escaping over walls and through openings. Sent, in hope of better results, to school, he "could not be taught anything, he used to swear at the teachers and throw missiles at them." At Hanwell his repeated escape attempts got him into a high-security ward, but he then won the confidence of an attendant with better behavior and a new interest in learning. The story ends with Philip's terrible death following an ingeniously deceptive nighttime escape attempt through a window opening that proved too small for his head. (The report gives precise measurements.)[39] The physicians digested stories like these into data on the causes and circumstances of mental illness among their patients.

Such patient narratives were extremely uncommon at the time. The accounts by the physician of the women's ward during this period are far less elaborate, though the medical data is more complete. It appears that health officers at the parish of origin compiled patient records prior to admission. A woman who entered in October 1846, Sophia, is described as having "had no Family" and as "living in a state of concubinage, for several years." The report describes bruises "indicating prior mistreatment or mechanical restraints" and gives details of her poor state of health at the time of admission. "The cause of the present attack is ascribed to Jealousy." Sophia attempted suicide when her lover announced that he would not marry her, and her loss of blood on this occasion added a corporeal dimension to the moral cause of her insanity. "She was bodily ill afterwards, but exhibited no Symptoms of Insanity until six weeks ago (unless indeed the act itself may be assumed, as it probably may, an act of madness) when she attempted to throw herself from her bedroom Window."[40]

Another report describes bruises on poor Susan, admitted the same year. Her assigned cause included pregnancy, birth, and lactation. "The disease is hereditary; her brother has been the subject of Insanity, and an Inmate of this Asylum." The admission records chronicle suicides as well as the insanity of family members, generally in a way that does not depend on an interview. But the case of Mary Ann, the mother of four children who lay down one night after dinner then awoke proclaiming "that Jesus Christ was looking at her thru' the window," seems to require the testimony of a family member. Indeed, one is soon mentioned: "No member of her family has been insane, and there is no cause known to her Husband to which the Malady can be ascribed."[41]

Asylum Statistics and the Economy of Information

In Germany, even before 1850, most asylum patients were admitted with a report from the district health officer, who usually filled out a form to certify insanity and to provide basic information on the patient's illness, history, present condition, and descent.[42] An undated pamphlet with rules issued for the Erlangen asylum listed the documents required for admission, including a proof of residence, a guarantee of payment, and a medical case history certifying mental illness. It then specified items of information to be included in this history, from name, age, religious confession, estate, and marital status of patient and parents to disease histories for the whole family, a presumed cause, and the family predisposition to disease.[43] These reports, the doctors understood, might not be reliable. An early asylum director in the southwestern German state of Württemberg complained of lay observers who confused the etiology of insanity with its initial appearance, so that the entry for cause might be nothing more than the first noticeable symptom.[44] William Hutcheson made the same point about his institution in Glasgow. When not innocently uninformed, relatives were often deceitful, substituting, from "false delicacy," some innocent cause for a true but embarrassing one. "[I]n a majority of cases, any hereditary predisposition is carefully concealed."[45]

Case narratives, which can be captivating, and data sheets, which are dull, provide indispensable points of access to the economy of information in the nineteenth-century asylum. The alienists labored to create, from scenes they never witnessed, data that could be lifted from all context and combined with results from other institutions. Especially on the question of causes of insanity, there was often no alternative, though a sufficiently energetic alienist might hope to get a sense of insane heredity by tracking down relatives.

CHAPTER 3

New Tools of Tabulation Point to Heredity as the Real Cause, 1840–1855

> No question is so frequently put to the physician of a Lunatic Asylum as that which calls upon him to state the most productive cause of insanity. Those who are most familiar with insanity, find themselves the least able to reply to this question.
>
> —Joseph Workman (1860)

The mission of insane asylums, to check and reverse the growth of insanity, was not limited to treatment. Right from the start, asylum superintendents took on the charge to block, somehow, the springs of insanity. Their tables of causes, based mainly on lay attributions, were featured in annual reports, but many of the causes assigned, especially "moral" ones such as disappointments in love or business, were never convincing to asylum doctors. They focused particularly on drink, masturbation, and heredity, with heredity gradually gaining the advantage over its rivals. Almost no one doubted the reality of hereditary transmission, for which the evidence came down to observations of similarly afflicted relatives.

Heredity had the feel of a deep cause and yet seemed to be supported by ordinary experience. It was simultaneously scientific and popular, and it worked well with statistics. The data of heredity provided a basis for international collaboration among alienists. The principal advantage in the investigation of causes was their access to records of their own institutions and to the data and conclusions of far-flung colleagues. Hence, the push for science and for the medical-administrative rationalization of asylums was never merely inward-looking. French, German, American, and British alienists all founded organizations or journals in the 1840s. They were not just willing but eager to share their data and findings.

With the support of ministries and boards of visitors, they cultivated habits of travel to keep up with the latest remedies, architecture, science, and techniques for keeping order. They emphasized statistics as their prime tool for assessing causes of insanity as well as results of asylum care. By 1840, they were beginning to discuss ways to improve hereditary data by tracking down the relatives of patients. While many alienists hesitated to merge the numbers of different institutions, they were keenly interested in enhancing their techniques for gathering and ordering data.

Tracking Heredity

The case books at the York Retreat document the origins of systematic hereditary investigation as something more than recording what informants told them. The first of these, extending from 1796 to 1818, is a small, unprepossessing volume, about 8 by 6 inches, meager as a storehouse of information. The records of these patients, though sometimes extending over decades, rarely fill as much as a single page. An 1802 watermark indicates that the first pages were filled in retrospectively. Many of the earliest patients had already been disturbed for years, and their prospects were bleak. Patient 1 returned home for a time but died finally in confinement. Patient 2, described as kindly and imaginative and noted for her vivid descriptions of "figures in the clouds & fire" and of strange remedies (these perhaps did not have to be imagined), lived out her life there. The third patient improved at first but soon turned melancholy and ended his stay as the Retreat's first suicide, grounds for the dismissal of a neglectful attendant.[1]

Case 9 is notable for the first appearance of spidery, slanted annotations in black ink. The patient had been marked "Hereditary" in 1813, but there is a later insertion: "& has had a son deranged since, see 599." Patient 599 was admitted about two decades later. For case 18, the spidery hand introduced "Hereditary?" as a hypothesis, adding "Cousin to Tho's and Benj'n Atkinson." Patient 19 was retrospectively cross-referenced to page 220 of Tuke's *Description of the Retreat*, published sixteen years after her admission.[2]

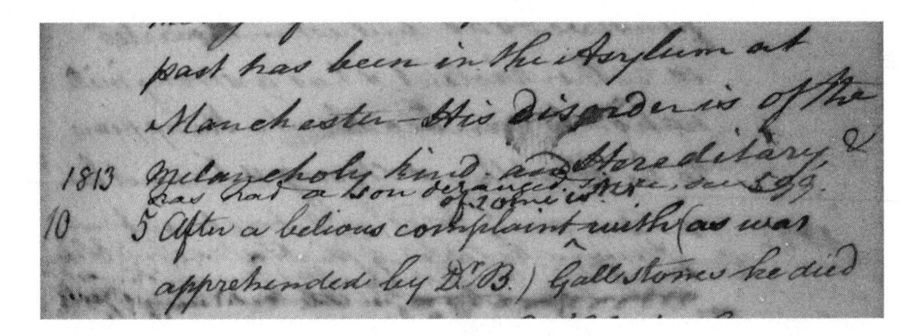
past has been in the Asylum at
Manchester — His disorder is of the
1813 Melancholy kind and Hereditary &
has had a son deranged, see 599.
10 5 After a bilious complaint with (as was
apprehended by Dr B.) Gallstones he died

FIGURE 3.1. Retreat at York, Case Book, Patient 9. About 1840, John Thurnam inserted, for a case already identified as hereditary: "& has had a son deranged, see 599." Such insertions demonstrate his active interest in patient heredity. Reprinted from an original in the Borthwick Institute, University of York, RET 6/5/1/1A Case Books.

These inscriptions haunt the case books. They speculate about hereditary origins, supply names of disturbed relatives, track the subsequent career of discharged patients, and record results of postmortem dissections. They tell of "strong hereditary decay" in patient 28, whose father was "very eccentric and he is nearly related to the Waring & Moxham families," the maternal cousins of James Moxham. Patient 83 "had a sister Martha Miller No. 454 here & it is believed the Mother destroyed herself. She was deranged and another sister, single, also manifested symptoms."[3]

Patient 236, after discharge, "remained, to say the least, very singular, carried a basket of wares & became a sort of hawker about the country. She is believed, 1840, to be still living." In the second case book, the record for a patient admitted in 1833 has inscribed in bold letters, then thin ones: "**Retrospective History**. Some hereditary tendency to insanity appears to exist especially on the Maternal (Wilson) side," followed by remarks on the patient's childhood, religious experiences, strange beliefs, and unconventional experiments.[4]

The hand that recorded these comments is easily surmised. John Thurnam was appointed medical superintendent of the Retreat in 1838. His scheme of standardized records, as we have seen, put special emphasis on tables of presumed causes of insanity, and his statistics point to an overwhelming concern with heredity. They show 70 patients with heredity as the only predisposing cause, 72 with

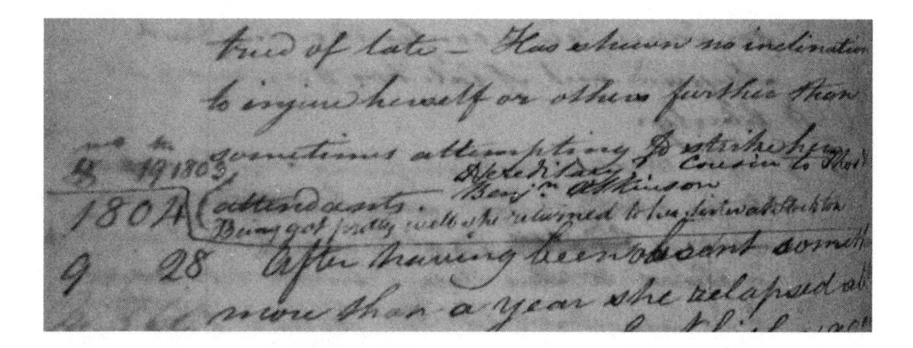

FIGURE 3.2. Retreat at York, Case Book, Patient 18. John Thurnam's insertion: "Hereditary? Cousin to Tho's and Benj'n A, . . . Being got pretty well, she returned to her Sister at Brockton." Reprinted from an original in the Borthwick Institute, University of York, RET 6/5/1/1A Case Books.

heredity in combination with some other cause, and 71 more involving collateral relatives rather than direct ancestors, constituting, in total, 51% of his patients. He insisted that even this number, based on incomplete knowledge, was by no means the limit. For practical as well as scientific reasons, he aspired to investigate still further. He was, by his own account, tireless in tracking down families, never accepting the mere declaration of a relative, always pursuing "more private information" and listening for what was well known in the community. The predisposing causes, he explained in a long introduction to his tables, "have attracted less attention than the exciting, and from their frequently insidious character, are often altogether overlooked" yet "they are frequently the most important" and "they are also, in many instances, more easily guarded against." Hereditary causation, in short, left room for effective intervention.[5]

Thurnam's investigations of insane heredity commenced two decades prior to Darwin's *Origin of Species* and a quarter century before Francis Galton became curious about hereditary genius. Although his family inquiries and his organization of data appear original, his hopes of checking the reproduction of human defects were already becoming familiar. George Chandler, who succeeded Woodward at the Worcester asylum, wrote in his annual report for 1846: "The prevention of insanity should be the aim of an enlight-

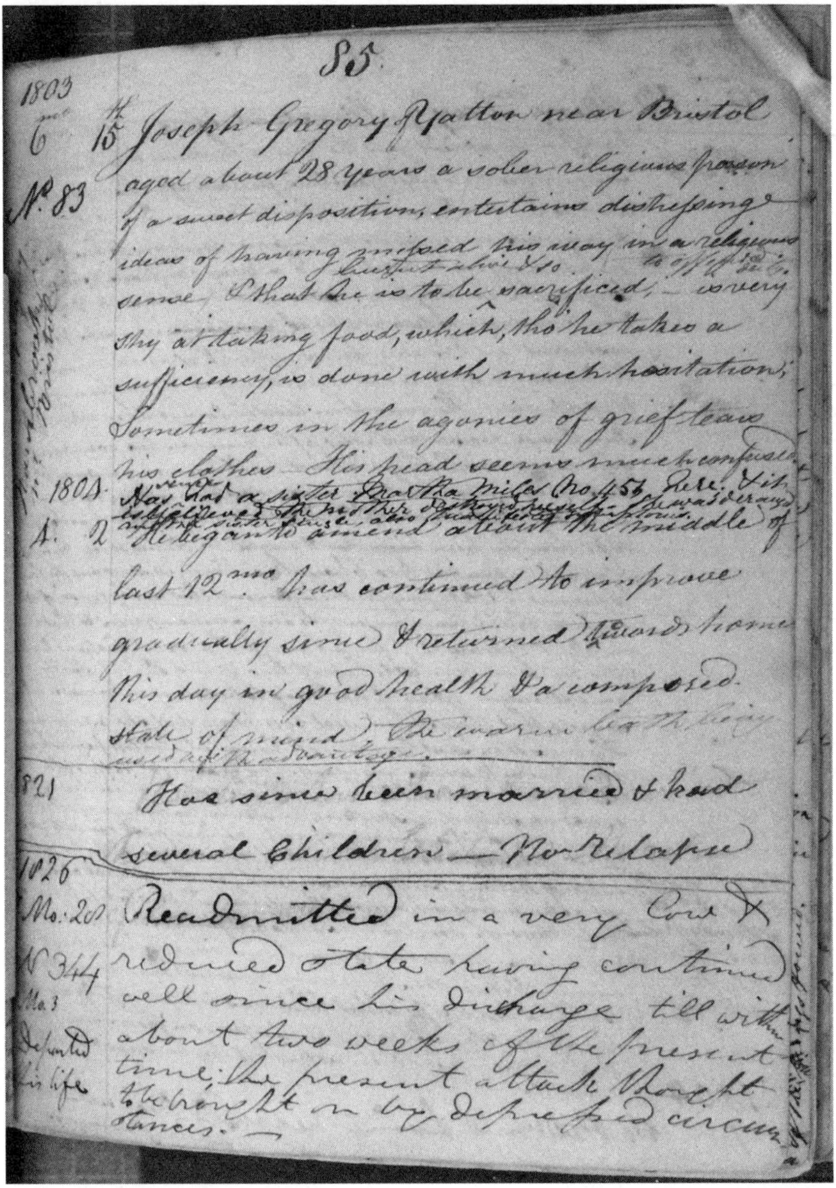

FIGURE 3.3. Retreat at York, Case Book, Patient 83, John Thurnam's insertion: "Has had a sister Martha Miller No. 454 here & it is believed the Mother destroyed herself. She was deranged and another sister, single, also manifested symptoms." Reprinted from an original in the Borthwick Institute, University of York, RET 6/5/1/1A Case Books.

FIGURE 3.4. Photograph of John Thurnam (1810–1873). Thurnam was among the very first medical superintendents of an asylum to investigate the behaviors and mental conditions of relatives of his patients. He was also an energetic statistician and advocate for uniform categories. Reprinted from an original in the Borthwick Institute, University of York, RET 1/8/7/12.

ened community as well as its cure. This means obeying the laws of health. . . . Hereditary predisposition to disease, which is either inherited from ancestors or acquired by the parents themselves, by abuse of their own physical systems, is transmitted to the lineal descendants."[6] In 1835, James Cowles Prichard, physician-ethnologist and outspoken opponent of the slave trade, summed up the import of his humane and learned book on insanity with this thought:

> In adverting to the inquiry, whether any means could be adopted that would tend to diminish the extent of this evil, we are struck by the obvious consideration that the numbers of deranged persons in the community might be very much lessened if it were possible to regulate or establish any surveillance over the marriages of the lower orders, or if some measures could be adopted to prevent the propagation of idiotism and an hereditary tendency to madness. Idiots who are at large wander about the country, and the females often bear children.[7]

Thurnam's originality was the fruit of his determination to track down the relatives of his patients and his skill in organizing the resulting data. He recognized the insane asylum as an ideal site for this kind of work. His advantage over others with similar ambitions owed much to the special population he treated. The Retreat, a Quaker institution, mainly served members of this close-knit religious community, which was his community too. He was thus able to trace family members across generations, whether they had been institutionalized or not, and to check out vague or uncertain claims. The concentration of mental illness in his institution enabled the families to be treated as model organisms for hereditary research.

While Thurnam carried out this research within the asylum he directed, his work coincided in time with the formation of national and international communities of asylum physicians engaged in research on patient heredity. If their backgrounds were mainly medical, their tools and methods were social and statistical. He provided an important model for their work and a basis for comparing and combining results.

Traveling Knowledge

"An interval of professional leisure, during the last summer, enabled me to gratify a long cherished wish of seeing a little of the Old World, and especially its institutions for the care of the Insane," wrote Isaac Ray in 1846. He had just been hired at the Butler Hospital for the Insane in Providence, Rhode Island.[8] Government ministries and boards of overseers authorized these alienists' holidays to familiarize new superintendents with the latest techniques of asylum management. Psychological medicine, if not quite a research specialty, formed a distinct branch of the medical profession. Pliny Earle, after taking his medical degree from the University of Pennsylvania, continued his studies in Paris, where he was won over to Pinel's moral treatment. Upon his appointment as superintendent of the Quaker asylum in Frankford, Pennsylvania, he set off on a European tour featuring some of its most notable institutions. While Bethlem, still wary of spectators, would not let him through the gates without an appointment, most welcomed him. In England, he visited asylums at Hanwell, York, the West Riding of Yorkshire, and the York Retreat, which he particularly admired. He proceeded to Amsterdam, Utrecht, and Antwerp, the great Paris hospitals, Milan, Venice, Malta, and Constantinople. His travel narrative bore, as subtitle, *An Essay on the Causes, Duration, Termination, and Moral Treatment of Insanity. With Copious Statistics*. Published in 1841, it included a mix of visual impressions, recollections of conversations, excerpts from other eyewitness accounts, and fragmentary statistics. He added a descriptive inventory of American institutions as well as Old World ones that he could not visit, extending as far as Cairo. In 1853, Earle again described an asylum tour, this one through seventeen German and Austrian institutions which, apart from Siegburg, near Cologne, were little known in America. He was impressed by German efforts to make asylum medicine a special field of study. He also commented on local censuses of the insane organized by institutions in Silesia and in Oldenburg.[9]

Medical tours, especially to France, were common for antebellum American physicians. Alienist visits proceeded in every direction. European alienists traveled often to neighboring countries, and sometimes to America, whose asylums, like its prisons, gained a degree of international renown. These New World institutions were adept at self-promotion. Behind the classical columns and pediments of a US state asylum building, however, the visitor might be confronted with dark stinking rooms and noisy tumult.

Everywhere, in this business, appearances could be deceiving. On a beautiful September day in 1869, the lead editor of the *Allgemeine Zeitschrift für Psychiatrie* (*AZP*), Heinrich Laehr, floated serenely on a gondola to the Venetian hospital of San Servolo. The island setting and Mediterranean climate would appear to his countrymen as an impossible dream, he remarked. Inside, it was a depressing place. The patients arrived despondent and fearful, most of them suffering from pellagra, whose treatment they had put off until it endangered their lives. The archives there confirm Laehr's description. The remedy for pellagra was simple: a better diet. Patients who did not die improved rapidly. And then, the doctors complained, they insisted on leaving before the cure was complete.[10]

Even in El Dorado, the alienist's paradise was elusive. In 1870, when California got serious about containing mental illness, it sent the newly appointed commissioner to visit forty-five American asylums and ninety more in Canada, Bavaria, Austria, the German states, Switzerland, France, Belgium, Holland, England, Scotland, and Ireland, in quest of medical and architectural instruction. His state would need "all the light that the wisdom and experience of the learned men in other States and countries could shed." His own statistics showed a disturbingly high rate of lunacy, already a California stereotype. In the 1880s, an Australian achieved the ne plus ultra of asylum tours, filling a 1,567-page volume with his observations "from all parts of the world" under the fitting title *Lunacy in Many Lands*.[11]

FIGURE 3.5. Artistic rendering of the Ohio Lunatic Asylum in Columbus, printed as frontispiece to the annual report from its opening in 1839. Most American states constructed new buildings, which could be quite grand, for their mental patients. The interior was typically less lovely than the façade. Courtesy US National Library of Medicine.

Printed Reports and Circulation of Data

The circulation of documents was equally central to professional and scientific exchange. Annual and multiyear reports provided information on practical topics such as building and garden layout, facilities, schedules, and work routines. They also supplied patient statistics in a familiar form, providing material for waves of excerpts, digests, and comparative tables. National journals of asylum medicine, three of which sprung up almost simultaneously in 1843 and 1844, watched each other closely. The networks of outreach and exchange can be traced through their contents, which reveal an intense, shared concern with the data of insanity and its causes. The *Annales médico-psychologiques* (*AMP*) called attention to patient heredity in Jules Baillarger's editorial introduction and made frequent reference to statistics, both at home and abroad.[12] The 1844 volumes include a news clip on the tripling of the burden of insanity in England in just twenty years and reviews of the statistics of Bethlem.[13] Having taken notice of the new *AZP* in 1844, the French journal printed an extensive summary of its first volume in 1845, with particular attention to asylums in the German states of Baden and Württemberg, across the Rhine from French Alsace. An alienist at La Salpêtrière contributed a sixty-page narrative of his visit to the well-known asylum of Illenau in Baden.[14] The first volume of the *American Journal of Insanity* (*AJI*) celebrated the launching of the French *AMP* and, a few months later, of the German *AZP*. It reported on hereditary investigations from Paris and statistics from the asylum of Saint-Yon in Rouen.[15] These are only a few examples, most from a single year, of news exchanges that went on for decades.

The Rouen institution was directed by Maximien Parchappe, whose devotion to asylum statistics in France may be compared to Thurnam's in England. In an 1839 study of the causes of insanity, Parchappe brought together reports from the United States, Italy, and Belgium, as well as France, for the sake of more inclusive numbers. He was already looking forward to a golden age of asylum statistics in consequence of all the new institutions required by the 1838 French asylum law.[16] In 1844, he published a data-filled annual report on Saint-Yon, followed the next year by a coauthored one with his colleague Lucien Deboutteville extending from 1825 to 1843. Thurnam, in a review, commended its cosmopolitan spirit of statistical emulation as well as its copious tables, which he described as a model for the multiyear report, demonstrating how the statistics of insanity could be raised above mere routines of recordkeeping.[17] A comment like this one seems to cast doubt on the value of annual reports, but Thurnam took the opposite view, applauding their recent advances both in number and quality, at home and abroad. The value of these international exchanges, he continued, would be enhanced by greater uniformity, especially of statistics of causes.[18]

On this issue of causes, Thurnam extended his praise to a recent Danish author, Peter Jessen, whose report on the territory of Schleswig made a compelling case for uniform classification. Jessen found little value in the data on moral causes supplied by impressionable family members. Hereditary causation, by contrast, could be reliably documented, since the evidence of sick family members was medically observable.[19] Against the background of whimsical, wavering moral causes, hereditary factors stood as a beacon of hope. Alienists agreed on the reality of hereditary causation, which was underreported, they argued, not on account of an inability of family members to recognize it, but because they had lost contact with their kin or were ashamed to admit an inherited family taint.

Doctors Know Best

Sorting out the true causes of insanity was a great international challenge. Apart from heredity, masturbation was the cause that families seemed most concerned to cover up. Perhaps the recognition of its role might explain away some of the more transitory and less credible moral causes, such as political events and religious excess. Often, the causes they recorded in the admission book were not what they really believed. It was especially hard to escape the roller coaster of spiritual and behavioral explanations.

Amariah Brigham mentioned in 1844 in his second annual report for the state asylum in Utica, New York, that while some insanity had no doubt been occasioned by the French Revolution, the American Revolution, Luther's Reformation, and South Sea speculation, passing events like these must be of relatively minor importance. He was much more impressed by predisposing causes, above all heredity.[20] However, his tables, based on what the families told him, highlighted religious enthusiasm. For example, Millerism, promising an imminent millennium, showed up as a distinct cause in 1844. It surfaced also in Connecticut, whose tables caught the eye of Bénédict Augustin Morel, soon to become famous for his theory of hereditary degeneration. Morel had a reputation for his statistical tables and reviews of reports from other institutions. He commented in *AMP* on the unhealthy religious doctrines that charlatans inflict on gullible Americans.[21] "Spiritual rappings" burst onto the insanity charts in 1851, just months after their first exhibition in Rochester and New York, sending 7 men and 11 women to the Utica asylum. The next year these rappings reached Ohio, where Elijah Kendrick, as superintendent, registered 13 male and 13 female cases "caused by the present popular delusion, 'Spirit Rappings.'" It promptly began to subside: 5 men and 6 women in 1853, then 3 men and 4 women, 1 man and 2 women, until, in 1856, it slipped beneath the dark liquid of lost memory. Kendrick now argued that alleged causes are not always real ones and that the attribution to religion was merely a mask for masturbation.[22]

Alienists, as respectable people, hated the idea that true religion caused insanity. Could they find a better explanation? Ohio, according to the statistics, was ravaged by onanistic lunacy, more so even than Massachusetts, where Woodward warned of its deadly effects and where Samuel Gridley Howe, writing on idiocy, grouped it with intemperance and heredity as a sin of the progenitors.[23] Samuel Hanbury Smith, Kendrick's predecessor in Ohio and an admiring disciple of Woodward, had already discerned this true basis for what was blamed on religion. "His mind, already weakened by the enervating effects of his pernicious practice, is quite unable to withstand the perturbating influences it is there subjected to, gives way, and fits of maniacal excitement, alternate with periods of dreadful prostration and agonies of despair. . . . During the continuance of the excitement, his distempered fancy prompts the outpourings of rhapsodies of devotion and blasphemy intermingled."[24] Always lurking and rarely acknowledged, masturbation also affected the woman, compromising her essential duty to the health of her children. Such a threat called for strong medicine: "cauterization of the urethra, and blistering the prepuce with cantharidal collodion, accompanied with the steady use of large doses of camphor and lupulin at bed time, and of the tincture of muriate of iron, much diluted with camphor water, during the day."[25]

Brigham had complained in the first Utica report that too few authorities "give anything more than the *supposed* or *probable* causes."[26] Upon his death in 1849, a new director elevated the charge from doubt to deceit. Relatives of the insane substituted "religious anxiety" for the deadly effects of masturbation. Alienists should ignore unreliable lay witnesses and report causes "as they are developed from the history and progress of the disease."[27] The

Glasgow asylum in Scotland emphasized careful recordkeeping as an alternative to reliance on ignorant or dissimulating relatives. He took Thurnam as his model. "From the accuracy, however, with which our records are kept, we are often enabled to trace the hereditary predisposition, in cases where it has been pointedly denied." The causal force of masturbation, however, was more elusive and tended to recreate those mad novellas from which the alienists longed to break free.[28]

In gold-mad California, where there often were no relatives to be found, many insane were brought in by sheriffs or constables who knew nothing about them.[29] Heredity was practically a closed book in a state populated by lone immigrants. California in the early 1850s, with its unattached young men dreaming of riches, seemed almost uniquely vulnerable to the deadliest moral causes of insanity. "It is fearful to contemplate the amount of mental excitement, the violent passions, the ungoverned tempers and continued turmoil prevailing throughout the entire population of the State," wrote asylum physician Robert K. Reid. The good side of California—its political and religious liberty—as well as the bad—its unsettled communities—each tended to increase lunacy.[30] Reid expected it to trail off as civilization took root, but in 1867 the resident physician G. A. Shurtleff conceded that his state had become notorious for insanity. The profile of causes diverged from the rest of America by its low figure for heredity and high one for masturbation, number seven among causes in the eastern states, and number one in California.[31] By this time, Shurtleff was beginning to think that masturbation might be more effect than cause of madness. In the territory, then state, of Washington, it lingered on as a leading cause into the twentieth century.[32]

Anglo-German Interactions

Thurnam's book provided an international model for hereditary investigation and, more generally, for organized recordkeeping and statistical calculations. The Retreat had become famous in Germany thanks to Maximilian Jacobi, pioneering director at Siegburg in the Rhine Province of Prussia. The York physicians returned the favor with their praise of Jacobi's asylum for its subtle effectiveness in inducing even the wealthy to undertake curative labor. They also had a role in the 1841 translation of Jacobi's book on the architectural layout of asylums. Samuel Tuke introduced that volume with an essay on the disconcerting increase of insanity, which he hoped might finally be approaching its limit. The same year, a German asylum doctor making a tour of British institutions was so captivated by the Retreat that he devoted more than half of his book to a translation with commentary of Tuke's essay and Thurnam's newly published *Statistics of the Retreat.*[33]

A few years later, Thurnam's work was featured by the founders of the *AZP*. In his editorial introduction, the lead editor, Heinrich Damerow, outlined a vision of psychiatry as a collective project to take shape on the pages of his journal. Reliable, unified statistics was to be central to this shared endeavor, a prerequisite for proper institutional planning. He proposed to compile and print annual returns from every German asylum and even foreign ones, so far as possible. A community of statistical investigation, he went on, needed to reach consensus on the forms for collecting and presenting data, a particular challenge in the disunited German state. "Objective" psychiatry required a collaboration of physicians and state administration. Together, they could achieve a unity of theory and praxis, word and deed, idea and execution, *Technik* and administration. "It appears therefore as one of the tasks of our journal that we German insanity doctors must unite and then work together on the principles and methods of statistical recording of lunatics in Germany." He called for a uniform, simultaneous census of madness throughout Germany, whose results could be laid down in his journal.[34]

Comprehensive, unified statistics was thus a key element of Damerow's vision for German psychiatry. Alienists should continue tracking patients after their release. Carl Friedrich Flemming, second in the editorial collective, foresaw that its pages might in the future supply a rich and useful archive, invaluable to investigators whose findings are otherwise dispersed among so many accounting statements of so many institutions. It could function as a data bank, not just of asylum patients, but extending to lunatics wherever they were found. Flemming, in his introductory mission statement, proposed a set of tables for general adoption featuring disease forms, treatment outcomes, and causes, both physical and psychical. He stressed the need for homogeneous numbers to enable dispersed observers to reason by comparison, and lauded Thurnam's *Statistics* for its "superb circumspection." Every science, he declared, runs through two phases with respect to its numbers: an era of scarcity followed by one of superabundance. As mental medicine entered this second phase, it required principles for gathering facts so that the enterprise would not dissolve into hopeless confusion. German doctors stressed that data are never self-organizing but become meaningful only when shaped by human intelligence. Uncertainties of measurement would clear up with time if only alienists would cast off the dubious practice of passive recording.[35]

Their vision even included a role for mathematics, provided it was given the right materials to work with. Dr. Karl Reinhold Bernhardi, another admirer of Thurnam, proudly called attention to the 1,151 observations he had accumulated at his asylum in Königsberg. Thurnam had described how cure rates could vary as an asylum filled up with patients, even if medical effectiveness was fixed. Bernhardi explained his reasoning and suggested an improved basis for asylum statistics to avoid spurious numbers.[36]

Finally, he moved beyond statistics to a logical argument for focusing on predisposing causes. Many moral causes, he explained, lack "objective validity." Disappointed affection cannot be a true cause of insanity, because most of the time it does not lead to mental illness at all. The passage from disappointment to insanity must therefore depend on a constitutional factor that enables some to withstand the hard blows of fate while others are toppled by a light push. "It is necessary to penetrate more deeply into the darkness and take the trouble to ascertain the sources of predispositions." What was this underlying factor? In such darkness, the dim light of heredity shone brightly. Heredity did not rest on the surface but lay deep within the organism. Superficial explanations reflecting family experience left questions that could only be resolved in terms of the inner forces of heredity and constitution.[37]

This sense of a need to get beneath appearances was not confined to German writing. In 1844, Pliny Earle was appointed as the physician at the Bloomingdale Asylum in New York City, one of the oldest American institutions of its kind. He set to work on a comprehensive statistical account of that hospital from 1821 up to his own arrival. This meant, in practice, mining the data for significant conclusions, particularly on the question of causes. He now challenged the old division between predisposing and proximate causes, arguing that it often was impossible to decide which was which. Only heredity, which acted from within, retained its clarity as "inevitably a remote or predisposing cause." The hereditarily weak individual "will retain the healthy action of his mind until he is subjected to some other influence, more immediate, more active, more potent, and the tendency of which is to derange the physical functions of the system as to impair the manifestation of the mental powers." Something like a theology lurked in this complex phrasing, which preserved the sanctity and coherence of mind even as the brain disintegrated. Heredity, as a physical cause, could be documented by looking into the health of family members, and sometimes from hospital records. This way of proceeding seemed more convincing than trying to specify the impact of moral disturbances.[38]

More Than a Predisposition: The *Anlage*

The first substantive article in the newly established *AZP*, immediately following Damerow's mission statement, was Ernst Albert von Zeller's seventy-nine-page report on the curative asylum of Winnenthal, in Württemberg. Zeller, who published poetry and hymns as well as asylum reports, was its founding director.[39] Before assuming his new duties, he set off on a tour of institutions in England, Scotland, and France as well as several German states. Right from the start, he was skeptical of causal claims based on official tables. In his second report, he said it was impossible to distinguish cures from mere recoveries. In his first, he announced that a "tabular overview of the causes of mental disturbance" seeks what is impossible. Rather than comprehending the true convolute of causes, it picks out as cause the first notable symptom.[40]

Zeller could not simply run roughshod over established practices. His reports included statistical tables that were hard to reconcile with his own avowed principles. He was seeking a deeper basis of understanding. In his third triennial report, the one that found its way into Damerow's ambitious new journal, he listed not causes but "impulses [*Momente*] that can be seen as nearer or more remote causes." The first cause on his list was *erbliche Anlage*, a historically important term with well-developed medical roots. The usual translation, "hereditary predisposition," merely skims the surface of its meaning.

The great nineteenth-century *Deutsches Wörterbuch* (German dictionary), by Jacob and Wilhelm Grimm identified the original sense of *Anlage* as a laying on (*Anlegen*), of taxes, for example. Even as this meaning was preserved, it developed an inverted sense. In terms of the irresistible horticultural metaphor, *Anlage* began to refer not just to the seed that is sown but to the garden bed receiving it. *Anlage*, in this sense, signifies a potential that may or may not be realized, the soil in which certain seeds will grow and flourish while others fail to sprout. The human *Anlage* could be purely personal, but hereditary (*erbliche*) *Anlagen* were, according to Zeller, the "most frequent and the most important developmental impulses [*genetische Momente*]." He added that mental illness rarely arises from such factors alone. Just one female and four male cases from his institution could be attributed to this *Anlage* by itself. However, 75 men and 60 women owed their insanity to heredity in combination with "other circumstances." Hereditary causation may be direct or remote, and hereditary predispositions are not always noticeable in the family.[41]

In the instructions for registering patients, drawn up in 1853 for the new asylum of Karthaus-Prüll in Regensburg, Bavaria, the authorities laid out a plan for comprehending the *Anlage*. Item 7 in a list of facts to be recorded for a new patient addressed the question of causation.

> One must, in particular, investigate and record the relationships in which the patient has lived from earliest childhood, the causes that once impressed themselves on his body and mind, and the dispositions that supposedly could have provided an *Anlage* to mental illness, and then finally the proximate causes of the actual outbreak of mental derangement. Then, if possible, one should determine through what concurrence of inner *Anlagen* and external instigations the illness may have been generated and developed.

These instructions specified that the *Anlage* could as well be mental as corporeal and that it referred to "capacities and powers" arising during maturation from childhood to the outbreak of illness, not only to "original or congenital ones." Finally, the text pointed specifically to the *erbliche Anlage*, which it categorized among mental (rather than bodily) causes that act from the earliest years.[42] A similar document from Munich, drawn up as an appendix to the printed rules when the institution opened there in 1859, called for information on the "*Anlage* for similar diseases as far as the grandparents" as well as "*Anlagen* and direction of the mind and disposition, temperament, education, condition, favorite pursuits, social intercourse, morality, and religiosity." The *Anlage* was an alternative to confusing, superficial causes arising from miscellaneous life experiences. Within a few decades, the hereditary *Anlage* would appear as something concrete, a factor or element rather than merely a precondition. The 1876 revision of the Munich rules mentioned simply "family *Anlage* to disease, especially brain and nerve diseases" and then "mental *Anlagen*." In the early twentieth century, the Mendelian factor was called, in German, an *Anlage* or *Erbanlage*, and in psychiatry, at least, this term remained more common than *Gen* (gene) in the 1930s.[43]

Zeller called for observation informed by a deep understanding that could grasp things in their complexity. He aspired to comprehend how heredity worked in the transmission of insanity. Flemming pointed out in 1838 that the recurrence of parental mental disturbance in children and grandchildren was a well-confirmed fact, and yet that with a correct upbringing, these descendants might be protected. The *erbliche Anlage* was no mysterious curse, he insisted, but "a pathological condition or a disposition to develop such a condition."[44] Zeller stipulated that an *Anlage* was strengthened by exercise. "Here is one more reason why inheritance of mental disturbance is so prominent, for the procreator is, as a rule, also the educator. In the same way, a dormant *Anlage* for mental disturbance may be awakened early and nurtured by the parent's unreason."[45] The complexity of the *Anlage* seems better suited to narrative than to counting, and Zeller put no faith in simple answers. To pronounce a patient recovered is a subtle determination, he insisted, and to attribute the recovery to treatment is another. The figures in tables are always somewhat fuzzy, since every institution categorizes in its own way. Tallies, then, can never replace thinking. "And yet in the eyes of the world there is no way to get a grip on the effectiveness of an institution but with such numbers."[46]

The eyes of the world, it appears here, were distinct from the eyes of science or wisdom. On questions of cures and of causes, worldly eyes peered through bureaucratic spectacles and were often satisfied by numbers. It was not only Germans who found that viewpoint frustrating. At the asylum of Fains in Lorraine, Dr. Emile Renaudin conceded that "medico-administrative" statistics of "isolated" asylums were far from ideal for science. He called them, instead, tools of humanitarianism, "with which we can defend our budgets." Since "no one is a prophet in his own land," the alienist requires data for "the struggle against the prejudices of this proverb." The advantage of numbers is to be accessible to everyone and to occupy the only terrain on which physicians can reach accord with "deliberating assemblies that are not animated by the sacred fire of philanthropy."[47] For Zeller, too, numbers were integral to campaigns of public outreach. He wielded them to show that problems of mental health were not confined to acutely affected individuals but would persist even if every insane person were placed in an institution. In an eight-year report for the years 1846 to 1854, he explained that while the revolutionary movements of 1848 and 1849 had "fantasies of the delusional at their head," they also demonstrated what turbulence the "organ of soul" can withstand. Insanity participated in the great events of the day. "The whole life of man is involved in mental illness."[48]

Thurnam's investigation of insane heredity, with its focus on Yorkshire Quakers, was a local one. His push for standardization was cosmopolitan. Work with asylum data presented many problems and subtleties. Yet even the *Anlage* was employed as a discrete variable attributed to certain patients and not to others. Only occasionally did conceptual subtleties get in the way of counting, and the study of insane heredity, in Germany as in England and America, remained a matter of numbers.

CHAPTER 4

The Census of Insanity Tests Its Status as a Disease of Civilization, 1807–1851

> And if it is not without some surprise that we will find proportionately more insane in Norway than in England or France, we gain at the same time a proof that the result derives from the same general laws that rule the development of minds among all peoples.
>
> — Étienne Esquirol (1830)

Nobody concerned with the treatment of mental illness believed that the number of people in institutions was a valid measure of insanity in the population at large. It was the work of the census to make such determinations, tallying the insane wherever they were found. The diagnosis of lunacy or idiocy was especially uncertain when disconnected from the medical-legal processes of commitment to an institution, even if a village idiot or lunatic was part of ordinary life. Worse, asylum statistics of admission were radically incompatible with those of discharge. At least until the mid-nineteenth century, many asylums showed cure rates of 50% or higher, sometimes, as we have seen, much higher. Yet it seemed that every new asylum rapidly filled to overflowing, and the numbers of known insane increased in lockstep with the capacity of the institutions. Some new patients appeared to be cured, but those who remained were lingering, often-hopeless cases who drove down cure rates, discrediting that noble, humane mission of mental medicine. Counts of the insane were as protean and ill-behaved as madness itself, refusing to converge to any credible number or even to expand or contract along with total population and other plausible variables.

Census counts both reflected and stimulated anxieties about the effects of modern life on the quality of populations and were invoked on both sides of debates on state action. Whatever its imperfections, a census appeared necessary for the very practical purpose of planning asylum facilities. Beyond that, census results bore specifically (if unreliably) on vital issues of public health. Was the apparent increase of insanity real, and what could explain it? Would the costs of care for the insane continue to grow without limit? Most crucially, what were the causes of insanity, and how were they evolving with the progress of civilization? Inevitably, and increasingly, heredity was at issue in these counts, which were invoked more and more to justify the selective control of human reproduction.

Civilization Causes Insanity

Some of the earliest counts of the mad arose from a 1773 British law requiring registration of persons admitted to lunatic houses of all sorts. These reports enabled the semiofficial *Literary Panorama* in 1807 to print and ponder the diverse ratios of insane to total population. The returns were too erratic to inspire confidence. They showed, for example, seven insane out of 440,000 inhabitants of four counties around Cambridge, smaller by a factor of 24 than the ratio in Lancashire. Yet they could not simply be ignored. "It passes uncontradicted that the lunatic affection is a disease increasing in its frequency in this country." The journal proceeded immediately to a list of possible causes, all of them "moral" or environmental rather than medical or physical, and some hinting at social critique. Should the growth of insanity be attributed

> to the increase, or *decrease*, of marriages? to the propagation of disorders destructive to generation, and to morals?—to the depreciation of money, and consequent difficulties of support?—to the introduction of foreign luxuries and a mode of life less conducive to general health, than that of former ages?—to a more diversified system of education which injures the body in *very early life*, before it is able to support the requisite exertion?—to prevalent glooms arising from the unhappiness of the times?[1]

Figures for "lunatics in each county held in gaols, houses of correction, and workhouses" were allied also to a medical-social vision. The journal printed these numbers in a table to advance its dream of an organized system of asylums, each to hold about 300 patients. The best authorities doubted that the variability indicated by these tallies was real, however. Dr. Andrew Halliday, the most prominent writer on insanity numbers of his day, assumed that mental illness was a simple fact of nature, proportional to population. In a letter printed with the 1807 report, he proposed to distribute asylums across the counties not according to the number of registered insane, but so that each asylum would serve the same total population. In 1814, Richard Powell of the Royal College of Physicians published a table of registered admissions in five-year intervals from 1773 and argued that all the numbers were too low to be credible.[2]

Halliday was a forceful advocate for public mental hospitals, not least for the inspections and statistics they would provide. He accused madhouse keepers of exploiting ignorance to their own advantage by abusing the insane until the disease became permanent. A decade before Farr, he insisted that asylums should be opened to public observation. In 1828, another moment of acute parliamentary concern, he pronounced it curable. Halliday traveled all over Europe, from Sweden to Spain, gathering data and regretted having to make do with printed reports on faraway India. It was time, he thought, for England to bring its register of admissions up to the French standard of accuracy, since correct tallies formed the indispensable basis for a rational system of asylum care.[3]

The urge to count the insane incorporated a new acceptance of public responsibility for them and for the health of populations more generally. At the same time, many were anxious that the apparent increase of insanity could be real. Pinel had emphasized revolutionary excess as a cause of insanity, and a chart he drew up at Bicêtre in 1793 attributed a third of his cases to "events connected with the revolution." Esquirol's table of moral causes for his private establishment in 1816 attributed 18% to political events, still a high number.[4] In his medical dissertation of 1805, he worked out an appealingly pessimistic explanation of the epidemic of madness. He spoke of *perfectibilité* as the curse as well as the glory of human life, since this capacity for improvement is attended by moral disorders linked to love, anger, vengeance, and ambition. Artificial needs and desires that do not conserve our bodies are the fruit of developing moral faculties. Fluctuations of commerce, enthusiasms of exaggerated religion, and the inflammation of love by theater and novel reading all chip away at mental balance. In combination with hereditary susceptibility, these can easily lead to lunacy.[5]

In 1816, Esquirol provided a numbered list of moral causes alongside the physical ones. So many persons gone mad due to failed love affairs, family deaths, business reversals, abuse of alcohol, masturbation, and the pressures of work and study moved him to reflect on the tumults of modern life. His statistics revealed a heightened vulnerability to insanity in sedentary occupations. The unstable lives of merchants and courtesans must be still worse. He credited Alexander von Humboldt for an observation on the rarity of insanity among the savages of America, and a book of travels by John Carr, *A Northern Summer*, for its rarity in Russia. In France, he declared, insanity occurs mainly in cities. He concluded, invoking Rousseau, that civilization as such was not the cause, but that by multiplying sensations and providing the means of excess, it occasioned an increase of mental disorder. His thesis became familiar in a less nuanced form: madness is a disease of civilization.[6]

Esquirol returned to these questions in 1830 in response to a Norwegian census organized by the physician Frederik Holst. Like so many in his profession, Holst had been granted state funding for an alienist *Wanderjahr*, on which he reported in 1820 and 1823. He was most impressed by the hospitals of Paris, where he followed Esquirol's course on mental illness. His count was part of the 1825 Norwegian census, but he delayed publication until 1828 because he did not accept the competence of priests to classify forms of insanity on the census sheet. Esquirol praised this work extravagantly in a review. There had been various incomplete or unsatisfactory enumerations in England, Bavaria, and France, he wrote, but here at last was a proper count, "the most complete statistics of insanity

K. Sygdommens Aarsager, sammenlignede.

Aarsager.	Maniaci.				Melancholici.				Dementes.				Sindssvage af alle tre Arter.			
	Mand-kjøn.	Qvinde-kjøn.	Af begge Kjøn.	Forhold.	Mand-kjøn.	Qvinde-kjøn.	Af begge Kjøn.	Forhold.	Mand-kjøn.	Qvinde-kjøn.	Af begge Kjøn.	Forhold.	Mand-kjøn.	Qvinde-kjøn.	Af begge Kjøn.	Forhold.
1. Physiske:																
Arveligt Anlæg	12	10	22	1: 8	8	6	14	1: 16	6	5	11	1: 15	26	21	47	1: 12
Barselseng	:	6	6	1: 31	:	3	3	1: 74	:	7	7	1: 23	:	16	16	1: 36
Epilepsie	:	1	1	1:185	:	:	:		:	:	:		:	1	1	1:569
Apoplexie	1	:	1	1:185	1	:	1	1:222	4	3	7	1: 23	6	3	9	1: 63
Feber	1	:	1	1:185	5	2	7	1: 32	12	5	17	1: 10	18	7	25	1: 23
Slag paa Hovedet	2	1	3	1: 62	2	2	4	1: 55	7	2	9	1: 18	11	5	16	1: 36
Vand i Hjernen	:	:	:		:	:	:		1	:	1	1:162	1	:	1	1:569
Byld i Hjernen	:	:	:		:	:	:		1	:	1	1:162	1	:	1	1:569
Forkjølelse	2	:	2	1: 92	3	:	3	1: 74	1	2	3	1: 54	6	2	8	1: 71
Engelsk Syge	:	1	1	1:185	:	:	:		:	:	:		:	1	1	1:569
Nervesvækkelse	1	1	2	1: 92	3	5	8	1: 28	2	:	2	1: 81	6	6	12	1: 47
Hysterie	:	:	:		:	5	5	1: 44	:	1	1	1:162	:	6	6	1: 95
Udsvævelser	2	1	3	1: 62	1	:	1	1:222	2	1	3	1: 54	5	2	7	1: 81
Drik	10	1	11	1: 17	10	:	10	1: 22	4	:	4	1: 40	24	1	25	1: 23
Onanie	2	:	2	1: 92	1	:	1	1:222	7	:	7	1: 23	10	:	10	1: 57
	33	22	55		34	23	57		47	26	73		114	71	185	
2. Psychiske:																
Forsømt Opdragelse	:	:	:		:	:	:		:	1	1	1:162	:	1	1	1:569
Streng Behandling	2	1	3	1: 62	2	2	4	1: 55	3	1	4	1: 40	7	4	11	1: 52
Feilslaget Haab	4	2	6	1: 31	1	:	1	1:222	:	1	1	1:162	5	3	8	1: 71
Haabløs Kjærlighed	13	34	47	1: 4	11	29	40	1: 6	5	21	26	1: 6	29	84	113	1: 5
Ulykkeligt Ægteskab	1	5	6	1: 31	1	2	3	1: 74	:	1	1	1:162	2	8	10	1: 57
Jalousie	2	2	4	1: 46	1	3	4	1: 55	1	2	3	1: 54	4	7	11	1: 52
Sorg	6	7	13	1: 14	33	18	51	1: 4	7	15	22	1: 7	46	40	86	1: 7
Skræk	18	13	31	1: 6	8	5	13	1: 17	10	10	20	1: 8	36	28	64	1: 9
Religionssværmerie	6	3	9	1: 21	6	12	18	1: 12	1	:	1	1:162	13	15	28	1: 20
Samvittighedsskrupler	2	2	4	1: 46	9	3	12	1: 18	1	1	2	1: 81	12	6	18	1: 32
Krænket Æresfølelse	2	:	2	1: 92	3	2	5	1: 44	2	:	2	1: 81	7	2	9	1: 63
Overdreven Studering	:	:	.		4	2	6	1: 37	1	:	1	1:162	5	2	7	1: 81
Arrighed	1	4	5	1: 37	2	6	8	1: 28	2	3	5	1: 32	5	13	18	1: 32
Psychiske	57	73	130		81	84	165		33	56	89		171	213	384	
Physiske	33	22	55		34	23	57		47	26	73		114	71	185	
Tilsammen	90	95	185		115	107	222		80	82	162		285	284	569	

FIGURE 4.1. Table of causes of insanity against disease form, from the 1828 report of the Norwegian census of the insane. The first cause on the list and the most common physical cause listed was *Arveligt Anlæg* (hereditary factor or tendency). The disease forms given are mania, melancholia, and dementia (followed by total). This cross table, or correlation table, with a different variable of interest on each axis, was still unusual in the 1820s. From Holst, *Beretning, Betankning og Indstilling* (1828), Table K (back of book).

that we have." The statistics required for rational administration of hospitals for the insane, and for a scientific reckoning of its modern increase, had to be based on a census like this one.[7]

Holst, rather exceptionally, combined two variables in his table: cause (beginning with heredity: *Arveligt Anlæg*) on the left axis and disease form (as well as sex) along the top. This arrangement required extra work to sort out the cases as well as the inspiration to seek statistical relationships of this kind.[8] He was, however, most concerned by the gross figures, concluding sadly that his numbers demonstrated an increase of insanity in step with the advance of civilization. Esquirol could scarcely disagree, and in his review, again invoked Humboldt, now for the categorical claim that insanity was *unknown* among the indigenes of South America. He also cited Philadelphia physician Benjamin Rush as a witness to the rarity of insanity among savages. Rush, in fact, claimed no personal experience among primitive peoples but relied on the testimony of travelers. "Baron Humbolt informed me, that he did not hear of a single instance of it among the uncivilized Indians in South America." Humboldt probably spoke with Rush in 1804 as he was returning from five years of travel in Latin America. Esquirol almost certainly extracted this comment, which became famous, as well as Carr's, from Rush's *Medical Inquiries and Observations*. Rush also referred to Dr. William Scott, a member of Lord Macartney's embassy to the Emperor Qianlong in 1793–1794, who had "heard of but a single instance of madness in China."[9]

The authority of these travelers, recited again and again, inspired a little ethnographic industry to give evidence on the question of whether madness really did arise from the denaturing processes of civilization. Non-Europeans figured here as authoritative witnesses. Amariah Brigham wrote in an 1845 asylum report that insanity was rare among Indians and Negroes, then, generalizing, declared it uncommon in China, Persia, and Hindostan as well as Turkey and Russia.[10] An eighty-year-old Cherokee chief, described by a missionary-physician as intelligent, was reported in the *AJI* as saying that he never witnessed a case of madness among his own people to compare with those he saw in a Philadelphia hospital.

Unnamed "Amistad Negroes" remarked after seeing the Connecticut Retreat in the early 1840s that insanity like this was very rare in their country. Their leader, Joseph Cinqué, with greater precision, said he had seen one case.[11]

The *Literary Panorama* had proposed already in 1807 to examine quantitatively some basic questions of causation by comparing lunacy rates in manufacturing and agricultural districts. "Though we acknowledge, freely, that more correct returns are wanting to justify inferences, yet we cannot refrain from directing the attention of the inquisitive and especially of medical practitioners to these queries." Esquirol, chiseling in the same vein, suggested a comparison with Halliday's work on England and Wales and T. R. Beck's on some US states. If others prepared statistics with the same care and detail as in Norway, they would be precious for "philosophical and medical study of mental maladies" and provide valuable comparative results for different climates and customs.[12]

In the event, Holst's ratio of insane to total population, 1:551, was uniquely high: higher than Halliday's most recent numbers for England and Wales, higher than the 1825 results for New York, higher even than an 1821 figure from Scotland. Only the department of the Seine showed a greater ratio, and this, Esquirol explained, was meaningless because so many migrants came to Paris for treatment. For France as a whole there were as yet only round, hence speculative numbers: 30,000 insane in a population of 30 million, or 1 in 1,000. He thought Norway's excess of insanity paradoxical. How could a land of mountains and fjords, of shepherds, fishermen, and tillers of the soil, outpace England and France in this disease of civilization? But Holst had anticipated the objection and hinted at the answer. An excess not of lunacy, but idiocy, accounted for the elevated Norwegian numbers, and idiocy, as Esquirol taught, is most common in mountains. The figures for Scotland, just a notch behind Norway's at 1:573, admitted the same interpretation.[13]

If the increase of these numbers over time reflected something real, the reasons for worry were not limited to budgetary ones. The idea of insanity as a disease of civilization hinted already at a threat of degeneration, the dark side of progress, and cast a shadow over faith in human reason. In 1831, the liberal Bonapartist and Montpellier-trained physician Claude-Charles Pierquin accused Esquirol of a fundamental error amounting to a logical contradiction. There must be something wrong with the evidence. Travelers had failed to notice insanity among the *hordes barbares* precisely because it is so ordinary, and only against a background of enlightenment does madness stand out. To counter Humboldt's authority on the mental stability of primitive peoples, Pierquin invoked the ethnographic experience of the Baron Jacques-François Roger, recently governor of Senegal and a member with Pierquin of the Société Universelle de Statistique. Roger had told him in a letter that furious madness was almost unknown in Senegal. This was not from any deficit of insane people but owed to their utter freedom from constraints, even of bulky clothing, and to the admirable tolerance of their generous nation. Since the mad were never tormented in Senegal, they were rarely violent. Monomania, always difficult to diagnose, was scarcely distinguishable there from ordinary religious practices. Dementia, Roger went on, appeared if anything to be more common in Senegal than in civilization, but this impression, too, was an illusion, a result of the charity bestowed on these unfortunates. People with dementia, entering a village, could expect a warm welcome and generous provision, in accordance with Islamic custom, so they had no reason to conceal their condition. Some, indeed, feigned madness in order to live well without working.[14]

Pierquin included an array of proofs that civilization is opposed to insanity. He noted, for example, that in counts of the institutionalized insane by profession, the highest numbers are given for domestic servants, seamstresses, day laborers, and the like, who make the least use of intelligence. (It seems he did not think to adjust the absolute numbers for relative frequencies of different occupations.) He argued that level of instruction was much more fundamental to insanity than all the little causes from Esquirol's table, and that the greater prevalence of insanity in Norway than Scotland could be simply explained by the superiority of Scottish schooling. Holst's numbers, in sum, reflected the commonness of idiocy and dementia in ignorant societies in contrast to mania and monomania in educated ones. It all made sense against the background of another, much hotter debate of the era, which obsessed Pierquin, on the relation of education to crime. He took the enlightened view that crime and insanity were not caused by education but cured by it and that citizens of Paris, the most enlightened in the universe, were the least insane or criminal. Its prisons were indeed filled to overflowing, but with foreigners.[15]

The work of recording demographic, commercial, agricultural, and judicial numbers intersected less than we might expect with asylum statistics, but censuses of the insane are another matter. Adolphe Quetelet took up the points of debate between Esquirol and Pierquin in his most important book, *Sur l'homme* (*Treatise on Man*), which appeared in 1835. For his chapter on mental alienation, Quetelet relied on Esquirol's review of the Norwegian census. He also thanked Esquirol for unpublished asylum data. He endorsed the typical result of asylum statistics, contested by Pierquin, that insanity tends to erupt at the age of most vigorous intellectual development, when persons are in their twenties and thirties. He rejected Pierquin's bundling of crime and insanity with ignorance, and he endorsed the distinction between idiocy and mental alienation as the explanation for high insanity figures in Norway. In short, he supported Esquirol on every point. Mental alienation is an attack on intelligence in its seat, provoked either by its too intense exercise or by an excess of passion and sorrow.[16]

Esquirol held resolutely to his views, reprinting his old papers and reports in 1838 as the summation of his life's work. Even in 1805, the link between madness and civilization was scarcely shocking. Insanity had long been associated with leisure and luxury, not least in England. This, as Andrew Scull has shown, was an understanding that mattered, since images of insanity as a tragic disease attacking respectable people were routinely mobilized in campaigns for public asylum systems. As soon as they succeeded, however, the experience of such systems began to undermine in a very concrete way the association of madness with luxury and cultivation. As these institutions filled to overflowing with pauper patients, families with resources began to avoid them. Under these conditions, insanity came to be interpreted as an affliction of the poor.[17] In 1872, one of the commissioners in lunacy for Scotland, Sir James Coxe, presented statistics showing a massive expansion of insanity, especially among institutionalized paupers, as a basis for reversing Esquirol's assessment. Far from being a disease of civilization, insanity was nurtured by poverty and ignorance. He put great stress on social conditions that led to the increase of asylum populations. Institutions for the insane, he thought, had very little power for good or for evil, and many so-called cures owed instead to the recuperative power of nature. He had seen many patients cured at home who would have rapidly become incurable in an asylum. Not purgatives, hypnotics, narcotics, and tonics, but hygiene and mental cultivation were the best that asylums could offer their patients.[18]

Counting and Being: Alienation as a Medical Specialty

The growth of asylum systems and of the knowledge that supported them appears haphazard until about 1840, when, in short order, the forces of professional and bureaucratic order marched onto the stage. Alongside the new journals of mental medicine, societies of asylum doctors took form, beginning with the (British) Association of Medical Officers of Asylums and Hospitals for the Insane in 1841. Still more significant were new laws requiring public asylum systems in much of Europe and North America. A simultaneous surge of census activity gives evidence of heightened public attention to the social problem of insanity. While the Norwegian census of insanity in 1828 preceded such legislation there, the counts of the early 1840s appear as efforts to gather systematic data for an accelerating institutional expansion. None of these counts, however, seemed quite satisfactory.[19]

The first American census of the insane, part of the 1840 US census, used a complex new paper form with an array of brave new questions. This was social politics on a census card, summed up in the entries for persons "at public charge" and for those over twenty who could not read or write. The card included a jumble of little spaces for the blind, the deaf and dumb, and insane and idiots, all divided between "white persons" and "colored persons." The deaf and dumb of the white race, but not the colored, were partitioned into three age categories. A space was set aside for black insane and idiots, and two more for white ones at public and at private charge. The tangled process of recording, combining, and copying to get totals for each state brought forth absurd results on race and insanity, and then a scandal.

Almost immediately upon the release of census results in 1841, its readers noticed massively higher rates of insanity among blacks living in the northern states than had ever been found for any population anywhere. In the free northern states as a whole, 1 in 163 blacks were shown as insane or idiotic, almost ten times higher than in the South. In Maine, with only a few black inhabitants, this ratio rose to the astonishing level of 1 in 14. By 1842, the Massachusetts asylum doctor and medical statistician Edward Jarvis had uncovered damning inconsistences, such as towns that tallied no colored residents at all and yet, in a separate entry, showed one or more insane ones. The entire population of the all-white Worcester asylum was entered as colored. Historians have reconstructed the mistakes, mainly of copying, that produced these implausible results. Each page of the printed census has about seventy-five lines, and it is barely possible to follow the entries for a town or county across facing pages. The discrepancies involving free blacks were not even on adjacent pages, and the population tables for all the states included over 100,000 numbers, excluding blank spaces. Checking for consistency could not have been easy, unless a reader was impressed or startled by the totals, and avoiding errors was not a high priority in this bungled operation.

Advocates of slavery eagerly embraced the rhetorical opportunities offered by this unexpected result. Even Jarvis briefly considered the possibility that high levels of black insanity might be a result of their sudden exposure to the pressures and choices of commercial society. His subsequent demand that the published numbers be corrected went nowhere, partly from a reluctance of the census officials to admit mistakes, but mainly because champions of the peculiar institution were so entranced by seeming evidence that the mind of the African could not bear up to freedom.[20]

A French naturalist and political anarchist of Spanish and Cuban background, Ramon de la Sagra, assayed these numbers and proposed some explanations in the first issue of the *AMP*. "Profiting from these tables that have just been published by the government of the United States on the population there, I have carried out a great number of comparisons and statistical calculations with proportional numbers that I have deduced from absolute numbers furnished by the census." The proportion of insane among the free colored, he observed, is without equal in Europe, "to the point that some expert statisticians in a famous academy have cast doubt on the exactitude of the document." That was the American Statistical Association, of which Jarvis was a founding member. While it would be no surprise, Sagra continued, if slave owners should downplay problems of mental health, the federal government had no reason to exaggerate the insanity of colored people in the North. The numbers, he concluded, must be correct, pointing to a serious problem of American civilization. The elevated rate of black lunacy must owe to the disdain and contempt of Northern whites, who appeared to him even more culpable in regard to race relations than slaveholders. He also got in a few jabs on the severity of religion in America, whose puritanism, lack of public amusements, and disorderly exaltation of Methodist meetings heightened madness in both races.[21] This was just the sort of thing that refined American critics liked to say about their own culture. John Butler of the Connecticut Retreat was moved by this census to deplore the circumstance "that in no section of the world is insanity more prevalent" than in New England and the upper Midwest. "We are too much obsessed with business, leaving too little time for recreation, social intercourse, literature, and science."[22]

The first French census of the insane, published in 1843 as a sixty-five-page section in the *Statistique de la France*, consisted mainly of tables on the operation of asylums. Its charge had been to count patients in private establishments or kept at home as well as the inhabitants of public institutions. A summary table at the end combined the departmental figures into national ones, giving mental alienation by profession, by presumed cause, and by condition: idiots, epileptics, and mad (*fous*). While the data was mainly bureaucratic, the issues at stake extended to civilization itself, as was made clear in debates at the Académie des Sciences. Alexandre Moreau de Jonnès, the first head of the revived French census, appeared there on 10 July 1843 to discuss its bearing on the great moral questions of the day. There were a lot of false numbers about, he intoned, mentioning first some very early British figures on insanity. The American census of 1840, with proportions mad as high as 1:14, was alarming for a different reason. He did not mention that this figure was only for a tiny racial minority in one state or that Jarvis had challenged it. Instead, he condemned in quasi-religious terms the terrible degradation of the human species implied by so much madness.

Moreau de Jonnès was concerned primarily with matters closer to home. The figures for insanity in France had risen steadily during recent decades to 32,000, implying about one insane per thousand inhabitants. That number, like every assertion that insanity must increase with the advance of civilization, was baseless. His office, he modestly declared, had at last solved the problem of counting insane persons outside as well as within institutions. The correct figure was 18,350, a ratio to population of 1:1,900 or 1:2,000, and it was not increasing.[23] But what was this methodological innovation of the new French census? A week later, the alienist Alexandre Brierre de Boismont offered proofs that the new numbers were purely institutional figures, by-products of the new French law for care of the insane. There had been no true census at all. The mad in *départements* with no asylum as yet would have been missed, as would those residing at large or in other kinds of institutions. A well-conducted recent census of Belgium had found 1.22 insane per 1,000 population, and Parchappe's authoritative count in the Lower Seine gave a ratio almost as high. The correct number for the insane in France could not be lower than 30,000.

Moreau de Jonnès denied everything. The Belgian census was flawed, and his critic did not understand the French one, which did not depend on the asylum law at all. It had been rigorously planned and executed using all the resources offered by a public authority, and it in no way supported the pretended increase of insanity with civilization. His tables showed a dominance of physical causes by a ratio of 7:3, while all the moral ones were such as could be found already in the Bible. In short, these causes of insanity included nothing specific to modern life.[24]

Moreau's tally, vigorously debated at home, also attracted attention abroad. Brigham declared such a count to be beyond the capabilities of the French census, since physicians alone were competent to draw up medical statistics.[25] To alienists in German lands, by contrast, it seemed, even if flawed, a daunting achievement of centralized authority, accenting the tragedy of poor, fragmented Germany. "The lunacy statistics of Germany are fractions of no whole," lamented Heinrich Damerow. German disunity implied the incoherence of its statistics and the impossibility of any coordinated asylum system, even within the Prussian state.[26]

Abandon Hope: Recording an Epidemic

In reality, there was as yet no established model for tracking and enumerating the insane at large. Thurnam had the advantage of a relatively homogeneous population and of access to informal networks of information. Relapsed patients were likely to return to the Retreat, whereas patients in state asylums might circulate through poorhouses and prisons as well as diverse asylums. Tallying the mentally ill in the context of a national census provided a basis for approaching big questions of social medicine. It appeared increasingly that state asylums were accelerating the increase of the recognized insane without relieving the pressure of mental illness in society. As Scull observed, "It remains perhaps the most paradoxical feature of the entire reform process that the adoption of a policy avowedly aimed at rehabilitation and the rise of a profession claiming expertise in this regard should have been accompanied by a startling and continuing rise in the proportion of the population officially recognized as insane."[27] At the same time, outcomes for these patients went into a tailspin. Physicians complained that so many incurable patients deprived them of space to house the very ones who could benefit from treatment.

Samuel Smith in Ohio, who had warned of just such an outcome, gave republican arguments for a ban on private institutions. Not only was luxury an obstacle to cures, but the availability of a private alternative tended to undermine support for medical facilities so desperately needed by the poor. Lacking state assistance, lunatics without means would drag their families into destitution or, if left at large, endanger the public. Finally, "it is the *duty of the Commonwealth jealously to guard the rights and liberties of her citizens*," which is only possible within "state Institutions under proper control." Smith vehemently opposed the custodial asylum on the ground that only the prospect of curing some patients can keep up morale among the staff and encourage decent care for the hopeless ones. His state, still close to the frontier, was, he said, mostly free of the destitution that plagued other societies. Yet insanity and its causes seemed to increase unremittingly, even in Ohio.[28]

A growing pessimism about cures undermined sympathy for the insane. At Worcester, George Chandler complained of Irish patients smuggled into the state and brought to the asylum with dubious claims that their long-festering illnesses were new. Laid low by drunkenness, their prospects of recovery were bleak. They were filling places needed by "our native population." His report for 1854 recalled fondly the days when Worcester was a model, drawing visitors from near and far to witness its excellence. While other asylums had experienced "changes and improvements, amounting to revolutions" in the pathology and treatment of the insane, Worcester was now packed to the gills with patients no longer drawn from an educated, intelligent class of yeomanry but from one without refinement or culture, "and not much civilization even." Such an institution must degenerate, bringing a revival of private hospitals and the evils that always come with them.[29]

As cure rates declined, asylum officers battled to preserve their status as places to treat and cure insanity rather than merely to warehouse it. Some recited Dante's line "All hope abandon, ye who enter" to protest the elimination of medical care for patients deemed hopeless. In his 1845 textbook, the noted Berlin alienist Wilhelm Griesinger worried that the segregation of chronic patients from those undergoing active treatment might create places "where 'Lasciate ogni speranza' is written on the brow."[30] In Quebec, supporters of a new asylum imputed to its predecessor institution the dismal condition described by Dante.[31] Critics of the proposed Willard Asylum in New York, designed to house chronic pauper patients cheaply by withholding treatment, complained in 1865: "Truly over the gateway to such institutions should Dante's inscription to the portals of hell be written." The next year in Toronto, it was again quoted as a protest against a custodial hospital.[32] It was impossible now to do without asylums, yet the dream of humanity and hope was giving way to darkest hell. Having set the train of generous asylum care in motion, it was hard to climb off as it careened out of control.

By the 1860s, the hope of cures was clearly receding, but this pessimism did not mean giving up on medical science. Ironically, the failure of medicine to control the growth of asylum populations brought new resources for recordkeeping and statistical study. Alienists labored to create a foundation of data for the investigation and relief of hereditary causes.

It is tempting, but mostly incorrect, to suppose that the growing fixation on heredity was an evasion of responsibility for medical failure. If anything, asylum doctors were moving in the contrary direction, reciting statistics to prove that inherited insanity was not less curable. Maximilian Jacobi's experience at Siegburg during the 1830s is revealing on this point. At first he referred to the high frequency of inherited insanity along with the poor condition of many patients as impediments to successful treatment. The responsible committee of the Rhine Province *Landtag* agreed, praising him for curing about 57 of 270 patients even though 69 were incurable and 50 more, "on account of hereditary *Anlage*, or for other reasons, presented at the time of admission greatly diminished hopes of a fortunate success." Jacobi's report for 1833–1836 emphasized the disadvantage of treating 183 patients "who, because of hereditary or congenital *Anlage,* the long duration of their illness, or earlier and as yet fruitless attempts at cure, offered as good as no hope . . . for a successful treatment." In 1840, he pointedly omitted heredity from the list of factors tending to reduce cure rates.[33] The first doubts that modern asylums could halt and reverse the growth of insanity coincided with the initial appearance of statistics on the curability of hereditary patients. The numbers were taken as showing that a hereditary *Anlage* increased vulnerability to recurrence or relapse but did not imply incurability. New tables showed hereditary insanity to be, if anything, more readily curable than nonhereditary. In 1847, *AZP*'s associate editor C.F.W. Roller referred scathingly to the ignorance of an author who had invoked a hereditary *Anlage* to excuse medical failure, in opposition to "the most rigorous observations."[34]

This view became, for a time, a consensus. Heredity is "among the most prevailing causes of insanity," was how William Malcolm summed up the statistics in his 1849 report for Perth, Scotland. "I by no means find the disease is less easily cured when this is the case."[35] B.-A. Morel, not yet a famous degenerationist, cited statistics to support his enduring faith in curability. He celebrated a new asylum at Maréville, with its great population of 760 patients, as an "inexhaustible mine of riches" for science. He even denied that expanding asylum populations reflected any real increase of insanity.[36]

Better Hospitals and Rising Insanity

Beginning about 1852, Edward Jarvis bravely undertook to separate social from medical sources of rising asylum numbers. Although the insane appear ever more numerous, he remarked, "it is impossible to demonstrate, whether lunacy is increasing, stationary, or diminishing, in proportion to the advancement of the population, for want of definite and reliable facts." A determination would require at least two accurate censuses of the same population. Such were nowhere to be found. The French census of 1843 was manifestly incomplete, while British counts were vitiated by their limitation to asylum patients. Jarvis respected the 1850 US census, but the one in 1840 had been a fiasco. An excellent new Belgian count would provide information on the growth of insanity when it had been repeated. He also praised the Norwegian census of 1828, which he knew from Esquirol's review, but he seemed unaware of subsequent Norwegian counts.[37]

In 1855, he reported on a census of the insane in Massachusetts, which had been authorized to assess the need for a third state asylum. In contrast to prior tallies in his own and other states, which relied on untrained officials, the commission for this one sent an inquiry to every physician in the state requesting medical information on all insane persons. There were enough physicians, Jarvis thought, to carry out a valid census: the "whole Commonwealth is, in detail, under the eye of the medical profession." The result, 2,632 insane and 1,087 idiots, would provide guidance in the future, once the exercise had been repeated.[38]

For now, he had no choice but to reason in reverse, from causes to statistics. Starting with a list of 176 physical and moral causes extracted mainly from the Worcester hospital reports, he surmised their direction of change. Civilization makes sensibilities more keen and passions more powerful and abiding, he reasoned, creating vulnerability but allowing the affections to become more permanent, providing stability. Religious enthusiasm waxes and wanes. Education, the greatest benefit of civilization, brings, alas, no sufficient increase of wisdom to guide cerebral action. The evidence from all sources, he concluded, supported Esquirol's assessment of a disease on the rise.[39]

In Britain, many informed commentators suspected that the increase of insanity was an artifact. According to a British statistical report from 1861, "the great increase which has taken place in the number of Patients in Asylums is limited almost entirely to Pauper and criminal Patients."[40] It appeared that the abundance, quality, and cheapness (to patients) of asylum care might provide the real explanation. Alternatively, the demands of modern life may have raised the bar for effective participation in society, driving the less able to asylums.

The British census first counted the insane in 1851, but only those in asylums.[41] Other numbers derived from registration processes at various sorts of institutions: county asylums, licensed houses, hospitals, poorhouses, and private homes. Since numbers from different sources remained distinct in the reports, they could be investigated separately. For example, pauper patients as tallied by the Commissioners in Lunacy were compared with those given in reports of the Poor Law Board to show that increasing pauperism, by itself, could not explain the alarming growth of pauper insanity. There were several ways for asylum patients to increase without any real increase of insanity. For example, life will be prolonged when "destitute and diseased persons" are placed in "well constructed, well regulated" establishments, "specially adapted for their protection and treatment." Asylum professionals, of course, liked to emphasize this one. The apparatus of data circulation could also have an impact. A simple example is the introduction of automatic registration of new patients, which of course made for higher patient numbers. A more consequential shift of recording practices was the requirement for inspectors at workhouses to include in the returns for lunacy "all persons receiving relief on account of mental infirmity." Some of these reclassified individuals were promptly sent to asylums.[42]

There was always a possibility that some cause, perhaps one as vague as "civilization," was stimulating the increase of a bona fide mental disease. "Degeneration," though less specifically medical, also implied a decline of human quality or resilience. The extraordinary growth of asylum populations was almost as alarming and just as expensive even without causes like these. The very tangible consequence of relentlessly growing populations of patients and prisoners drove these painful debates about causes. A series of reports from 1875 to 1877 of an expert commission set up by the *Lancet*, a medical journal, held that the problem was not really a medical one at all. "If the moment a new asylum is opened, with

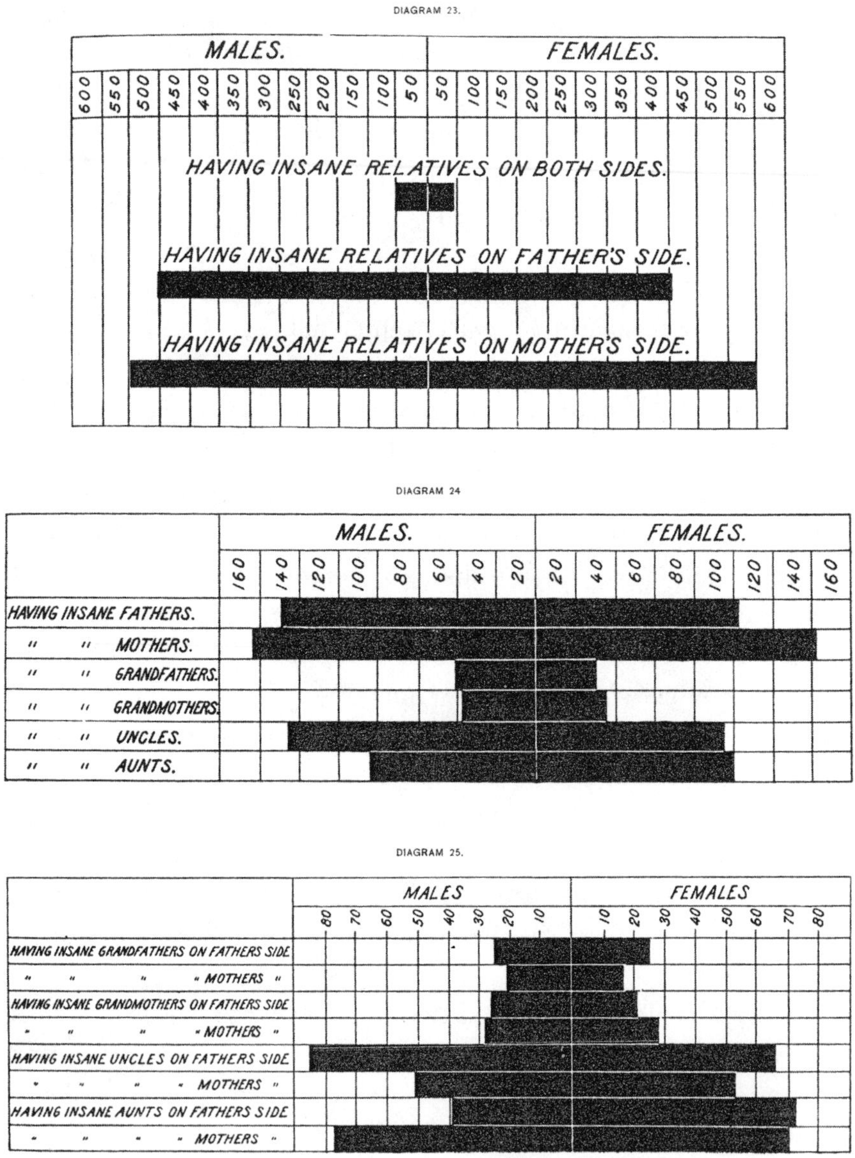

DIAGRAM 23.

MALES. FEMALES.

600 550 500 450 400 350 300 250 200 150 100 50 50 100 150 200 250 300 350 400 450 500 550 600

HAVING INSANE RELATIVES ON BOTH SIDES.

HAVING INSANE RELATIVES ON FATHER'S SIDE.

HAVING INSANE RELATIVES ON MOTHER'S SIDE.

DIAGRAM 24

MALES. FEMALES.

160 140 120 100 80 60 40 20 20 40 60 80 100 120 140 160

HAVING INSANE FATHERS.

" " MOTHERS.

" " GRANDFATHERS.

" " GRANDMOTHERS.

" " UNCLES.

" " AUNTS.

DIAGRAM 25.

MALES FEMALES

80 70 60 50 40 30 20 10 10 20 30 40 50 60 70 80

HAVING INSANE GRANDFATHERS ON FATHERS SIDE

" " " " MOTHERS "

HAVING INSANE GRANDMOTHERS ON FATHERS SIDE

" " " " MOTHERS "

HAVING INSANE UNCLES ON FATHERS SIDE

" " " " MOTHERS "

HAVING INSANE AUNTS ON FATHERS SIDE

" " " " MOTHERS "

FIGURE 4.2. From a report of the 1880 US census (1888), an early example of graphical representation of asylum statistics, here comparing the prevalence of hereditary taint of women to that of men and indicating the importance of different relatives as sources of insanity. From Frederick Howard Wines, US Census Office, *Report on the Defective, Dependent, and Delinquent Classes of the Population of the United States as Returned at the Tenth Census (June 1, 1880)*, vol. 21 of 1880 census (Washington, DC: Government Printing Office, 1888), Table 17.

all the best modern appliances, it be filled with patients withdrawn from the licensed houses, and treated as an almshouse for the aged and infirm paupers who happen to be eccentric and troublesome in the neighbouring workhouses, it will be necessary to go on building asylums until no inconsiderable portion of the pauper population is returned as lunatic."[43]

The insane asylum as a restful, ordered place grew up in part as the remedy for a disease linked to modern hustle and bustle. It was, in a way, a backward-looking remedy in an age of industry and progress.[44] It had much in common with communitarian utopian visions of this period, and it did not seem to be working. Hence, the problem had to be confronted outside the walls of institutions. Asylum doctors were already aware of this, and census investigations revealed more fully the social dimensions of insanity. Its medical character remained elusive, and no one could say if civilization was its deepest cause. It was, in any case, expensive and deeply disturbing. The work of the census intensified this frustration. The terrible increase of the insane was a sink for public expenditures, one that medicine seemed powerless to reverse. "*Hereditary predisposition* doubtless exists in a far greater number of cases than is generally supposed," wrote Richard Dunglison in 1860.[45] The census might even be mobilized to explore and to depict this power of hereditary causation.

PART II

Tabular Reason

> It would seem as if results like these could not be otherwise than correct, because they are but the general expression of the facts themselves. It is this very appearance of certainty which sometimes, as in the present case, blinds us to the actual fallacy, and we go on accumulating and hugging our treasures of knowledge as we fancy them, until we find at last that we have been ingeniously deceiving ourselves with an empty show, while the substance has completely escaped us.
>
> —Isaac Ray (1849)

The expectation that statistics must be simple and transparent, then as now, was bound to be disappointed. Early in 1858, Edward Jarvis prepared a set of cumulative tables for the twenty-fifth annual report of the State Lunatic Hospital of Worcester, Massachusetts. These were printed alongside the tables for the current year. Although Jarvis was the most distinguished American medical statistician of the day, the tables were a flop. The categories, never systematized, had clearly evolved over the decades, with the result that many assigned causes from early years had withered away while new ones kept springing up. John Gray of the Utica asylum mocked this alphabetized mishmash in his annual review of American asylum reports. Could there be any value in a list with "death of a brother" entered as a cause distinct from deaths of sisters, nieces, and cousins? What the report defended as the "partial light" of this table was, according to Gray, "inefficient," the flickering deception of an ignis fatuus and typical of the confusion sown when causal attribution was left to friends and relatives of the insane who knew nothing of medicine.[1] Even Hercules would have strained to clean up these Augean tables.

The obvious remedy was to work toward uniform categories. Thurnam was pushing in this direction by 1840, and Baillarger's data card for patient heredity, with which chapter 5 begins, had a similar purpose. He looked to rigorous, standardized data entry as a frame for collaborative research. Not everyone believed this to be possible. Luther Bell, director of a private asylum in Boston, complained in 1849 that ostensibly simple facts will often be superficial, leaving truth veiled in "mystery and darkness."[2] Isaac Ray in Rhode Island insisted that since patients committed to asylums are disparate, cure rates and other statistical measures must be misleading unless treated discerningly. "It is a very common saying that figures will not lie, but it is very certain that in the hands of the ignorant, the careless, the undiscriminating, they may become most potent instruments of falsehood."[3] In France, opposition to statistics went much deeper. Baillarger was sharply criticized by colleagues who understood heredity as active and ineffable, a tendency or process that could never submit to statistics. In Germany, by contrast, his work inspired statistical studies of differences between the sexes in the force of hereditary transmission. One admirer was Wilhelm Jung, author of an influential study of heredity based on the thirty years of data from his institution in Silesia. Jung made his tables as empirical as possible, and like Baillarger's, they permitted statistical associations between heredity and almost any variable for which the institution gathered data. His most consequential result, however, was simply the finding of strong inheritance of mental illness, suggesting that restrictions on reproduction would be more effective than medical treatment in the battle against insanity.

The biggest story in this period was of a series of moves to standardize national or regional statistics, leading by 1869 to an extravagantly ambitious proposal for homogeneous asylum statistics, the topic of chapter 7. It was a model of administrative and statistical organization, in principle the work of an international committee and in practice an expression of centralized French administration. Even so, many alienists from other nations welcomed it at first, until they discerned the burdensome foreign system of classifying and recording that came with it. Their resistance culminated in Friedrich Wilhelm Hagen's mocking critique of the push to consolidate and centralize. It was, however, no attack on statistics, but a defense of local culture and experience in shaping statistics whose validity was not merely on the surface. In the event, the fate of uniform standards did not depend on philosophical arguments but was sealed by the outbreak of war between France and Prussia. Even apart from international conflict, uniformity faced the insuperable obstacle of divergent institutions, regulations, and legal structures in the different states. The real achievement of the campaign for international standards was the establishment of shared categories for mental illness within the newly unified German nation.

Just as the French standardization campaign was starting up, the German alienist Wilhelm Tigges issued forth with a sharp critique of the new French theory of *dégénérescence*, or hereditary degeneration. His weapon, as we see in chapter 8, was asylum tables. These might seem too ponderous to snare such an ethereal target, yet his skillful deployment of data arrays made his reputation. A decade later, Hagen and then Tigges published tabular methods of hereditary prediction, urgently needed, they said, to demonstrate what selective breeding could contribute to the battle against insanity.

The episode of hereditary research discussed in chapter 6 was only loosely connected to table-making or to standardization efforts. Ludvig Dahl relied on Norwegian census records to identify parishes with exceptional levels of insanity and then to track down individuals who had been classified as insane. He published his first pedigree charts of mental illness in 1859, four decades before the eugenics movement began compiling them by hundreds and thousands. Dahl's story has that much in common with Gregor Mendel's, though Dahl's work was received with admiration right from the start, especially in Germany, and never disappeared from view. While he never achieved anything like Mendel's status as retrospective founder of a whole new science, the tables were widely reproduced and imitated. Dahl was interested especially in tracking hereditary factors to account for the greater or lesser prevalence of mental illness in different places and populations, another aspect of his work that was taken up eagerly by the eugenics movement.

The final chapter in this section returns to practices of gathering and recording data, mainly through institutional registration.

We have at first the impression of uncooperative sources and inadequate professional insight, or, worse, of barren, purposeless routines, especially in the 1850s and 1860s. Yet institutional recordkeeping scaled up sharply from about 1870, and many believed that better as well as more copious data might hold the key to improved medical outcomes or to healthier human reproduction, or quite possibly to both. By the 1880s and 1890s, when a new statistics began to take shape, the data techniques of asylums had reached the point that both sides, doctors as well as statisticians, could see their way to active collaboration.

CHAPTER 5

French Alienists Call Heredity Too Deep for Statistics While German Ones Build a Database, 1844–1866

> There is one great cause of insanity, a primordial cause, the cause of causes, *heredity*, which fixes the disease in families and makes it transmissible from generation to generation.
>
> —Ulysse Trélat (1856)

> On these unconscious laws [of heredity] rests the secret of breeding and crossing in the realm of animals and plants, and the possibility of attending in a similar way to human society. *Its necessity lies in our numbers.* [Ihre Nothwendigkeit liegt in unseren Zahlen.]
>
> —Wilhelm Jung (1866)

Asylums began using printed sheets and admission books to record information on new patients about 1820. At first they posed open-ended questions that invited discursive responses. By 1840, many relied on detailed forms to be filled in with nuggets of information, yet often still without specifying the field of acceptable answers. Before long, these pioneers of the information society were learning to suppress ambiguity by narrowing the alternatives.[1] Previously, when asked the cause of a father's or a wife's insanity, people had told stories. Now they might be instructed to choose from a list before spelling out details. If heredity was involved, the doctors wanted to know which relatives had been affected and by what kind of physical or moral disorder. For about half a century beginning in the 1840s, such data offered hope for shaping a science of hereditary defect.

In 1846, Jules Baillarger, editor of the *AMP*, drew up a standard form for data entry designed to structure a collaborative program of hereditary research. His scientific goal was to measure the comparative hereditary influence of father and mother in determining patterns of mental illness in male and female offspring. This work inspired a shift away from the old, one-dimensional asylum tables, collected simply as part of the registration process, to a statistics that tried to pose scientific questions. The idea was to make the table into a tool for identifying and measuring relationships. Statistical tables emerged in these years as the characteristic data technology of their era, designed to make connections visible. Often, the key variable on these tables was heredity.

In France, Baillarger's initiative failed. The ambition there to use empirical data to find patterns of human inheritance was never strong. The critical reaction was animated by a sense of biological heredity as immanent process bound up with forces of bodily development, too deep to be plumbed by mere statistics. French statistics of insanity, while plentiful, were focused mainly on administrative issues. Baillarger's endeavor to gather statistics specifically for the purpose of a scientific study of heredity ran up against this barrier. A similar effort achieved better success in Germany as a program to measure statistically the import of biological sex for hereditary transmission. It was articulated most influentially by the alienist Wilhelm Jung based on more than three decades of data from the asylum at Leubus in the Prussian province of Silesia.

Yet the problems of making data useful never went away. Jung's conclusions on sex and heredity were a bit miscellaneous and had no clear impact. What mattered most, including to Jung, was the hereditary reproduction of mental illness. It was, he declared, plainly evident in his numbers and made a compelling case for better breeding.

The Problems with Statistics

Statistics was central to the ambitions of the newly-founded *AMP*. Like their German peers, the French editors described their journal as a hub of integrated or collective investigation.

Baillarger laid out the terms of this project in an unsigned introductory essay for its first number. Statistics, he declared, might advance on the basis of clear observations despite the protean character of mental alienation. In the second year of the journal, 1844, he published his own study of the inheritance of insanity, based on 600 observations, including 440 cases of direct heredity and 160 more involving siblings and collateral relatives.[2] In medical writing, the word "observation" referred to a case or an episode of illness, which, especially for the insane, might be observed in multiple ways over a considerable period.[3] He regarded the research as Herculean in its scale, a point his medical critics conceded. Yet he modestly declined to assert that any single study could be definitive. The prospects for mental medicine would be more favorable if a project like his could enlist a score of physician-researchers in different parts of France and beyond, laboring collaboratively to identify patterns of hereditary transmission.

To this end he created a tabular form, divided between paternal and maternal sides, each with columns for parent, grandfather and grandmother, uncles and aunts, great uncles and great aunts, male and female cousins, and brothers and sisters. The data form had rows for up to ten insane children in any family, a generous allotment. It included a blank space in the upper left on which a doctor choosing to participate could enter his name. Baillarger found, as he seems to have expected, that fathers transmitted insanity preferentially to their sons, mothers to their daughters. The maternal effect was the stronger, appearing in 271 cases, while paternal influence was limited to 182.[4]

On the basis of this work, Baillarger presented himself for a position in anatomy and physiology at the Paris Academy of Medicine.[5] Although he was, in the end, elected, the assessment of his candidacy laid bare a French understanding of medical heredity as too deep or too innate to yield to empirical investigation, and thus far beyond the scope of statistics. The academy appointed a commission to assess his research, with the physician Hippolyte-Louis Royer-Collard as reporter. His report was printed in the academy's par M. le docteur

Numéros.	Initiales des noms.	Age.	Côté maternel. Mère.	Grand'mère.	Grand-père.	Oncles.	Tantes.	Gr.-oncles.	Gr.-tantes.	Cousins.	Cousines.	Frères.	Sœurs.	Fils et neveux.	Filles et nièces.	Total.	Côté paternel. Père.	Grand-père.	Grand'mère.	Oncles.	Tantes.	Gr.-oncles.	Gr.-tantes.	Cousins.	Cousines.	Frères.	Sœurs.	Fils et neveux.	Filles et nièces.	Total.
1	D	37	0	0	0	0	0	0	0	0	0	0	0	0	0	0	1	0	0	0	1	0	0	0	0	0	0	0	0	2
2	G	62	1	0	0	0	0	0	0	0	0	0	1	0	0	2	0	0	0	0	0	0	0	0	0	0	0	0	0	0
3																														
4																														
5																														
6																														
7																														
8																														
9																														
10																														
			Total des cas de folie transmis par la mère sur 10.														Total des cas de folie transmis par le père sur 10.													

NOTA. Pour les fils et les neveux ou les filles et les nièces on ajoutera les lettres F ou N après le chiffre.

FIGURE 5.1. This tabular form was designed to facilitate cooperative research on the importance of paternal versus maternal hereditary influence by making the data of different doctors interchangeable. There is a space in the upper-left corner to enter the name of the doctor. From Jules Baillarger, "Recherches statistiques," 338.

Bulletin in 1847 then excerpted at length in German translation in the *AZP*. Royer-Collard, son of a doctor and nephew of an influential statesman and philosopher, professed a deep respect for Baillarger, which, if sincere, did not extend to his methods or conclusions. The inheritance of insanity, he declared, is really a question for philosophy, "I would also say, for administration; I mean, at least, for those men who occupy themselves seriously and disinterestedly with ways of improving the conditions of social life in civilized states." The problem was not Baillarger's observations but statistics itself, "from which men have tried, so improperly, to make a science." An idolatrous cult, he complained, has grown up around the image of "what are called facts." Too often, statistics has been described as the most positive and most certain of all methods of scientific investigation. He endorsed numbers as an administrative tool. Scientific truth, however, lies at depths that statistics cannot plumb.

> Figures, for some men, even highly enlightened ones, are an irrefragable expression of truth. But . . . figures, like words, have only a representative value; statistics, which collects them and shows their direction is, by itself, but a blind woman [*une aveugle*] who does not reason: and, to get at the truth she promises, one must necessarily cut through all this drapery and go straight to the things represented, bypassing the signs that represent them.[6]

Royer-Collard bypassed mere statistics to divine a more essential truth of heredity lying at the crucial node of reproduction. The newborn is not merely produced by its father and mother but extends their existence, including their moral and intellectual faculties, into the succeeding generation. Triggering factors are of very little account. It is not the *maladie*, the illness, that is perpetuated, but the *malade*, the sick person, whose predispositions extend to health, constitution, and temperament. It is enough to grasp this principle to comprehend the crucial role of inheritance of insanity, occurring as an immanent process that could scarcely depend on empirical evidence.[7]

And that was fortunate, because Royer-Collard dismissed the evidence as utterly mendacious. He told how his father, the Charenton alienist Antoine-Athanase Royer-Collard, had accumulated information on causes for twenty years. For fifteen of those years he was assisted by Louis-Florentin Calmeil, his student, who subsequently confided to the son that the data was rubbish. Despite taking all possible care in his interviews with family members on the question of hereditary influences, Calmeil had discovered afterward that "eight times out of ten" they had deceived him. Every lie and every uncertainty had tended to understate the power of heredity, which must remain hidden when, for example, an early death prevents the expression of inherited mental illness. The blame for the faults in his report, Royer-Collard concluded, belonged not to Baillarger, but to statistics itself. Facts so complex as those of this living economy, double- and triple-sided facts, fused and entangled, cannot be added up as if they were independent units.[8]

Hard Facts and the Allure of Narrative

Ulysse Trélat found equally memorable reasons to reject testimonies on the causes of insanity, and with them, every kind of hereditary data. He had had a prominent role in public health during the revolutions of 1830 and 1848, then served during the Second Empire as a physician at La Salpêtrière, treating "a vast clientele of insane women," whose lives he featured in an 1856 essay on the causes of insanity. He had learned to be wary of deceptive evidence, so readily conjured into mesmerizing narratives. His case stories exhibit this sweet seduction while proving his own masterful capacity to resist it.

He tells, for example, of a woman of "thirty years and some," Madame J., who married while not yet thirty an active septuagenarian laborer. The dear husband is indispensable to his employer, and she thinks of nothing while he is at work but to keep up their little abode and await his return. One day, a young clockmaker, arriving to repair the pendulum movement, shocks her by declaring his love. She virtuously rejects these avowals, and he is polite and respectful when he brings back the clock, yet the flame in his eyes burns her soul. Tormented beyond her powers of endurance, she throws herself into the Seine then is rescued and sent to the asylum. "Assuredly we have here a sufficient explanation of the despair of this young woman.... Was there not in this little story enough misfortune, enough poetry to seize the entire soul?" Later, Trélat discovered that she had long suffered attacks of nerves, as had her sister. The cause of her madness was heredity, and the clockmaker with his long hair and black eyes achieved no more than to precipitate what would otherwise have developed later with a different provocation.

It was the same with a general's widow who went mad after falling into destitution, and with an English working woman taken in by a scheming dentist who, promising marriage, used her resources to set up his practice and then abandoned her. Other cases that the doctors had explained in terms of physical causes such as childhood falls or blows to the head also gave up their secrets when he discovered a deranged family member.[9]

"We pass to observations that are still more striking":

Lise R . . . was a young and very beautiful girl in the valley of Montmorency. A rich man, married, but of dissolute morals, noticed her, seduced her, and brought her into an utterly different world from that in which she had been raised. A simple life in the fields, of hard work during the day and deep sleep at night, gave way abruptly to excesses of the table, gambling, orgies, dances, spectacles, stays at the baths, journeys abroad. R . . . had from her seducer two beautiful girls, and these had scarcely begun to grow up when another household began taking shape amidst hers. R . . . and the new woman, who promptly became mother of a son, linked up and shared the same evenings of pleasure. One of the two little girls died. The children were raised in this disorder by the two concubines of the debauched old man, who still maintained his legal household in Paris, half a league from his harem.

But, after these joyous dissipations of every kind came the bad days. When you consume above your revenues, you devour the capital. They had to flee their creditors and transport the debris of lost luxury into an abode that was barely modest, living in shame, privation, and misery. Poor R . . . soon came into our asylum, where she remained the most beautiful, the most furious, and the most formidable of our patients. She was often visited there by the *other lady*, who would say to her child: "Come see Lise!" The girl had the same beauty as her mother. We didn't see her for several years. What could become of her with such models?

For a long time we never looked for the causes of the insanity of Lise R . . . except in the life her seducer made for her. A complete change of life, pleasure, exhilaration, excess and exhaustion day and night, all the seductions and the satiety born of sumptuous habits of opulence, and then all at once suffering, tiredness, fears, quarrels and affronts engendered by ruin with no honorable recollections to redeem the abasement of such conditions: quite enough causes to trouble the

> soul and to shatter it. Well! All this wasn't what distracted the reason of poor Lise. It is less poetic, but it is more truthful: she went mad because there were several insane in her family.[10]

Layered over these operatic tales, recounted with evident pleasure, there is always another story featuring a tough alienist who refuses to be taken in by their flimflam. The sinuous story line strikes a barrier and is completed, or annihilated, by facts and laws that are wholly external to it. Heredity arose here as an unyielding force, the irreducible continuity of life. Trélat had the steel to resist the tawdry allurements of surface explanations. "Error is attractive by nature, because it is always full of consolation and hope, while the truth, in most cases, is sad, inexorable."[11] The blows and misfortunes of life can be painful, but they cannot explain insanity. Those who keep looking will discover, behind the fog of appearances,

> the true cause, the living cause, imperishable and transmissible. . . . The germ was there, and sooner or later it had to grow.
>
> There is one great cause of insanity, a primordial cause, the cause of causes, *heredity*, which fixes the disease in families and makes it transmissible from generation to generation.
>
> This is a law.[12]

Troubles with Tables: Some International Exchanges

The mood of Royer-Collard's and Trélat's hereditary doctrines was distinctively, if not quite uniquely, French. Carl Hohnbaum, Royer-Collard's German translator, rejected his claim for hereditary continuity from parents to offspring. The propagator, he declared, must be distinct from the propagated. Although much concerned with causal mechanisms, he refused to allow that theoretical claims could ever supplant empirical investigation. Nature, retaining its creative power, brings forth mysteries but also reveals patterns, including great talents that run in families and

FIGURE 5.2. Asylum at Stéphansfeld (near Strasbourg). This lithograph from 1841 of the first asylum in Alsace shows a tranquil, rural site enclosed by walls. The institution was a key site of French statistics of insanity and, along with some nearby institutions, an important conduit for translation and data communication between French and German alienists. It remained so after 1871, when it was incorporated into the new German empire. From *Renseignements sur l'asile départementale d'aliénés de Stephansfeld (Bas-Rhin)* (Strasbourg: Vve Berger-Levrault, 1841), plate 1. In public domain.

hereditary sources for a range of diseases. Royer-Collard went too far in dismissing statistics, which should be combined with close study of particular cases. Without taking a position on Baillarger's scheme to coordinate research with a printed form, Hohnbaum endorsed his effort to infer causal mechanisms from the statistics of medical experience.[13]

Royer-Collard's critique of statistical research rested not only on his sense of the deep forces of heredity but also on an understanding of statistics as a bureaucratic endeavor, distinct from science. Morel's career similarly exemplifies the possibility of treating heredity as an ineffable force while cultivating statistics for practical, administrative purposes. Baillarger, looking to statistics as a basis for science, dismissed asylum reports as hemmed in by their reliance on administrative categories. For the researcher taking an original approach to a serious question, the numbers from his own institution will rarely be sufficiently copious for reliable statistical conclusions. His solution was collective data, compiled and distributed by a well-managed journal. In this way, medical statistics of insanity could be freed from the inflexibility of institutional compilations.[14]

This scheme of original research coordinated by a printed form does not seem to have gone anywhere. There was no getting around the separate institutions as principal loci of inquiry. Although Baillarger drew German supporters, in France he was caught between two poles. What seemed aimlessly empirical to philosophical authors like Royer-Collard was criticized by others as detaching statistics from the essential needs of state administration.

Emile Renaudin spent most of his career in Alsace, which functioned as a French-German borderland. His four-year report on the asylum of Stéphansfeld, near Strasbourg, printed in 1840, included an exemplary combination of administrative and medical data, including an exploration of heredity as predisposing cause of insanity.[15] Five years later, he inserted into the *AMP* a multipart essay, really a book, on the administration of insane asylums. Here the focus was more narrowly practical. Beyond his duty to the state to operate an effective institution, he wrote, the asylum director needs to educate the public. He should print annually a "moral and administrative account" according to a uniform plan. This would be mainly a catalog of observations: patient data in the medical part and budget figures in the administrative. The state should collect these documents each year from all over France and publish them in May. They would provide a statistical basis for comprehending how insanity varies in relation to geography and to legal regimes. This sounds like the work of a census, but he preferred to rely on institutional records. They would require, however, to be reformed and standardized, which emerged here as his mission.[16]

Baillarger, who did not welcome a state publication competing with his *Annales*, responded with a critique of official statistics. To ward off this rival, he announced a new statistical yearbook, the *Annuaire historique et statistique des établissements d'aliénés*, to be published in April (!) as a number of the *AMP*. He does not appear ever to have acted on this intention. He pursued numerical uniformity by a different route, outlined in an open letter to Renaudin on the application of statistics to mental illness. Amariah Brigham promptly had Baillarger's letter translated by "an Inmate of the New York Asylum" and published in the *AJI*. It begins: "Statistical researches have doubtless rendered great service to the study of mental diseases; but confined of late years in a circle consecrated by habit, they have ceased in my opinion to be so useful." The true need was for collective work on a uniform basis by associations of physicians, who should fix in advance their research questions, such as the nature of inherited insanity.[17]

Brigham, it appears, was moved by Baillarger's vision of uniform statistics to initiate his own series of articles on the statistics of insanity. While praising Baillarger, he seemed instead to follow Renaudin's lead. Annual asylum reports provide the public with irreplaceable evidence of the curability of insanity, in this way encouraging legislators to fund these valuable institutions. Their principal defect was a lack of uniformity, and he held up the proposed English registers of 1842, modified from Thurnam's proposal, as a proper basis for harmonized statistics in America. For his next installment, Brigham promised to answer objections to statistics. The series was, however, cut off by the sickness and death of this author.[18]

Renaudin, meanwhile, had answered Baillarger's letter. The men agreed on the need for large-scale statistical research and agreed also on the desirability of standard forms and tables. But Baillarger had in mind a project of independent research designed to yield results of scientific interest for his journal. Renaudin, like Brigham, valued routine data collection according to rules set by the state. His praise for Baillarger's "wise commentaries" served mainly to accent his doubts as to the value of Baillarger's data. Statistics is no panacea, Renaudin wrote, least of all for science. Asylums require numbers, but mainly for rhetorical and administrative purposes. They can be understood by everyone; indeed, nobody listens to reasoning unless it is flanked by figures. Armed by data from his own and other institutions, he would challenge error and prejudice to win the battle for humane treatment of the mad.[19]

By 1848, Renaudin had begun reporting for the *AMP* on German institutions, which he now brought into the debate. Perhaps this reviewing heightened his sense of disparities of meaning among institutions for categories that, on the surface, appeared homogeneous. "Everywhere, and especially in Germany, the need is felt to unify and coordinate the numerous facts observed in asylums for the insane," he wrote. The German *AZP* editors had just published a *Normal-schema* for asylum data, embodying scientific goals that he dismissed as worthless. It would contribute nothing, he declared, to knowledge of the causes and geographical history of insanity.[20]

The German report was preoccupied with statistics of heredity. It has the look of a compromise, an adaptation of Baillarger's ideal to the shaping of scientific statistics for official reports. The editors were pushing an all-German project of compatible, if not yet collaborative, measurements, the data to be gathered up and diffused through publication in the *AZP*. Damerow, as editor, nurtured high hopes for this empirical undertaking, so different in spirit, he said, from the "dogmatic" psychiatry of the prior generation in Germany. He had just released an exuberant editorial introduction to his 1846 volume, vowing that statistics would now at last fulfill its promise to bring evidence to bear on the great questions of psychiatry. This, he let on, was to be achieved through a redesign of asylum and census reports.[21]

While Renaudin did not specify his objections, he could not have approved the ambition to set aside routine reporting and to redefine insanity, with its urgency for public health, as a scientific problem. Most of the sixteen tables proposed in Flemming's schema involved converting raw numbers into ratios or percentages, sometimes in relation to a larger average of patients, sometimes as a fraction of the whole population. The latter would facilitate comparison across regional and national boundaries. These German alienists may have overreached, however. There are indications of discord. For some events, such as relapses, they were unable to specify any breakdown or typology, and on the crucial question of how to tabulate "etiological factors," they came to a stalemate. The statistical problem of causation was thus to be left to the wise discretion of local alienists: "The correspondents will be allowed to compose a tabular overview according to their best insight and careful consideration of the respective causes." This was scarcely a basis for standardization! At least they reached a consensus that hereditary predisposition (*erbliche Anlage*) should be a particular focus of attention.[22] Heredity, in Germany, was the exception to midcentury doubts about data on causes. Baillarger's work suited them nicely.

Methods of Tabular Statistics: Wilhelm Jung at Leubus

Paul August Wilhelm Jung followed Baillarger both in his commitment to statistical research on heredity, which stood above official routines, and in structuring his analysis according to the variable of sex. To Germans, at least, his two long papers in the *AZP* in 1864 and 1866 were a compelling model of what could be achieved with asylum data for research on inheritance of mental illness. They were regularly cited for half a century.

A range of German commentaries shaped Jung's understanding of Baillarger's work, in particular a close analysis of his text by Rudolf Leubuscher. Leubuscher had grown up in Breslau (Wrocław), a few miles from Leubus (Lubiąż), the source of his family name. The asylum at Leubus, converted from a Cistercian monastery after the Napoleonic wars, was where Jung, as assistant physician, gathered up thousands of patient observations. Leubuscher took his doctorate in Berlin with a thesis on religious mania and then served from 1845 to 1847 as Damerow's assistant physician in Halle. His paper on inherited madness appeared in 1847 in the *AZP* and a year later in English translation. He subsequently made his reputation in Berlin as a public health reformer and ally of Rudolf Virchow.

Leubuscher emphasized the shaping of individuals by heredity, calling it "the best established and least doubtful" among the causes of mental illness. He acknowledged its obscurity as a medical concept, and even the inconsistency of measures of its effects. He had

FIGURE 5.3. Asylum at Leubus, formerly a Cistercian abbey. A visitor wrote in 1852: "The building, which is of vast size, had originally been erected as a palace, was afterwards used as a convent, and finally, in 1830, converted to a lunatic asylum" (W. F. Cumming, *Lunatic Asylums in German and Other Parts of Europe* [London: John Churchill, 1852], 45). Most early German asylums were converted religious buildings. In public domain.

no truck with Royer-Collard's theory of heredity as a continuation of the life of the parents. The instantaneous combination of maternal and paternal elements in the fertilized egg, a "middle thing" between father and mother, confers a type on the offspring, leaving no further role for the pregnant mother except to nourish her fetus. Mental dullness (*Blödsinn*), often congenital, seemed to him a more convincingly hereditary condition than insanity, which appears in adolescence or adulthood. He applauded Baillarger's use of systematic data gathering to clarify the processes of insanity and to reconcile discrepant measures of its inheritance. On this basis, it might become possible to intervene effectively and to reduce mental illness through better practices of breeding and child-rearing.[23]

The faith of German alienists in statistics was at a high in the 1840s. Wilhelm Griesinger, leader of a Berlin group, explained in his 1845 textbook of mental illness how multiple causes create complex variability both within and among institutions.[24] Flemming echoed the thought in 1852, concluding that averages alone could stabilize this Proteus.[25] In 1854, the physiologist Georg Schweig discussed three possible ways to apply mathematics to insanity and concluded in favor of the simplest, the method of classification into groups, requiring only that the groups be mutually exclusive. Physicians too often neglected this rule, he continued, citing the wildly discrepant causes in Esquirol's tables. A proper statistical categorization would distinguish heredity without other causes from heredity with other causes and from other causes without heredity. Such reasoning privileged the causal import of heredity. It also called attention, as Jung later would, to interactions of disparate causes.[26]

The institution at Leubus had been founded about 1830 by the physician Moritz Martini, who remained as director until Jung replaced him in 1873. On the basis of a census of the insane, and supposing the continuation of high initial cure rates for fresh patients, Martini had reckoned that one hundred beds would suffice to care for the local population.[27] The first report in 1832 did not take up causes. But he must have collected causal data from the beginning, because Jung's essay on the inheritance of mental disturbance, appearing in 1864, five years after his arrival at Leubus, included the results of thirty-three years. From discussions and abstracts of Leubus reports in the *AZP*, it appears that German alienists esteemed them as among the very best of the genre. The tables for this institution, appearing every few years, were always accompanied by results of a census of insanity in the province of Silesia. Heinrich Laehr described the four-year Leubus report for 1860 as excellent. Other provinces now prepared similar reports, he remarked in 1865, but none as yet had quite matched Leubus. It would be desirable to print up reports like these in enough copies to distribute to all readers of the journal.[28]

Jung's paper became a model for the next generation of asylum studies of heredity. Its scope was monumental: 3,606 patients filling 130 pages, or 177 if we include the 1866 continuation. German research on psychiatric heredity, as on so many topics, was now very serious. Jung made no effort to probe bodily mechanisms of hereditary transmission. This was a data project, focused on family relationships among those diagnosed as mentally ill. "The task of this work is to be purely statistical, and it should therefore hold at a distance all *Raisonnements*, with their openness to ad hoc considerations." He would "leave it to the naked facts alone to speak and from them to draw naked conclusions."[29] Jung gave no ground to those troubled souls who insisted on interpretation to ascend to sound conclusions, quantitative or not. This meant ignoring subtle issues of temperament and its diverse tendencies, since these would lead to speculative and inductive conclusions rather than statistical ones. His tables relied strictly on diagnosed mental disorders.

But subtle reasoning, barred at the door, came in through the window. The detection of patterns of inheritance depended on satisfactory descriptions of the phenomena of insanity, and these, he had to concede, were far from transparent. He supposed that physical damage or organic disease could lead to mental disturbance and that such cases could become hereditary. It would be valuable to identify the circumstances under which an *Anlage* goes dormant in the direct line while popping up among collateral relatives. What is transmitted from parent to offspring, he reasoned, cannot be the disease itself but rather a bodily order, form, or condition, an "organic *Anlage*." Martini spoke in those years of "the inherited *Anlage*, the transmitted seed." Often what really matters for disease, wrote Jung, is the "material substratum," which can be stimulated to develop by any kind of "accidental impulse," whether internal or external.[30] It begins to sound like *Raisonnements*. Jung struggled to comprehend not just transmission as such, but how it interacted with other factors recorded at the time of admission, including age, sex, disease form, and the delay from disease onset to treatment. He hoped to determine how results of treatment depended on these factors, alone and in combination. He acknowledged that much about insanity, even the demarcation of sickness from health, was clouded by ambiguity. In many patients the illness evolved over time, even to the point of requiring a revised diagnosis. This kind of information could not just be transcribed unthinkingly from the medical reports that served as his sources.[31]

Above all, he sought out variables that determined the probability of hereditary transmission, and with them, a better measure of the percentage of mental illness attributable to heredity. Like many writers on causes of insanity, he believed that the role of family *Anlage* was almost always underestimated. In the public section of his hospital, filled with paupers, many knew almost nothing of relatives beyond their own parents. The paying patients, though better informed of their lineages, tried to conceal evidence of family disease. Sometimes, what the families hid might be learned from reports of the recommending physicians, who also could provide information about the medical condition of these relatives. Many families, unfortunately, were so dispersed that nothing could be learned about them. Still, the institution at Leubus held more than thirty years of records. It is pertinent that German institutions in this period kept their patient records in files rather than the unwieldy case books used by British and American asylums. From the scale of Jung's investigation, it is evident that Dr. Martini was willing to dedicate quite a lot of the time of his third physician to this investigation of insane heredity. Martini, indeed, had initiated the study and had printed results in 1860 for the first thirty years of the institution. He was keenly interested in the force of heredity and how it differed by religion as well as sex. In one report, Protestant insanity proved to be hereditary in 20% of cases, Catholic in 17%, and Jewish in 35%.[32]

Jung gave much thought to the problem of organizing data and appears proud of his solution: the compilation of lists. Hereditary patients were to be categorized by their relationship to another sick person, such as mother to daughter or uncle to nephew. Like Baillarger, he carefully distinguished the paternal from the maternal side. Vague references in the records—for example, to a family *Anlage* or a mentally ill grandfather—required another set of more provisional lists. Some persons would appear twice, both as sources and as recipients of hereditary influence, a duplication that required numerical correction to avoid double-counting.

Jung was as systematic as possible, using a hierarchy of letters and numbers to sort out heredity by sex and by nearness of relationship. It is all a bit numbing. Letters distinguished male from female patients; roman numerals indicated inheritance on father's side, mother's side, both sides, or between siblings; arabic numerals distinguished insanity of a parent from other relatives. He proceeded to a still finer categorization. For

	A. I.		A. II.		B. I.		B. II.	
	A. I.	A. III.	A. II.	A. III.	B. I.	B. III.	B. II.	B. III.
unter 1. . .	37,17	—	43,70	—	42,45	—	58,86	—
unter 2. . .	33,96	—	32,34	—	41,00	—	35,29	—
überhaupt aufgeführten	36,14	—	41,17	—	42,06	—	54,28	—
mit Angabe der Form								
unter 1. .	70,83	71,18	65,21	68,83	77,08	78,70	81,60	81,63
unter 2. .	80,95	73,07	75,00	64,28	69,56	68,00	70,00	70,58
überhaupt aufgeführten	73,90	—	66,66	—	74,64	—	77,57	—
überhaupt als geisteskrank aufgeführten	41,97	45,16	47,00	49,53	53,62	55,81	61,73	60,60

FIGURE 5.4. Wilhelm Jung offered this tabular scheme as a compact and precise way to record data on hereditary transmission and to facilitate analysis. On the horizontal axis, he used *A* for male and *B* for female insane. *I* indicates mentally ill relatives on the father's side, *II* on the mother's, *III* if on both sides, *IV* for siblings. On the vertical axis, he used *1* if the affected relative was a parent, and *2* if it was some other relative. The table continues down the page with finer distinctions and percentages. From Jung, "Untersuchungen," 622.

example, "A.I.1" designated insane men whose nearest affected ancestor was the father. There were 48 such patients, in 17 of whom the *Anlage* was reinforced by at least one additional sick relative, and of these, 11 belonged to families with only a single form of illness. After splitting his cases as far as possible, he descended to the level of individuals. Of the male patients with an intensified *Anlage* whose fathers suffered the same form of illness, there was one case of melancholia in which the paternal grandfather and brother of the patient also suffered melancholia, two cases of melancholia in which just the paternal grandmother also had melancholia, and so on. Jung continued in this vein for about thirty pages. The family characteristics extended beyond diagnosable mental illness to include drunkenness, apoplexy, epilepsy, and singularities or unusual behavior. Would Jung have been shocked to be told that he was compiling a database? Like Baillarger, he wanted his technique of disciplined data entry to permit any researcher to add or recombine, to compare and even (granted a few simplifying assumptions) measure the strength of hereditary effects.[33]

Jung's conclusions, as in so many asylum reports, lead us back into the world of naked, unexplained, and, for this reason, strikingly miscellaneous facts. Insanity breaks out mainly with the onset of adulthood, affects Protestants more than Catholics, and is more frequent yet more easily cured for women. Cases within a family group are often similar. Women have a stronger *Anlage* for inherited mental disturbance than men. Martini's result on differential inheritance by religion, however, he brushed aside as an artifact: it is merely that the most inbred (and hence degenerate) groups are the best informed about their kin. Jung's data confirmed the familiar result that hereditary insanity was more curable than nonhereditary, but they surprised him by suggesting that inherited insanity had the best outcomes when the affected ancestor suffered mental illness rather than a less severe condition such as epilepsy, hysteria, or peculiarities of character.[34]

Soon after publishing his study, Jung was drawn back to this topic by a closely related paper from Scotland. Since he lacked access to the *Journal of Mental Science*, he consulted the French translation, which was published within months in the *AMP*. The author, Hugh Grainger Stewart, had compiled a report based on hereditary data from twenty-five years of experience at the Crichton Royal Institution, where he served as medical assistant. These records, Stewart explained, had been kept by three successive physicians and relied on information from friends and medical advisors of the patients. The questions posed to them were extremely loose: "Is the patient, or his relatives, subject to any hereditary, nervous, or periodical disease, and what? Or have they manifested any peculiarity, eccentricity, or prominent propensity, or tendency to crime?" Jung was especially interested in results concerning

4. Empfänglichkeit des Geschlechts für das erbliche Irresein.

	Männer	pCt.	Frauen	pCt.
Hood	121	8,58	240	10,62
Thurnam	65	32,82	77	35,48
Crichton-institution	253	48,56	194	51,05
Baillarger	271	49,72	274	50,27
Leubus	492	48,71	518	51,29

FIGURE 5.5. Table of male and female "receptivity" to hereditary influence, as measured in five studies in different institutions. Jung stressed that while the differences of receptivity between the sexes were small, all pointed in the same direction. The comparative roles of male and female in hereditary transmission were of great interest to asylum researchers. From Jung, "Noch einige Untersuchungen," 220.

hereditary influence as a function of sex.[35] Stewart's findings were in good accord with his own results, and with many others: a solid, repeatable finding of asylum-based statistical research. He was moved to what we can call meta-analysis, based on a meta-table. Even though the absolute measures varied greatly, from Charles Hood's 8.58% and 10.62% at Bethlem to figures around 49% and 51% in Scotland, Paris, and Silesia, the differences, however slight, were all in the same direction, all showing greater female "receptivity" to hereditary madness.[36]

For decades, beginning in the 1840s, the breakdown of hereditary results according to sex appeared the most promising way to get beyond mere numerical summaries to some kind of insight into mechanisms of heredity. By 1865, the range of numerical conclusions on inherited insanity was expanding. Another such study, a twenty-year report, appeared in 1866 from the noted institution at Illenau in Baden, directed by Christian Friedrich Wilhelm Roller, third in the triumvirate of editors of the *AZP*. He charged his younger brother Robert to prepare the statistics, which he then mobilized to shed light on causes and on the effectiveness of treatment.[37] The Illenau weekly newsletter, written for personnel at the asylum and perhaps for higher officials, commented on the cheering conclusion of these statistics: that patients with a hereditary *Anlage* are not thereby condemned to confinement until death but are actually more likely to recover. Less encouragingly, Roller appealed to an "inner conviction" of alienists that the increase of insanity was no statistical illusion but a genuine consequence of excessive demands on the modern brain, stormy politics, extremes of urban wealth and poverty, and the dangerous use of stimulants.[38] Like Jung, the Rollers relied where possible on a cutting-edge technology of data analysis, the correlation table. They were particularly interested in co-related variables, including disease form and percentage of cures; assigned cause and disease form; and cause and percentage cured. They vowed to avoid "excessive subjectivity" and "to strive for all possible objectivity."[39]

Damerow also weighed in at almost the same time with a twenty-year report on the Halle asylum. He focused less on tabular technologies than on a troubling rise in the percentage of hereditary patients. Through marriages of relatives, it seemed, unnoticed personal *Anlagen* were infecting whole families. It was, he said, like the statistics of suicide and crime in a well-known book on statistical laws by the economist Adolph Wagner, an accumulation of separate causes from which emerged a disconcerting pattern of regularity. Wagner, the son and grandson of doctors, helped set off a German debate on free will by advertising the uncanny statistical stability of suicide, a central concern of alienists. Damerow, who knew Wagner's family, looked for an explanation in patterns of heredity.[40]

Selective Breeding as Social Medicine

By 1865, the statistical investigation of causes was entering a new phase. Jung, too, set out on the path of the correlation table. It took a lot more work to process figures in this way, all the more so when thousands of cases were involved. The asylum doctors were moving into an era of arduous data management.[41]

Jung's willingness to make this effort reflected his firm conviction of the importance, practical as well as theoretical, of hereditary causation of insanity. Dismayed by English historian Henry Thomas Buckle's disregard of the power of heredity, he appealed to Trélat's pronouncement that heredity was "the cause of causes." "Heredity has an essential role in the genesis of mental illness," especially in the case of marriages between branches of the same family. Mental weakness arises in the spoiled marriage bed. Weakness and disease are not only inherited by children and grandchildren but can "leap over whole generations, 'remaining there hidden within and in this hidden form be transmitted again.'" This concluding phrase Jung copied from Dr. Friedrich Rolle, an early German expositor of Darwin's theory with a particular interest in biological inheritance. The biology reinforced hereditary concerns that alienists had been discussing for decades. Damerow too was now crediting Darwin as having shown how propagation of weak and inferior types leads in biology to species ruin. For humans, this applies most urgently to hereditary forms of psychical illness. Jung, following Rolle and suspending his ideal of naked, unexplained facts, posited hereditary elements, borne by the cells and susceptible to environmental influence, that could act as a tendency or receptivity.[42]

Again we are converging on eugenics. Jung concluded that "unconscious laws" of heredity explain "the secret of breeding and crossing in the realm of animals and plants and the possibility of attending in a similar way to human society. *Its necessity lies in our numbers.* If these are convincing, there will be no need for prohibitions, no external coercion." Jung quoted here the words of another medical Darwinian, Hermann Eberhard Richter, who had caught a vision of *Zukunfts-Medicin*, or "medicine of the future," and pointed to a new and higher duty, "to lighten the burden of the struggle for existence" and to promote the "improvement of human capacities through rational breeding." The advisability of controlling the reproduction of persons of defective heredity seemed clear. The "rational necessity" of Jung's numbers might permit breeding decisions to remain voluntary. He added the happy thought that medicine was now on the path to "calculating in advance the success or failure of any marital bond," and the darker one of looming "degeneration and the most frightful proletariat."[43]

CHAPTER 6

Dahl Surveys Family Madness in Norway, and Darwin Scrutinizes His Own Family through the Lens of Asylum Data, 1859–1875

> The "medicine of the future," in many of its departments likely to become almost wholly preventive, and thus of a public and general character, must owe more and more of its progress to statistical science.
>
> —John Gray (1861)

> Such a factor [*Anlæg*] having once arisen—from unknown causes—in a small and isolated population, may easily increase and disseminate in the course of time, notably by continued marriages in kin already affected.
>
> —Ludvig Dahl (1862)

Norway, on the European periphery, had attracted a degree of interest from alienists for its thorough census and—no mere coincidence—its uniquely high measured rates of insanity. In 1857, when Ludvig Dahl quietly began work on his *Contribution to the Knowledge of Insanity in Norway*, it had only recently set up its first dedicated institutions for the mad. Yet Norway possessed an unusual capacity to survey its population. That infrastructure enabled Dahl to study mental illness outside of asylums at an unprecedented level of detail and to assemble family pedigrees of insanity decades in advance of other countries. He reasoned upward from families to the level of census, putting forward an explanation of geographical differences in the prevalence of insanity in terms of the transmission of hereditary factors. Despite the local specificity of its subject matter and its linguistic inaccessibility, the book attained an international reputation as a key resource for investigating and representing the inheritance of mental illness.

How Much Madness?

In 1860, the *American Journal of Insanity* printed a list of the proportions insane in different places. The highest ratios were for Norway (1:550), Scotland (1:513), and the canton of Geneva (1:446). John Gray, *AJI* editor and superintendent of the Utica asylum, lifted these figures from B. A. Morel's just-published *Treatise on Mental Maladies*, which followed by three years his initial articulation of the theory of hereditary degeneration in 1857. The new book, focusing more specifically on phenomena of mental illness, reemphasized his commitment to statistics. Morel kept his distance from cultural pessimism, insisting that civilization could no more cause insanity than true religion. Yet he recognized that asylum patients had increased alarmingly, along with the systems that housed them, and he, like so many of his professional colleagues, wondered anxiously where it would end. New counts almost always showed bigger numbers, as in Scotland, for which Morel and his reviewer mentioned a more recent and still higher figure recently published by an Italian doctor, 1:417.[1] He had somehow missed an enumeration by a Scottish commission in 1855, distinct from the decennial census, incorporating data from sheriffs, boards of supervisors for the relief of the poor, ministers, and rural constables as well as its own detailed investigations. The figure they eventually released in 1857 implied a ratio in Scotland of 1:390.[2]

Yet even this was not the limit, as Gray already knew. A still higher ratio appeared six pages up in the same issue of the *AJI* in a reprinted extract from the Irish medical press, "Notice of Dr. Dahl's Report Respecting the Insane in Norway." Since 1825, the Norwegians had counted the insane as part of their decennial census. Morel knew only of the first, probably from Esquirol's French review. Each successive count revealed some growth of population and a disproportionate increase of insanity. Holst's figure (1:550), so startling in 1828, had been surpassed in 1835 and again in 1845.

The most recent census, made known outside Scandinavia by reviews and discussions of Dahl's report, found 5,071 insane, including idiots, in a population of 1,490,047. This meant 1 insane per 293.8 of population, a number, as Dahl admitted, without parallel among nations. He warned against exaggerating its significance, and especially against any Norwegian exceptionalism.[3] He might as well have commanded the incoming tide to freeze in place. Alienists wanted to know what conditions in Norway could account for this unparalleled scourge.

Gray introduced the next year's volume with a historical meditation on statistics. "Looking over the annual reports of American Asylums, we need not say how almost entirely they are filled by numerical tables. . . . With a certain license—as pardonable perhaps in the specialist as in the poet, who coerces to the demands of measure or rhyme the noblest and most comprehensive words—these are called 'statistics of insanity.'"[4] Gray spoke of license because he knew that statistics, according to its etymology, should be about the state. He explained, incorrectly, that the numerical method in medicine, made famous by Pierre Louis's test of the efficacy of bleeding, was the model for Esquirol's (earlier) tables of insanity. Gray declared that statistics had failed utterly in pathology and provided only negative proofs in therapeutics. Now it lingered on mainly in epidemiology and public hygiene, areas of medicine devoted not to individualized treatment but to principles and to masses, where he thought it appropriate.

On this matter he invoked another French source, the 1857 census of establishments for the insane, which the *AJI* had recently reviewed. He endorsed the claim of the French census director that the greatest predisposing cause of insanity was heredity, yet he attributed the explosive growth of the numbers (from 10,539 in 1835 to 24,524 in 1854) to the magnetic appeal of an expanded asylum system.[5] There was no real increase of the insane, he wrote, but only of institutionalized patients. Although Gray's opening remarks on the tyranny of tables sound cynical, by page 6 he was praising the unified French census for its "thorough system of records" providing indispensable guidance to the legislator. Even as he acknowledged the force of Isaac Ray's critique of statistics and the obstacles to tallying a thing as fuzzy as insanity, he called not for strategic retreat but for a still bolder advance. Statistics so far had merely recorded the results of insanity, but Morel's clarified taxonomy opened the door to statistical scrutiny of its pathological development. According to the new French category of "hereditary insanity," inheritance was no longer a mere symptom but the disease itself. Morel had cleared the way for the "medicine of the future," which would focus on prevention rather than treatment.[6] The case is similar with Norway.

Tracking the Elusive Hereditary Anlæg

Dahl brought research on insane heredity down to the level of minute studies of particular communities. Norway, which in those years was subject to the Swedish king, enacted legislation in 1848 authorizing construction of dedicated insane asylums. The first of these opened at Gaustad, outside Christiania (Oslo), in 1855. Dahl returned from two years of medical study in France, Austria, and Germany to take up an appointment there as assistant physician. He then received a stipend from the Crown of 250 Norwegian *Speciedaler* to undertake a tour of institutions for the insane in Holland, Belgium, and Great Britain. Adding Denmark to his itinerary, he set off in March 1856, visiting thirty-three asylums in about three months. His report, published in the Norwegian journal of medical sciences, met the usual expectations of the genre, including a basic sketch of the history and laws of each institution, its architecture and layout, sanitary arrangements, medical treatments, daily schedules, and rules of behavior.[7]

Other nations seemed eager to learn from Norway, whose new asylum attracted laudatory press coverage right from the start. I find no evidence that Dahl visited James Murray's Royal Asylum at Perth during his visit to Scotland. But when Lauder Lindsay, its superintendent, chanced to travel to Norway on holiday in 1857, he could not resist the temptations of a visit to Gaustad. He wrote a glowing report for the *Journal of Psychological Medicine*, which was then picked up and debated in Paris at a meeting of the *Société médico-psychologique*. That discussion formed the basis for a report in the *AMP*. The building, all agreed, was wonderful. It is perhaps even excessive, wrote Lindsay, who contrasted the intelligence and charity of the Norwegians to the small-mindedness apparent in the designs of so many county asylums in Britain. Brierre de Boismont, in the French report, spoke of a picturesque site and a "veritable palace" whose construction cost six times more per patient than what his colleague Parchappe had described as acceptable for these hospitals. Lindsay derived the same multiplier, six, by comparing its building costs per patient to those for the Scottish asylum in Montrose. Gaustad had rooms enough to do things right: a ward for quiet patients of good social class, another for quiet pauper patients, a third for the noisy and turbulent, a fourth for the excited and destructive, and a fifth for patients who were dirty.[8] It went almost without saying that men were kept separate from women.

Brierre de Boismont took great interest in a therapy employed by Ole Sandberg, medical superintendent at Gaustad: a prolonged warm bath combined with cold trickles spilling onto the head. The treatment, he explained, was his own, introduced a decade earlier. A footnote leads us to a paper he read to the Academy of Medicine, with seventy-two patient histories ("observations") and tables of results according to diverse variables. His remedy had spread to Britain, notably Scotland, after he demonstrated it to David Skae of the Royal Edinburgh Asylum during Skae's visit to Paris. What a triumph to learn from Lindsay's review that his baths were in use now in Gaustad as well as Edinburgh and were yielding excellent results! Dahl's report indicates that he spoke with Skae in Edinburgh, and it is possible that he learned of the head-chilling Brierre bath in this way.[9] It is a dense web of alienist connections we confront in these documents.

The Norwegians adopted the English model of annual reports, and Lindsay was almost as enthusiastic about the cornucopia of statistical tables on Gaustad as he was about its lavish facilities. His review of the asylum drew also on discussions in French and German medical literature. The table of causes, according to its

FIGURE 6.1. Ludvig Dahl (1826–1890) combined demographic and historical research with family investigation of the insane in selected parishes of Norway. Image from a painted portrait held by the Norwegian Medical Society, with special thanks to Øivind Larsen. See Øivind Larsen and Magna Nylenna, "Profiler og portretter i norsk medisin," *Michael Quarterly*, supplement 11 (2012), 43.

heading, was based on information from doctors and other relevant sources. In the first report, for 1856–1857, hereditary predisposition (*arveligt Anlæg*) was the leading cause,[10] just ahead of masturbation and drink, both overwhelmingly male, and grief, worry, and disappointments in love, all characteristically female. In subsequent years the order varied, partly due to changes in the terms used and partly to recombinations of categories. But the initial report, which received the most international attention, supported the expectation, originating with Holst's 1828 census, of the exceptional role of hereditary insanity in a nation of matchless statistics. It went on to list specific family relationships among the patients at Gaustad: father-daughter, uncle-nephew, brother-sister, and the like.[11]

The most detailed and influential response to Dahl's book was in Germany. By 1859, much Norwegian scientific and medical research was conducted within the German orbit. The capacity of Scandinavian countries to census or to survey their own populations inspired admiration and a touch of envy from German alienists.[12] Although the book was not translated, a long essay in the *AZP* by Gerhard von dem Busch may have been even more effective. The reviewer, a Bremen physician and translator, was convinced of the profound importance of Dahl's empirical study for the understanding of causes, especially hereditary ones, and he provided a full translation of the sections concerned with heredity. He also was impressed by the role of the Norwegian state, which was beginning to be recognized for pioneering statistical studies of its own population. It had supported Dahl's research with funds for seven months of travel and with indispensable infrastructure in the form of census data. Dahl relied on comparative results at the parish level, made visible with shaded maps that he included at the end of his volume, to identify sites of study. He also used census lists to pick out families for closer investigation, tracking down 270 out of the 283 mentally ill recorded by schoolteachers in Trondheim, Bergen, Christiania, and Christiansund. Following a common practice that Holst had made explicit, he took care to distinguish "acquired" (*erhvervet*) insanity, meaning insanity that arose during or after adolescence, from congenital weakness of intellect, appearing in infancy. Especially for "idiots," who often died young, he recognized the need to take age distributions into account to sort out misleading frequency differences. His research confirmed Norway's stand-

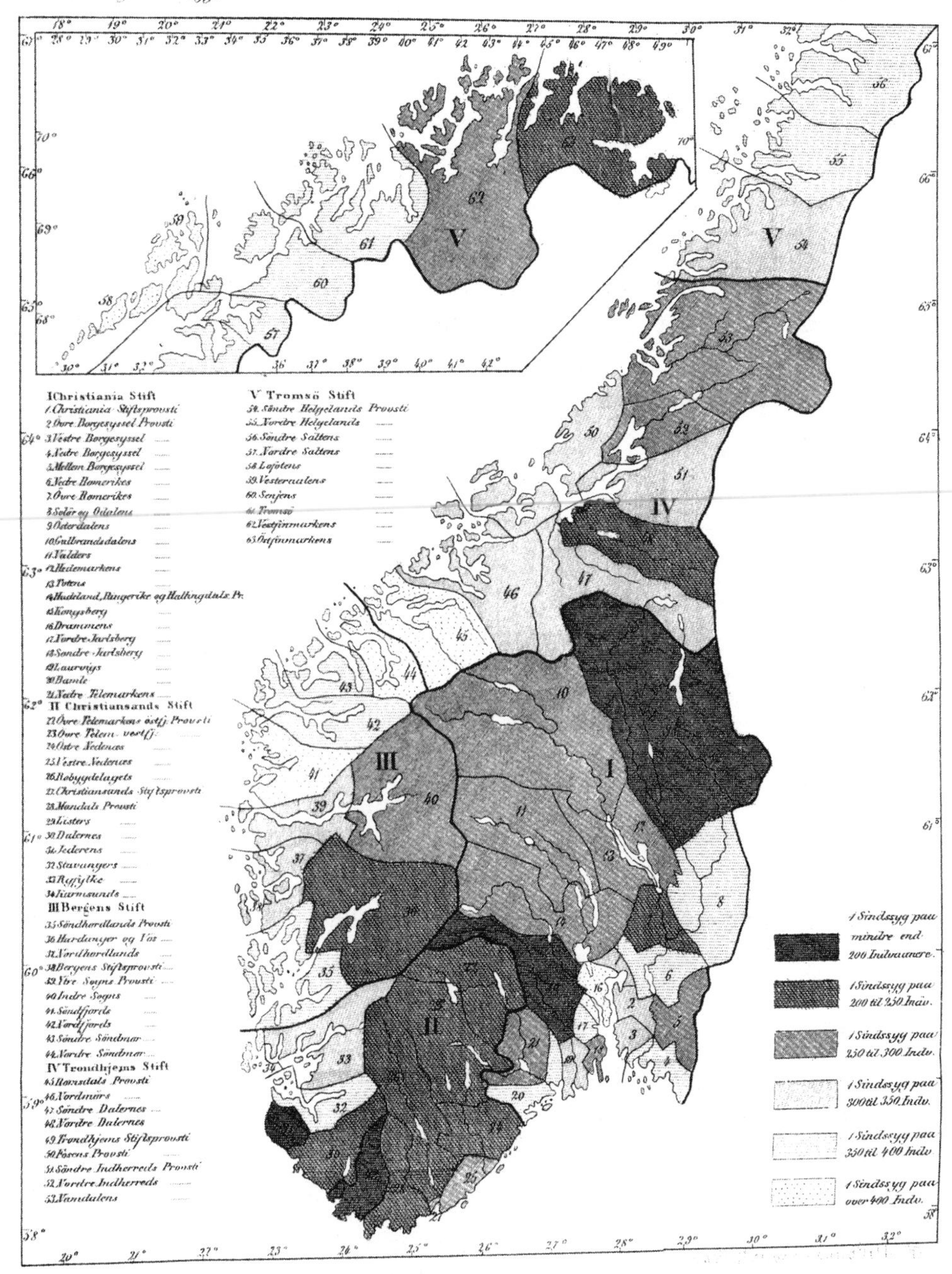

FIGURE 6.2. Norwegian shaded map showing relative frequencies of insanity, based on census data. Ludvig Dahl used these maps to pick out regions deserving detailed genealogical study, hoping to identify sources as well as paths of transmission for hereditary factors. From Dahl, *Bidrag*.

ing as a test case (we might say model system) for the examination of hereditary insanity.[13]

Dahl's hierarchy of causes was in line with Norwegian census results and with tables from the Gaustad asylum. He identified hereditary predisposition and marriages among close relatives as the leading cause of insanity. Merely having mentally ill relatives, he warned, even if these were direct ancestors and even if they suffered the same form of illness, did not prove the existence of a hereditary *Anlæg*. Proof of heredity had to be statistical. It depended on showing that a disease or disease class appeared much more commonly within a particular family than outside of it and on the absence of any other known causes that could explain it. Possible alternative explanations for childhood idiocy in one or more children included excessive use of brandy by the father and scaring or mistreating the mother.[14]

He aimed to uncover families whose bad heredity was beyond doubt. By focusing on regions with relatively stable populations, high rates of mental illness, and frequent marriage within communities, he was able to secure data for pedigree tables of inherited conditions, apparently the first such tables ever printed.[15] Much later, about 1900, they became an indispensable template of information for eugenic research, the raw material for investigating the laws of hereditary transmission and for keeping records of good and bad lineages. I am not aware of any others of their kind for two decades after 1859. Dahl's book included tables of defects he took to be hereditary for eight kin groups, or *Slægter*. His *Slægttavler*, kinship tables, became, in von dem Busch's translation, *Geschlechtstafeln*, which preserved etymology, or *Stammtafeln*, both terms used at the time mainly for royal and noble lineages, an ascending line of male ancestors. Dahl's kinship tables, by contrast, recorded patterns of transmission of an ancestral *Anlæg* to the descendants. In 1877, when the superintendent of the Scottish National Institution for the Education of Imbecile Children, William Ireland, needed to illustrate the inheritance of mental weakness for a book on idiocy, he reprinted Dahl's tables with translated captions. Evidently he knew of no suitable English tables of this kind. Ireland

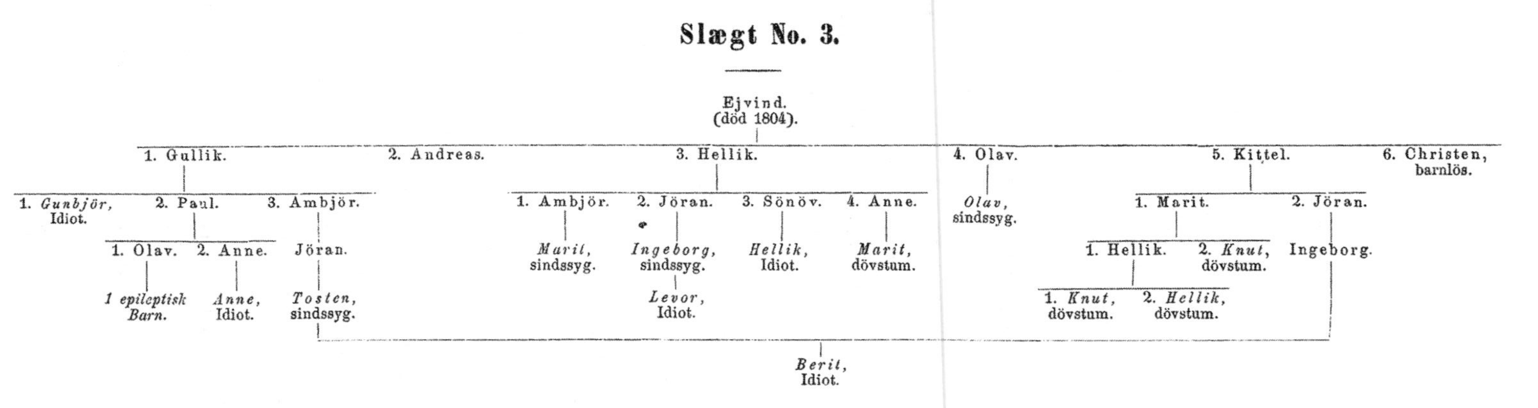

FIGURE 6.3. Pedigree chart of mental illness for kin group number 3. These tables attracted international attention and were often reprinted. Dahl's charts remained of great interest to hereditary researchers on account of the wide range of disabilities he found within kin groups. I have not found pedigree charts of mental illness from other countries until the 1890s. From Dahl, *Bidrag*.

called them *family trees*, here again pouring Dahl's new wine into a very old bottle.[16]

Dahl's most certain result, as he called it, and the one that most impressed Ireland, came out of the town of Flesberg, west of Oslo in the parish of Numedal, where he gathered data for his kinship group number 3. The table included 9 insane and idiots, of whom 8 were still living, plus 4 deaf-mutes and one epileptic. In the entire parish he found 24 insane (including idiots) and 6 deaf-mutes. This result was consistent with the 24 insane and 11 deaf-mutes given by the 1855 census, since 5 of the deaf-mutes were also idiots. The census had included them in both categories. In all, he determined, 12 of these 30 insane or deaf-mutes were members of a single extended family, descendants of Ejvind, who had died in 1804. Ireland spoke of "scattered branches of a common ancestor," every one the victim of "ancestral taint." A son of Ejvind's son and a few other relatives assisted Dahl in identifying all living members of this lineage, who numbered 126. Putting this kin group to the side, he observed, would reduce considerably the proportion of idiots and deaf-mutes in the parish of Numedal, population 2,922.[17]

Dahl was pursuing an alternative to the explanation of rates of mental illness in terms of physical environment, diet, or social milieu, interpreting them instead as an accumulation of diseased family lines. Yet the hereditary *Anlæg* was an elusive quarry, one that often skipped generations or appeared in collateral relatives rather than in the direct ancestral line. It also varied its form, appearing now as insanity or idiocy, now as epilepsy, now as deaf-mutism or albinism. Sometimes it seemed to be present but not fully formed, as in the case of two sane sisters of dark and violent character who each bore mentally ill children. Although a family *Anlæg* had real consequences, it was by no means a sufficient cause. A strong *Anlæg* in the parent, he explained, could increase the frequency of hereditary illness in the children by a factor of 15. The arithmetic seems to refer specifically to Ejvind's descendants, with 12 insane out of 126 (1:10.5), by comparison to 18 out of 2,796 (1:155) among the remaining inhabitants.[18]

A year after the publication of his book, Dahl was commissioned to resume the research, first in an area around Trondheim and then in the far northern regions of Nordland and Finnmark. He published these results in the Norwegian journal of medical science as two continuation chapters of his book, which von dem Busch again summarized at length in German in the *AZP*. The state, having taken responsibility for the health of its population, was pressing forward to gauge the medical needs of the insane in these remote northern regions. Although there were no asylums so far north, Dahl again could rely on the census, now six or seven years out of date, to pick locations for study. Tracking down what information he could, he ended up with a smaller proportion of hereditary cases than in his prior studies. He attributed this difference to incomplete information. Once again, his most startling conclusions involved patterns of inheritance in specific families. None of his far-northern kin groups were so extensive as Ejvind's, but he printed tables for five of them, each beginning with a male progenitor—an identified or unknown father. These tables omitted the healthy, recording only those lines involving conditions that he thought hereditary, including lameness and epilepsy as well as blindness, deafness, idiocy, and insanity. Some of the children had badly deformed skulls, which he duly measured, and multiple bodily, sensory, or mental problems. Several were products of cousin marriages.[19]

Dahl chose one small region in Nordland for a model study, the basis for his most striking results of all. This was the parish of Saltdal, running north from the Arctic Circle, which he picked for special investigation on the basis of disproportionately high figures for insanity in the 1855 census. The census there had recorded 11 mentally ill in a population of 1,690, a ratio of about 1:150. When he arrived, the priest told him of 35 insane in the parish. Investigating the matter himself, Dahl found 24 mentally ill and five idiots plus six doubtful cases. He could round this to 30, he proposed, and use 1:60 for the proportion, or he could exclude recent outbreaks involving intermittent illness and call it 15 plus five idiots, a ratio of 1:90. Dahl's willingness to work with so much uncertainty in his numbers was unusual. Either way, this was undoubtedly a very high number by comparison to census figures for Nordland as a whole, whose ratio was about 1:400. It would be interesting, he thought, to compare Saltdal with the next parish west, Bejeren (or Beiarn), where mental illness was much lower despite similarities of climate and landscape and a diet that, if anything, favored Saltdal. As before, he chose to track the *Anlæg* rather than to pursue explanations based on geography, diet, or differences of civilization.

In pursuit of his quarry, Dahl inquired locally about relatives of the mentally ill and pored over parish records. A book on the natural and human history of Saltdal, under the title *Physisk-oeconomisk Beskrivelse over Saltdalen* (Physical-Economic Description of Saltdal), opened up new resources for tracking the insane *Anlæg*. It was written by Søren Christian Sommerfelt between 1824 and 1827, just after he resigned as parish priest there. He is now best known as a Linnaean botanist who compiled long lists of local plant species. He also inventoried animals, described the geography and climate of the region as well as the character of its inhabitants, and tried to reconstruct patterns of settlement going back to medieval times. Since he had left Saltdal in 1824, he could not very well have been directly involved in Holst's census of insanity of 1825–1828. But he had already noticed that mental debility in the parish appeared disproportionate to its population, and he proceeded to tally and describe the insane. At least eight current residents, Sommerfelt found, had for some period of time been so violent that they had had to be bound. He added, consolingly, that seven had recovered, leaving only one who, for many years, has been "like a wild beast." Even her condition was not hopeless, he suggested, for Sommerfelt had found in the church register a still less promising case of a woman who was absolutely crazy for a very long time, and then, when she was almost seventy, suddenly recovered her senses.[20]

Sommerfelt wrote nothing about heredity, but Dahl relied heavily on his discussion of population movements in the parish. There were, according to Sommerfelt, 1,049 residents at the end of 1823. Examining church registers all the way back to 1730, he counted 2,453 births and 1,777 deaths, implying, he calculated, a population of just 373 at the beginning of the period. He acknowledged that the completeness of the record might be challenged, but in view of the impressive growth of population since his own arrival in 1818 and the much healthier habits of former times, he supposed that the increase since 1730 could easily have been still greater, and, accordingly, the population in 1730 even smaller. Extrapolating backward, he conceived that inhabitants of this valley must have been virtually wiped out by the Black Death in the fourteenth century. On the basis of linguistic patterns as well as the evidence of its oldest surviving houses, he surmised the origins of the new settlers. They must have come in very slowly, beginning with Finns, Lapps, and Swedes from across the Swedish border.[21]

Dahl accepted this idea of a near extinction and resettlement, including the surprisingly low population number in 1730, and he quoted approvingly Sommerfelt's remarks on the prevalence of insanity in 1823. He also learned, apparently from the priest in Ranen, the adjacent parish to the south, that a "couple hundred" settlers from Ranen had recently settled in Saltdal and that they were largely free of mental illness. The *Anlæg* for insanity, Dahl inferred, "seems instead to belong to a tribe originating somewhere other than the neighboring villages." In Saltdal, as in Flesberg, insanity appeared to be mainly a "family sickness." He proceeded to scour church records for the identity of Sommerfelt's "caged girl" (perhaps the one who behaved "like a wild beast"). He learned that she had three brothers, one of whom was "completely insane" and another who was mentally weak to a high degree. Each of these sick brothers, in turn, had fathered two insane children. Dahl prepared no family tables for Saltdal, but he gave a full list of infected families: two families with three mad children each, one family with four, and two more families in which madness affected the mother and two children. He provided no balance sheet, but there seem to have been, in total, from these families alone, 23 mentally ill and one doubtful.[22]

That would be a solid majority of the 29 confirmed mentally ill in Saltdal. Could the story be simplified still further? Dahl hinted at hopes to comprehend all or most of the mental illness in Saltdal as resulting from the migration of one single *Anlæg*, perhaps in a few interlinked families. Regrettably, since many of the insane siblings were very old, he was impeded by a lack of reliable information. He read through the church records from 1828 to 1845 without turning up a single marriage of close relatives. There had been a few such marriages more recently, and he reasoned that in a parish so sparsely populated, there must be many marriages between more-distant relatives. "In this way, a sickness-*Anlæg* that is already present may be strengthened." He did not overlook other causes. As in his book, and in so many asylum reports from many lands, he put great emphasis on alcohol and masturbation. Other dangers included religious enthusiasm (*Svarmerie*), spooky folk tales, and dangerously copious bleeding by rough doctors in the far north. Yet he treated all these as secondary, and their bearing on the problem as indeterminate. His crucial finding was "that such an *Anlæg*, having once arisen—from unknown causes—in a small and isolated population, may easily increase and disseminate in the course of time, notably by continued marriages in kinship groups already affected."[23]

It was, in its time, a brilliant research program. Dahl's hereditary *Anlæg* seems very like a gene. The appeal of explanations like these has endured into the present era. In the 1990s, for example, geneticists in Arizona proposed to investigate whether (rumors of) an extremely high incidence of schizophrenia among the Havasupai might owe to descent from a single shaman who had lived more than a century earlier.[24]

Reshaping Heredity

Dahl moved back and forth from the level of large regions, where the methods of the census were appropriate, to towns and parishes, where he could look into houses, talk with the priest, and examine documents pertaining to individuals. In some respects, his investigations were fitted to the template of the asylum report, and he devoted most of his professional life to asylum medicine. Like so many alienists, he understood the main causes he invoked as acting principally outside the asylum and as highly pertinent to public health. This was especially true for heredity, whose dynamics he tracked with unmatched precision. His book "found its way into the libraries of many European neurologists," wrote Ireland in an obituary notice. "It deals mainly with the causes and distribution of insanity, and especially attracted attention by the careful way in which he traced the descent of hereditary insanity in families dwelling in the quiet valleys of Norway."[25] Dahl focused on marriages among those with a hereditary *Anlæg*, itself a vital object of study, for the perpetuation of mental and nervous illnesses. This was a theme of growing importance, much discussed by 1859.

For example, in 1857, as we have seen, Lauder Lindsay journeyed to Norway and on a lark, or so he implied, visited Dahl's asylum at Gaustad. Lindsay's comments in the annual report of James Murray's Royal Asylum for 1858 took a very strong line on the role of heredity in the perpetuation of insanity. The official figures for causes, he there declared, are, as usual, valueless, owing to the "imperfect and unsatisfactory data on which they are founded; it were profitless, therefore, minutely to analyse them." Just why he allowed the records to be kept in so slovenly a fashion is not clear, unless of course his role was simply to record what the families told him. "Excitement in connection with celebration of Burns' centenary," an entry from 1859, shows little promise as a generalizable statistical category. There was a clear gap there between causes as recorded and what the alienists believed. His predecessor, William Malcolm, had mentioned hereditary predisposition in 1849 as the principal cause of insanity, even though his table of causes listed not a single man and only one woman under that heading.

Lindsay's confidence in the causal power of heredity, similarly, did not depend on statistical information from his own institution. His 1858 report is almost shrill in its call for restrictions on the reproduction of insanity. It is difficult to interfere with civil liberty, he said, but perhaps there should be legal restrictions on marriages of the insane. "The propagation of insanity by means of fatuous and facile female paupers is now amenable to civil law." This language we recognize from the 1857 report of the Royal Lunacy Commission for Scotland, and it may have been reinforced by his discussions at Gaustad. Lindsay's relationship to statistics changed completely in 1859. Although he complained in the report issued in June 1860 of his dependence on unreliable entries from the schedule of admission, it seems that he was now seizing the initiative. The new report provided full hereditary statistics including, when he had it, specific information on which relatives were affected.[26]

1859 and All That

To historians of biology, 1859 is Darwin's year. Francis Galton lauded his cousin's bold theory for demonstrating the analogies between biological species change and purposeful breeding. Almost everyone since has understood eugenics as what Galton made of Darwin's legacy. But we have already seen how much more water flowed into this river than can be attributed to the theory of evolution. Darwin, it turns out, was moved by family concerns to take a personal interest in inherited disease. He quickly found his way to medical writings on insane heredity.

By 1859, alienists everywhere recognized heredity as a key cause of the insanity that had filled to overflowing a crowd of new asylums. "Now, in relation to this malady, two important facts have been clearly established," wrote the Scottish Commissioners in Lunacy in 1859, and proceeded to a discussion of the key role of hereditary predisposition.[27] In 1858, a year before Darwin's *Origin*, John Gray was provoked by Buckle's *History of Civilization* to reassert its importance. "The large amount of statistical evidence in its favor, which Mr. Buckle sees fit to depreciate, is good at least until the first adverse generalizations are brought forward. In the entire medical profession there is almost no dispute of the law of heredity in bodily and mental disease."[28]

On 9 September 1859, the festivities organized by the University of Freiburg to celebrate the thirty-third birthday of the grand duke of Baden were capped by a lecture on the inheritance of ethical *Anlagen*. Alban Stolz, prorector and professor of pastoral theology and pedagogy, explained how free will had been overvalued through inattention to heredity. Many spiritual gifts and much "derangement of spirit and mental illness" are passed on by inheritance from parents to children. A thought experiment will convince us that the effect cannot be reduced to home environment. Imagine removing many children to a house of education where pupils are all treated identically, as if in a factory. They would assuredly yet diverge, each showing characteristics of its parents. Probably he knew nothing of Dahl's new book, but he called for research like Dahl's: "more exact observation" of heredity, "carried out statistically," and collections of exact and well-confirmed cases, such as criminal histories. What was needed was a new specialty in exact science "to serve as a norm of comparison, namely the inheritance of psychical diseases."[29]

In 1863, Isaac Ray joined in denouncing the historian's dismissal of hereditary causation of mental qualities. Buckle's extreme standard of causal efficacy, if applied more generally and not only to heredity, would block progress in every department of medicine. "The causes of insanity which spring up around us, are of far less potency than those which we bring into the world with us." Ray's insistence on hereditary stability of type was so emphatic that it would be impossible to regard him as following Darwin's lead. He was already writing in a eugenic way five years before the *Origin of Species* and without a thought of biological evolution. As a specialist in medical jurisprudence and asylum medicine, he could not ignore the hereditary predisposition for criminality and insanity. "None but they who have a professional acquaintance with the subject can conceive of the amount of wretchedness in the world produced by this single cause." The physician's wisdom had specific practical bearing for those who thought of marrying, and he warned them against "disregarding a law which carries with it such fearful penalties."[30] At least half the cases in institutions owe to a hereditary tendency. Breeders of animals know enough to insist on a lineage free of hereditary blemish, he wrote in 1853, so why do human families "go on forming alliances for life as if it were a fanciful speculation instead of a very serious fact?" Although there were other deadly causes, notably masturbation, which he saw as the chief causal link between insanity and civilization, heredity seemed a more promising focus of intervention. It was "at the same time the most prolific and the most easily avoided."[31]

Ray's reports from Rhode Island inspired the Lower Canada Lunatic Asylum in Quebec, in 1858, to endorse the analogy between breeding practices of domestic animals and humans. It had already enacted a requirement to report on hereditary taint in every new patient's application. Of the first 36, 12 certified "that insanity had manifested itself in the parents or in the immediate blood relations."[32]

By 1859, eugenics, in a broad sense, was old hat. Health officials may already have put their hereditarian doctrines into practice, as a consideration in decisions to intern insane or imbecile paupers. Dahl's great admirer, Ireland, remarked in 1877 that "unhappily there are too many instances" on record where imbecile or idiotic women have had children." He then quoted a reference to the transmission of idiocy for five consecutive generations from an 1857 report of the Royal Commissioners on Lunatic Asylums in Scotland. They announced "alarming figures" from the latest returns, showing no fewer than 126 idiotic women were shown to "have borne illegitimate children and whose mental defect is frequently manifested in their offspring." The commissioners considered that a full investigation would greatly increase this number and followed with specific instances from certain parishes.

> It thus becomes a matter of very serious import, whether for the sake of public morality and civic policy, all fatuous females should not be restricted in their liberty, and be gathered together in poorhouses. If it were possible to place all those who were at a child-bearing age, in circumstances where illicit intercourse would be impossible, much would be done to arrest an evil which has already entailed great misery and heavy burdens upon the community.[33]

Asylum Data, the Darwin Family, and Family Defect

Darwin first confronted human heredity as a personal issue in 1839, the year of Alexander Walker's *Intermarriage: Or the Mode in Which, and the Causes Why, Beauty, Health, and Intellect, Result from Certain Unions, and Deformity, Disease, and Insanity, from Others.* It was also the year he married his first cousin, Emma Wedgwood, and this big, diffuse book, treating domestic animals as well as human ones, made him anxious about the hereditary health of his children. By the year of Dahl's book, the question whether parents of new patients were blood relations was printed on many asylum admission forms. In consequence, the asylums became indispensable sources of (inconclusive) data on this vital issue. Darwin was greatly interested in 1865 by Galton's first paper on hereditary talent, but another paper that year on human heredity made an equal impression.

That one, on consanguinity, was written by the deputy commissioner of lunacy for Scotland, Arthur Mitchell. There he described pedigrees that pointed to ghastly hereditary damage. For example: "A married B, his full cousin, and had five children by her." Child 1 was sound in mind and body; 2 was imbecile; 3 died, age at death and mental condition not known; 4 was imbecile; 5 became insane. In the next generation, child 1 had by first wife four children: 1 was sane; 2 was sound in mind and body; 3 became insane in adult life; 4 died in early infancy; by second wife, five children, and so on. "If it be possible to conceive a family history more melancholy than that presented in the foregoing diagram, we shall find it in the cases which follow." But a conclusive answer, he acknowledged, required statistics, not just cases. Even collecting every episode that happened to present itself could prove nothing, since they would not be representative. Mitchell proposed two strategies for ascertaining the truth on marriage of relatives. His first was to identify all cases of some defect commonly attributed to consanguinity and see what percentage involved cousin marriage. Although his position as a Lunacy Commissioner gave him privileged access to data on families with mental defect, there were no authoritative figures for the frequency of cousin marriages at large. His second idea was to perform complete counts in several towns to see if idiocy was more common among the children of cousin marriages in these towns. Again his position provided advantages. He chose some locations to visit then "placed a schedule of queries in the hands of willing and competent persons." The research proved more difficult than he had anticipated, and the results were contradictory. He concluded on a balance of the evidence that cousin marriage does harm the offspring, though perhaps not the children of prosperous and well-nourished families that reside in healthful surroundings.[34]

While Mitchell referred also to the experience of cattle breeders, Darwin's attention was most engaged by the human question. He began exploring ways to procure the numbers that Mitchell had called for. By 1868, he was corresponding about the problem with William Farr, who told him in May that the International Statistical Congress, meeting in Florence the previous summer, had passed a resolution in favor of adding a column to census schedules to indicate the relatedness of father and mother of any family. On 17 July 1870, just as Parliament was about to take up legislation authorizing the 1871 census, Darwin asked his friend John Lubbock, an anthropologist and member of Parliament, to propose a new question on cousin marriages. He also wrote that day to Farr, arguing that such data would furnish "a standard by which to judge whether the proportion, (already tabulated in some cases) of persons in asylums for the dumb & deaf the blind & insane who are the offspring of cousins is in excess of the proportion of cousin offspring in the whole population." Lubbock organized what support he could and put the question to Parliament at 1:30 in the morning on 23 July. It was in vain.[35]

Darwin then passed the project to his son George, a mathematician. George Darwin made a selection of marriage announcements in the *Pall Mall Gazette* and of marriages registered at Somerset House and collected genealogies from *Burke's Peerage.* He also studied the works of noted alienists and surveyed asylum statistics in consultation with men like Lindsay, Henry Maudsley, Crichton Browne, and George Shuttleworth of the Royal Albert Asylum for feebleminded children. His father wrote to Shuttleworth in 1874 asking him to help George by querying his patients. On the basis of all this data, George reached the Scottish verdict of not proven, but when he read his paper to the (London) Statistical Society,

Galton stood up to say that he was too cautious. His paper had swept away an "exaggerated opinion" of the weakness produced by cousin marriage, and Galton knew of populations with much intermarriage that were magnificent. He had great hopes for schemes of inbreeding to generate castes with specific talents. George Darwin and his father, in contrast, were drawn to eugenics more by fear than by hope. These investigations left George deeply disturbed by the inexorable increase of insanity, and in 1873 he wrote a paper for the *Contemporary Review* advocating restrictions on the liberty to marry.[36]

Asylum doctors did what they could to turn back the increase of insanity, embracing the public-health role of alienist medicine and stressing prevention over cures. For doctors and patients, the "medicine of the future" might include homeopathy and clean living, but the focus of scientific attention was first of all on the inheritance of mental disease and the reproductive choices or restrictions that could stall or even reverse its inexorable growth.

CHAPTER 7

A Standardizing Project out of France Yields to German Systems of Census Cards, 1855–1874

> No other class of illness has been the object of such broad and persistent statistical study [as mental alienation].
>
> —Maximien Parchappe,
> International Statistical Congress, Paris (1855)

> Numbers have become a world power even in science. Every doctrine hastens to draw the greatest possible benefit of applying them. Even psychiatry is pulled along by this train of time.
>
> —Ludwig Wille (1872)

In July 1863, at a meeting in Stockholm, the statistical section of the Scandinavian Naturalist Society took up the question of uniform methods for tallying insanity. In a paper published just before the congress, Ludvig Dahl argued for harmonization through the adoption of a common form, with questions and definitions simple enough to be administered by nonmedical census takers.[1] Five years later, he gave a lecture on insanity counts to the same organization, now meeting in Christiania. The problem of harmonization still awaited a solution. The 1855 census in Norway gave one insane per 294 residents, which was close enough to the Danish result ten years earlier of 1:316. The Swedish ratio for 1855, however, was radically discrepant, 1:935. So great a difference between peoples so similar was scarcely credible. "I remarked that the great advantage of Sweden in this matter must in part at least be only appearance, arising from a less inclusive count." One obvious difference was that the Swedish count did not include idiots. Just after the 1863 congress, Sweden released the results of its 1860 census, showing many more insane and a ratio of 1:512. The Danes, for the sake of harmonization, had just advanced their decennial census from 1865 to 1860. They simultaneously began recording the insane by name, and that, according to Dahl, made everyone more cautious. The proportion insane in Denmark dropped almost to the new Swedish level, 1:493. It was a heartening result.[2]

Norway, while putting off the switch to a new census cycle, also began recording names in its 1865 census, but the figures, in contrast to the Danish ones, scarcely budged. The new ratio was 1:327. Dahl racked his brain for an explanation. As a country of small towns, Norway might have attained more exact figures through minute oversight. There is much discussion of insanity as a condition of civilization, he remarked, but perhaps the real source of discrepancies was bad census administration in Sweden and Denmark. Certainly that was his explanation for low numbers in most countries outside of Scandinavia. He acknowledged that his countrymen were prone to alcoholic excess and kin marriages. But the campaign against giving brandy to mothers and infants, one of his pet causes, seemed to be paying off in lower levels of rural idiocy. So the question remained: what confidence could the alienist invest in national differences in measured rates of insanity? Even for meticulous Scandinavians, the evidence was inscrutable.[3]

Since about 1850, standardization and statistics have seemed to go hand in hand. If nations could be compared on the basis of accurate counts, it would facilitate the identification of causes of insanity, enabling more effective interventions to improve mental health. Still better, it might allow the pooling of data from different sources, something like Baillarger's ambition but using routinely collected public statistics. Quetelet was thinking along these lines when he organized the first International Statistical Congress in 1853. Alienists, too, dreamed of a vast international reserve of reliable statistics, the results, as Quetelet had put it, of experiments already performed, promising solutions to the great problems of public health. This was scientific statistics, yet relying on bureaucratic means and directed, in part at least, to bureaucratic ends.

The Scandinavian deliberations appeared to be positioned on the scientific side of this continuum. The French, German, and

Swiss efforts that provide the main subject matter for this chapter involved a range of strategies for reconciling the bureaucratic and the scientific. Their scientific goals could never be separated from the structures and conditions of state administration. It is largely for this reason that international standardization, and not only of asylum statistics, has always been so difficult. The Franco-Prussian War in 1870 and 1871 abruptly converted the universal French project into a German national one, whose relative success would depend on the political transformation of Germany. Unification bolstered the country's status as an exemplary site of science and scholarship, a standing that extended also to mental medicine.

Administrative Ideals

In England, the outcome of standardization efforts had more to do with regularizing institutional practices than coordinating medical knowledge. At least a few, most notably Thurnam, conceived uniform statistics as a tool of science. Charles Hood, an assiduous researcher, focused his efforts on a ten- and then fifteen-year study of Bethlem, so as to get beyond the limitations of the annual report. He entered information on causes of insanity, but not other data, in his own hand.[4] John Conolly put forward his new tables and improved admission form at Hanwell as a model, boasting that they enabled him to record a cause in the vast majority of cases. This success, however, was perhaps due to the inclusion of separate columns for physical, moral, and hereditary causes, any one of which would meet this requirement. By 1842, his sense of triumph was receding: "No information is obtained with more difficulty than that which relates to causes." And in 1844: "Every year . . . increases the doubt with which I receive the Reports given in relation to the causes and the duration of their malady. The assigned causes are often but the first symptoms, and sometimes little more than conjectural."[5]

English alienists, like most, cared a good deal about reaching the public with their reports and statistics. The usual reason given was to encourage prompt action when a family member began acting peculiarly and to combat unhealthy practices, in particular what John Charles Bucknill called the "reckless" transmission of hereditary predisposition.[6] One of the first acts of the association of asylum officers, in 1842, was to devise uniform tables. This effort, however, was pushed forward mainly by the Metropolitan Commissioners in Lunacy, a regulatory body, rather than the asylum doctors themselves. The secretary of the Medical-Psychological Society, C. Lockhart Robinson, tried to take a loftier view. He complained that lack of uniformity reduced their tables to "a mass of labour and good printing rendered useless from a want of system." The fruit of his labor was just six standard tables summing up the operation of the institutions, with little pretense of science.[7]

Intermittent reports on this effort in the *Journal of Mental Science* (*JMS*) applauded the gradual increase of asylums that had signed on to this inconsequential reform. Then, suddenly, in 1867, a notice arrived from France of plans for a "General and Universal Medical Congress," organized by the Medical-Psychological Society of Paris, to be held on 10, 11, and 14 August. English hearts, too, beat a little faster. The letter listed four topics for the congress, one of which was the harmonization of statistics. Bucknill, editor of the *JMS*, added to this announcement his hope that England would be represented. Eventually he went there himself, with two countrymen. The project unfolded as a dream of data-driven medical administration.[8]

Ludger Lunier: Harmonizing Lunacy

The organizer of this standardization campaign was the French alienist Ludger Lunier, whose devotion to statistics was very special. The obstacles to international standardization were legion, to the point, perhaps, that it never had a realistic chance of succeeding. Yet within the borders of politically unified territories, the push for uniform statistics was hard to resist. Asylum doctors typically preferred to hold on to what they could of their autonomy, including control of their own archives. Friedrich Wilhelm Hagen emerged about 1870 as Lunier's most eloquent opponent. Their contrasting careers as well as their intellectual exchanges reveal what was at stake for definitions and deployment of hereditary data.

The centralized character of French asylum administration is readily apparent in its records. The Archives Nationales hold fragmentary files of reports from the *départements* in 1849 and extensive files for 1859 and 1860. The 1849 forms were created to document admission into the national asylum system and to indicate how many insane remained in jails and hospices or as vagabonds.[9] The files for 1859 and 1860 include a thick pile of preprinted booklets that had been filled out and returned by every institution. Labeled *Modèle No. 9,—Tableau B.*, the booklets include 28 pages for 29 tables plus a Form 30 for miscellaneous comments. They were to be transmitted to the minister on 1 July of each year, after all the numbers on each page had been checked to confirm their mutual consistency. Most tables are just lists breaking down some category of patients by age, occupation, disease form, time spent in the institution, outcome of treatment, or cause of death. The main exceptions are tables of causes: Table 7 for continuing patients and Table 8 for newly admitted ones. The causes are divided between predisposing and determining and include five choices for predisposing cause: father tainted by insanity (*père atteint d'aliénation*), mother tainted by insanity, father and mother tainted, neither, insufficient information.

It is a lot of data, and we can easily imagine asylum directors grumbling over the heavy burden of recordkeeping and form-filling. But a few surpassed expectations. The asylum of La Roche-Gandon in Mayenne attached a half-page to Table 7. The physician, suspecting that his numbers understated the importance of heredity, added material on siblings, uncles, aunts, and cousins: one case of three brothers insane; four cases of two brothers, three of a brother and sister, one of three sisters, four of two sisters, one each of a paternal uncle, two aunts, two male cousins, and a male and female cousin, plus three probable instances of consanguineous marriage. There is a clear family resemblance to Baillarger's investigation of inheritance of insanity by sex. Another site of irrepressible statistical energy was the asylum of Saint-Yon in Rouen, where Lucien

Deboutteville and Maximien Parchappe had carried out their well-known statistical work, now directed by B. A. Morel, who submitted additional data on heredity and on treatments. Morel's former institution at Maréville, directed now by Emile Renaudin, attached supplemental information on curative methods.[10] But the prize for statistical voluntarism must certainly go to Lunier, chief physician at the asylum of Blois on the Loire River, who, in 1859, worked out the additional categories required for an adequate table of predisposing causes, then drew up a revised form and filled it out for his patients. He insisted on a line for collateral heredity, divided into brothers, sisters, uncles, aunts, nephews, and nieces, plus one for combinations of direct and collateral heredity and a last for grandparental insanity when the parents were healthy. The next year, after the ministry neglected to revise the tables in accordance with his suggestions, Lunier repeated his observations and again included the supplement.[11]

An *éloge* in the *AMP*, two decades after Lunier's death, explained his statistical bent as ancestral: "This taste, in a way innate, . . . seems to have been for him a product of heredity, and even of an overcharged heredity." His father and grandfather were accountants and financial functionaries, and his own passion for figures arose early and never left him.[12] This faith in numbers assumed a variety of forms. In 1853, as chief physician of the asylum at Niort, he organized his report around a shaded map showing relative frequencies of insanity in the cantons of Deux-Sèvres (Poitou). The tables were suited to a mixed medical-moral-geographical explanation in terms of climate and soil, alcohol consumption, and distance from the asylum. He was especially keen to gauge the impact of asylums built under the 1838 law on the number of registered insane. These researches, as he indicated in a dossier, were interrupted that year by his move to Blois.[13] There, he introduced a "reasoned statistical report," to be published annually, he said, as every director should, so as to heighten awareness of the benefits of asylum care for local people. While these reports combined medical and administrative information, he brought only medical data to the 1867 congress of alienists. In his application for membership in

FIGURE 7.1. Photograph of Ludger Lunier (1822–1897), who was unsurpassed in his devotion to asylum statistics. He set in motion a movement to create universal standards on the French model, and it seemed for a time as if he might succeed. Portrait © Bibliothèque de l'Académie de Médecine (Paris), with permission.

the Academy of Medicine two years after that, he boasted that his recommendations had been received, not piece by piece, but as a package. It is no surprise that his tables for Blois closely resembled the ones he had sent to the ministry. Since 1864, at least, he had been obsessed with the question of causes, and especially of hereditary predisposition.

> Knowledge of the causes of the illness of the insane, which are confided to us, has so much importance for the insane who are put in our care, Monsieur le Préfet, from the double perspective of prognostication and of treatment. . . . Hence we make every effort to obtain from family and friends the most detailed information on the prior history of the insane patients admitted to the asylum.[14]

Alas, since patients from outside the *département* received few family visits, his access to information was limited to 185 of the 580 patients. Of these, 113 had relatives who were insane, epileptic, or tainted by some other cerebral condition.[15]

Lunier's statistical work soon extended beyond the asylum. His appointment in 1864 as inspector general of asylums and prisons enlarged his responsibility and opened up a wider field of data. After the war with Germany, he took up the vexed question of political disturbance as a cause of insanity. He tried out a novel statistical design, grouping the *départements* of France according to whether and for how long German troops occupied them, then looking for an association with the percentage of new cases whose cause involved war events. Interpreting the results was not simple. How should he understand the overall decrease in asylum admissions during the period of turbulence? War, he determined, enhances some causes of insanity while reducing others, including heredity.[16]

In 1878, in collaboration with the two fellow inspectors, he published a uniquely ambitious report on madness in France. The numbers seemed to go up and up, reaching 87,698 in the 1872 census. Setting aside idiots, imbeciles, and cretins, it was still 52,835. He could no longer discount this explosion of numbers, as he had in 1869, as merely an artifact of Moreau de Jonnès's undercount in 1843. Maybe Esquirol was right in claiming that insanity was a disease of civilization. "The century is nervous," he wrote. Beyond the effects of politics, religion, and war, the report found ugly effects of degeneration. For many illnesses, and especially insanity, it now appeared that the condition of the descendants is worse than that of the ancestors.[17]

Alienists of All Nations!

Lunier's push for international standards was not without precedents. The second International Statistical Congress, held in Paris in 1855, included medicine as a featured topic. Insanity had come up at the Brussels Congress in 1853 in debates on criminal statistics. A Danish representative in 1855 asked how best to gather data on the number and situation of the insane, given the tendency of family members to dissemble. On the second day, Parchappe took the floor to proclaim that insanity was the most statistical topic in all of medicine. The delegates understood that the appeal of asylum systems depended greatly on favorable numbers. Now they were becoming conscious of their problems. Tens of thousands were cured, and yet they continued somehow to multiply. Was the measured increase of insanity real or factitious? What were its true causes, not just in individuals, but from a social and statistical perspective? Parchappe stuck by the old tenets: that insanity had been proven curable and that heredity is a predisposing cause only, never an inexorable fate. Others called for comparative statistics to provide more reliable answers. Comparison required comparability, which meant standardization. How to standardize was another issue on the agenda, one this congress did not advance very far.[18]

Perhaps there was some uncertainty, even in Paris during that week in September 1855, whether a medically useful classification of forms and causes of insanity could ever arise from negotiations of officials at a congress. Parchappe's paean to statistics, reprinted in the *AMP*, was followed on the very next page by Ulysse Trélat's artful patient narratives, privileging any medical evidence of heredity over unreliable stories and useless statistics. Renaudin declared a few months later in the *AMP* that it would be better to abandon general statistics and to focus on particular causes of insanity. "On the first line, I would put heredity," he said, which he claimed to recognize from its mode of operation.[19] Heinrich Damerow, long an enthusiast, now complained that all the "numbers and ratios and sums and general sums" of statistics were shedding no light on causal relations or the effectiveness of treatment. "Conclusions from mere number-giving mad-statistical tables are again just numbers," which can only be useful when the cases are homogeneous.[20]

No concrete proposals came out of the Paris statistical congress. The vague hope to draw valuable lessons from comparisons on the basis of more uniform statistics persisted as a low hum in the journals of asylum medicine. In his 1859 report, Dr. Joseph Workman of Toronto complained of all the wasted statistical effort that could be made useful with a uniform system for recording and compiling numbers.[21] Seven years later, a new society of Swiss alienists had just taken up a project of standardization when word came in from the *Société médico-psychologique* of plans for an international alienist congress.[22] Lunier had first aired the idea at a meeting of this society on Christmas Eve, 1866. A report on his proposal in the next issue of the *AMP* emphasized the advantages of a closer association with foreign doctors. Lunier, meanwhile, used the occasion of a nomination for membership to reemphasize the key role of statistics in asylum reports. The members agreed that they would have to organize their own congress, since medical congresses permitted no sections, and since the exclusion of non-physicians was inappropriate for medical psychology. The alienists worked out their own agenda in a series of biweekly sessions beginning in mid-August 1867. The invitations they soon dispatched to *aliénistes de tous les pays* (alienists of all nations) listed four topics as especially timely: (1) legislation; (2) public and private education; (3) "Basis of a general system of asylum statistics"; and (4) pathological changes of the nervous system. Lunier's visionary statistical scheme was nested within a mélange of topics involving medicine, law, and publicity.[23]

The internationalism of this congress included an aspect of Potemkin, providing cover for a campaign, in the end unsuccessful, to let other nations capitulate to tables devised in France. The *AMP* printed a list of its eleven distinguished visitors, including three British, three Swiss, two Germans, and one each from Italy, Spain, and Moravia. Eight of them joined with four French alienists to form a commission. After three long sessions, we are told, it authorized Lunier to report on their discussions. His report, including specimen tables, was to be sent out for comments to societies of psychiatry and statistics and to governments. Lunier, reporting for the committee in the *AMP*, explained that the proposal was based on the suggestions of French asylum inspectors, while taking into account some unspecified English asylum documents, the statistics of Illenau in the German state of Baden, and the most recent report of his own asylum at Blois. There had been a lively discussion, he said, followed by gracious concessions, and almost the whole project was adopted unanimously.[24]

Two years later, Lunier described the agreement as unanimous without qualification. He gave credit to eminent practitioners from several European states and acknowledged a helpful impulse from all the leading alienist journals. He tried not to claim too much for the standards, acknowledging that expert, dispassionate assessments of statistical results would always be needed. When patient populations are dissimilar, numbers cannot be straightforwardly compared. But mental medicine is powerless without comparative statistics, since so many puzzles can never yield to observations by any single individual. His report was to provide a basis for harmonizing data. In contrast to prior efforts, this one entered into details, specifying, for example, a list of permitted occupational categories, the criteria for labeling a patient incurable, the formula for calculating a cure rate, and specific time intervals for duration of illness prior to admission. In all, it included twenty-nine patient tables plus two administrative ones. Classification of the noninstitutionalized insane was left as a project for the future.[25]

The most urgent and difficult problem they confronted was the classification of mental illness and its causes. Although the report listed seven categories of disease, just one of these, "la folie simple," corresponded to "acquired insanity" (meaning "madness" or "lunacy"). The others were epilepsy, general paralysis, senile dementia, organic dementia (identified by a brain lesion), idiocy (or mental weakness), and cretinism. Most important of all was the question of causes, for which the committee introduced a cascade of distinctions. Lack of information should never be confused with absence of a cause. Causes may be predisposing or effective, and moral, physical, or mixed. Since they are often unknown or plural, the sum of causes will not in general equal the number of patients. The most important predisposing cause, heredity, can be direct, collateral, or mixed. Direct heredity can be paternal, maternal, or a combination of both; collateral can involve brothers, sisters, or both; and the "mixed" category opened up to every combination. It made for a complex table in very small print, full of brackets and insets. The table was compound to facilitate detection of correlations, with disease forms laid out horizontally at the top of the table and causes arrayed down its left side.[26]

There must have been much discussion of this complex table. Certainly the causes and disease forms inspired animated debate when the report was circulated. But on the evidence of printed forms, the international committee was either mute or highly compliant. A comparison of the proposed standard table of causes with a filled-out table from the 1863 report of Lunier's asylum at Blois reveals modest differences of detail but an almost identical layout and many of the same categories. The most obvious difference is the inclusion in the international table of two classes of dementia among disease forms. We have seen what strenuous objections had been raised in recent decades to almost all effective or precipitating causes, and especially to medically unqualified friends and relatives who provided this information. Perhaps for that very reason it seemed pointless to try very hard to reconceptualize them. Disease form was a different kind of problem, a medical one, whose solution drew from therapeutic experience. The challenge to Lunier's scheme began with details then developed into a root-and-branch rejection of impersonal, standardized statistics. This escalation was

TABLEAU SIXIÈME. — CAUSES PRÉSUMÉES DE L'ALIÉNATION DES MALADES EXISTANT DANS L'ÉTABLISSEMENT LE 1er JANVIER 1863.

DÉSIGNATION DES CAUSES.	FOLIE Simple. H.	FOLIE Simple. F.	FOLIE Épileptique. H.	FOLIE Épileptique. F.	FOLIE Paralytique. H.	FOLIE Paralytique. F.	IDIOTIE. H.	IDIOTIE. F.	CRÉTINISME H.	CRÉTINISME F.	TOTAL GÉNÉRAL. H.	TOTAL GÉNÉRAL. F.	TOTAL GÉNÉRAL. 2 S.	OBSERVATIONS.
1° CAUSES PRÉDISPOSANTES.														
HÉRÉDITÉ.... Directe.. Paternelle (père, grand-père, oncle, tante)..	12	10	2	4	1	1	»	1	»	»	15	16	31	
Directe.. Maternelle (id.)..	8	14	»	1	»	»	2	1	»	»	10	16	26	
Directe.. Paternelle et maternelle..........	»	4	»	»	1	»	»	»	»	»	1	4	5	
Collatérale..........	12	10	1	»	3	1	1	»	»	»	17	11	28	
Mixte... Collatérale et paternelle (frères, sœurs, neveux, nièces).	7	2	2	»	»	»	»	3	»	»	9	5	14	
Mixte... Collatérale et maternelle (id.).	4	3	»	»	»	1	»	»	»	»	4	4	8	
Mixte... Collatérale, paternelle et maternelle.......	»	2	1	»	»	1	1	»	»	»	2	3	5	
Consanguinité des parents..........	»	2	»	»	»	1	»	»	»	»	»	3	3	
Grande différence d'âge entre les parents (plus de 20 ans)..........	»	1	»	1	»	»	»	»	»	»	»	2	2	
Influence du sol, du milieu ambiant..........	»	»	»	»	»	»	2	1	»	»	2	1	3	
Convulsions ou émotions de la mère pendant la gestation..........	»	1	»	1	»	1	1	»	»	»	1	3	4	
Grossesse, suites de couches, âge critique, période menstruelle.....	»	6	»	»	»	1	»	1	»	»	»	8	8	
Epilepsie et autres névroses..........	»	1	26	21	»	»	»	»	»	»	26	22	48	
Ivrognerie (1)..........	1	1	1	»	»	»	»	»	»	»	2	1	3	(1) Excès de boissons habituels datant de loin.
Excès de travail intellectuel prolongé..........	»	1	»	»	»	»	»	»	»	»	»	1	1	
Accès antérieurs d'aliénation mentale..........	15	18	»	1	2	»	»	»	»	»	17	19	36	
Autres (intoxication saturnine, onanisme)..........	1	»	1	»	»	»	»	»	»	»	2	»	2	
Nulles ou inconnues..........	14	11	1	»	3	3	1	2	»	»	19	16	35	
Renseignements nuls ou insuffisants..........	105	185	»	»	6	16	28	24	»	1	139	226	365	
TOTAUX..........	179	272	35	29	16	26	36	33	»	1	266	361	627	
2° CAUSES DÉTERMINANTES. Causes physiques, ou plutôt causes externes et organiques.														
CÉRÉBRALES.... Déformations du crâne..........	»	»	»	»	»	»	1	»	»	»	1	»	1	
Convulsions de l'enfance, dentition..........	»	»	3	3	»	»	3	3	»	»	6	»	6	
Congestions cérébrales..........	2	»	»	»	»	»	»	»	»	»	2	»	2	
Affections organiques du cerveau..........	»	»	2	»	»	»	1	»	»	»	3	»	3	
Sénilité..........	1	»	»	»	»	»	»	»	»	»	1	»	1	
NON CÉRÉBRALES. Diverses... Dermatose, pellagre..........	»	»	»	»	»	»	»	»	»	»	»	»	»	
Syphilis constitutionnelle..........	»	»	»	»	»	»	»	»	»	»	»	»	»	
Fièvre intermittente..........	»	1	»	»	»	»	»	»	»	»	»	1	1	
Fièvre typhoïde, fièvres éruptives..........	»	2	»	»	»	»	»	»	»	»	»	2	2	
Goutte, rhumatisme..........	»	1	»	»	»	»	»	»	»	»	»	1	1	
Vers intestinaux..........	»	»	»	»	»	»	»	»	»	»	»	»	»	
Suppression du flux hémorrhoïdal..........	»	»	»	»	»	»	»	»	»	»	»	»	»	
— menstruel..........	»	»	»	»	»	»	»	»	»	»	»	»	»	
— d'épistaxis..........	»	»	1	»	»	»	»	»	»	»	1	»	1	
— de la transpiration..........	»	»	»	»	»	»	»	»	»	»	»	»	»	
Métastase, goutte supprimée..........	»	1	1	»	1	»	»	»	»	»	2	1	3	
Poisons.... Boissons fermentées..........	8	1	1	»	4	»	»	»	»	»	13	1	14	
Poisons végétaux..........	»	»	»	»	»	»	»	»	»	»	»	»	»	
Poisons minéraux (mercure, plomb, cuivre)...	»	»	1	»	»	»	»	»	»	»	1	»	1	
Externes....... Insolation, chaleur intense..........	1	»	»	»	»	»	»	»	»	»	1	»	1	
Externes....... Coups, chutes graves..........	1	1	2	»	»	»	»	»	»	»	3	1	4	
Autres causes physiques (abstinence prolongée)..........	1	»	»	»	»	»	»	»	»	»	1	»	1	
CAUSES MORALES... Dévotion exaltée, fanatisme religieux, confession intempestive.......	1	5	»	»	»	1	»	»	»	»	1	6	7	
Remords..........	»	»	»	»	»	»	»	»	»	»	»	»	»	
Superstition, sort jeté, bonne aventure.......	»	1	»	»	»	»	»	»	»	»	»	1	1	
Education mal dirigée, mauvaises lectures....	»	»	»	»	»	»	»	»	»	»	»	»	»	
Amour contrarié, trompé, mariage manqué...	8	4	»	»	1	»	»	»	»	»	9	4	13	
Jalousie..........	»	1	»	»	»	»	»	»	»	»	»	1	1	
Maladie, éloignement ou perte d'une personne chère.	2	18	»	»	»	»	»	»	»	»	2	18	20	
Chagrins domestiques..........	3	14	1	»	1	2	»	»	»	»	5	16	21	
Discussion d'intérêts..........	3	3	»	»	1	1	»	»	»	»	4	4	8	
Revers de fortune, embarras d'affaires.......	7	4	»	»	»	»	»	»	»	»	7	4	11	
Vocation contrariée, tirage au sort..........	1	»	»	»	»	»	»	»	»	»	1	»	1	
Orgueil, amour-propre blessé, ambition déçue.	1	1	»	»	»	1	»	»	»	»	1	2	3	
Frayeur, saisissement..........	4	1	4	5	»	2	»	»	»	»	8	8	16	
Imitation (vue d'une personne en convulsions).	»	»	»	1	»	»	»	»	»	»	»	1	1	
Colère..........	»	»	»	»	»	»	»	»	»	»	»	»	»	
Pudeur blessée..........	»	»	»	»	»	»	»	»	»	»	»	»	»	
Crainte relative à sa santé..........	1	»	»	»	»	»	»	»	»	»	1	»	1	
Chagrin de ne pouvoir gagner sa vie..........	1	»	»	»	»	»	»	»	»	»	1	»	1	
Evénements politiques..........	1	»	»	»	»	»	»	»	»	»	1	»	1	
Nostalgie..........	»	»	»	»	»	»	»	»	»	»	»	»	»	
Misanthropie, tædium vitæ..........	»	»	»	»	»	»	»	»	»	»	»	»	»	
Joie..........	»	»	»	»	»	»	»	»	»	»	»	»	»	
Emprisonnement simple..........	»	»	»	»	»	»	»	»	»	»	»	»	»	
— cellulaire..........	»	»	»	»	»	»	»	»	»	»	»	»	»	
Autres causes morales..........	2	»	»	»	»	»	»	»	»	»	2	»	2	
CAUSES MIXTES..... Excès de travail intellectuel ou de lecture....	1	1	»	»	»	»	»	»	»	»	1	1	2	
Inconduite, libertinage..........	1	»	»	»	»	»	»	»	»	»	1	»	1	
Onanisme, abus vénériens..........	1	»	»	»	»	»	»	»	»	»	1	»	1	
Dénûment et misère..........	»	2	»	»	»	»	»	»	»	»	»	2	2	
Mauvais traitement..........	»	»	»	»	»	»	»	1	»	»	»	1	1	
Passage subit d'une vie active à une vie inactive, et vice versâ.....	1	»	»	»	»	»	»	»	»	»	1	»	1	
Nulles ou inconnues..........	9	6	»	1	1	1	1	1	»	»	11	9	20	
Renseignements nuls ou insuffisants..........	103	190	10	11	6	16	30	28	»	1	149	246	395	
TOTAUX..........	165	258	26	21	16	24	36	33	»	1	243	337	580	

FIGURE 7.2. Ludger Lunier's table of causes of insanity from the asylum at Blois including heredity as the first entry, subdivided according to relative affected, with disease form on the other axis. His tables reflected strenuous French requirements for the collection and reporting of asylum data, yet, as usual, his own efforts to document patient heredity surpassed expectations. See Lunier, "Asile départemental d'aliénés de Blois (Loir et Cher)," 8–9.

ALIÉNÉS ADMIS POUR LA 1re FOIS DANS UN ASILE.

TABLEAU IX — *Causes présumées de l'aliénation.*

Hommes	
Femmes	
Total	

Causes présumées.	Folie simple		Folie épileptique		Folie paralytique		Démence sénile		Démence organique		Idiotie		Crétinisme		Autres formes		Total			Observations.
	H.	F.	H.	F.	H.	F.	H.	F.	H.	F.	H.	F.	H.	F.	H.	F.	H.	F.	D.S.	
Malades sur lesquels on a obtenu des renseignements — nuls ou insuffisants																				
Malades sur lesquels on a obtenu des renseignements — suffisants																				
Total des malades admis																				
Causes prédisposantes. — Hérédité — directe — Paternelle (père, grand-père, oncle, tante)																				
Hérédité — directe — Maternelle (id.)																				
Hérédité — directe — Paternelle et Maternelle																				(1) Les excès alcooliques et l'onanisme agissent tantôt comme prédisposition en débilitant lentement l'organisme, tantôt comme cause déterminante en produisant un ébranlement dans les fonctions de l'encéphale. (2) Considérées comme causes et non point comme symptômes.
Hérédité — collatérale — Frères et Sœurs																				
Hérédité — mixte — collatérale et paternelle																				
Hérédité — mixte — ——— maternelle																				
Hérédité — mixte — ——— paternelle et maternelle																				
Consanguinité pure																				
Grande différence d'âge entre les parents (plus de 20 ans)																				
Influence du sol, du milieu ambiant																				
Convulsions et émotions de la mère pendant la gestation																				
Épilepsie																				
Autres névroses																				
Grossesse																				
Allaitement																				
Époque menstruelle																				
Âge critique																				
Puberté																				
Ivrognerie (excès habituels datant de loin) (1)																				
Excès vénériens, onanisme. (1)																				
Autres causes prédisposantes																				
Absence probable de causes prédisposantes																				
Causes déterminantes. — Causes physiques, ou externes et organiques. — Cérébrales — Déformations artificielles du crâne																				
Cérébrales — Convulsions de l'enfance; dentition																				
Cérébrales — Congestions cérébrales. (1)																				
Cérébrales — Aff. organiques du cerveau																				
Cérébrales — Sénilité																				
non cérébrales. — Maladies aigues et chroniques — Pellagre																				
Maladies aigues et chroniques — Anémie																				
Maladies aigues et chroniques — Syphilis constitutionnelle																				
Maladies aigues et chroniques — Fièvre intermittente																				
Maladies aigues et chroniques — Fièvre typhoïde																				
Maladies aigues et chroniques — Fièvres éruptives																				
Maladies aigues et chroniques — Rhumatisme aigu																				
Maladies aigues et chroniques — Goutte et Rhumatisme chronique																				
Maladies aigues et chroniques — Affections organiques du cœur																				
Maladies aigues et chroniques — Phthisie pulmonaire																				
Maladies aigues et chroniques — Vers intestinaux																				
Maladies aigues et chroniques — Autres maladies aigues																				
Maladies aigues et chroniques — Autres maladies chroniques																				
Suppressions et Métastases — Suppression du flux hémorrhoïdal																				
Suppressions et Métastases — Troubles menstruels																				
Suppressions et Métastases — Métastases																				
Poisons — Boissons alcooliques. (1)																				
Poisons — Abus du tabac																				
Poisons — Autres poisons végétaux																				
Poisons — Poisons minéraux — Plomb																				
Poisons — Poisons minéraux — Mercure																				
Poisons — Poisons minéraux — Cuivre																				
Poisons — Poisons minéraux — Autres																				
externes — Insolation																				
externes — Chaleur intense																				
externes — Froid intense																				
externes — Coups, chutes sur la tête																				
externes — Autres causes traumatiques																				
Autres causes physiques																				
Causes mixtes — Excès de travail intellectuel																				
Causes mixtes — Veilles prolongées, inconduite, libertinage																				
Causes mixtes — Onanisme (1)																				
Causes mixtes — Troubles des fonctions génitales																				
Causes mixtes — Dénuement et misère																				
Causes mixtes — Mauvais traitements																				
Causes mixtes — Passage subit d'une vie active à l'oisiveté & vice versâ																				
Causes mixtes — Perte d'un ou plusieurs sens																				
Causes morales — concernant — la religion																				
Causes morales — concernant — l'éducation																				
Causes morales — concernant — l'amour — amour contrarié																				
Causes morales — concernant — l'amour — jalousie																				
Causes morales — concernant — les affections de famille																				
Causes morales — concernant — la fortune																				
Causes morales — Chagrins domestiques																				
Causes morales — Orgueil, ambition déçue																				
Causes morales — Frayeur																				
Causes morales — Imitation																				
Causes morales — Colère																				
Causes morales — Pudeur blessée																				
Causes morales — Événements politiques																				
Causes morales — Nostalgie																				
Causes morales — Ennui; misanthropie																				
Causes morales — Joie subite																				
Causes morales — Emprisonnement simple																				
Causes morales — ——— cellulaire																				
Causes morales — autres causes morales																				
Absence probable de causes déterminantes																				

FIGURE 7.3. Proposed international standard table of causes from 1869, supposedly the outcome of international deliberations but closely resembling Lunier's table at Blois five years earlier (fig. 7.2). The failed international effort provided a starting point for deliberations on uniform asylum data in Germany. From Lunier, "Projet," 42.

respectfully noted, then ignored. In the end, as so often happens, criticism, however well-grounded, had less impact than the emerging contradictions of a grand rationalistic project.

Too Many Tables

The Swiss were first to take up Lunier's proposal as it emerged from the 1867 congress. Already in September, at their fourth annual conference, the society of Swiss alienists sketched out a plan for international statistics in the light of the Paris discussions. Ludwig Wille, asylum director at Rheinau, took the lead. "Strict scientific treatment, fully objective research from a single viewpoint, to serve the truth, are the conditions required above all," he declared. There should be no compromise merely to ease the labor or to preserve conventional categories. Wille was prepared to push statistical comparability even further. Since many of the insane were not in institutions, scientific statistics had to extend beyond them to census figures. Swiss alienists should take the lead in a push for "exhaustive statistics," even at the cost of increasing Lunier's thirty-one bulky tables to forty-four.[27]

Prussian alienists were deep in discussion of how best to census the insane when they got word of the French initiative. Mental medicine in Germany was acquiring an increasingly elaborate professional structure. Alienists associated with the *AZP* laid the groundwork for a national society at a meeting in 1860 in Eisenach, eventually adopting statutes at the fifth meeting in Frankfurt in 1864. In the next few years, regional sections and local societies were formed in southwest Germany, Lower Saxony and Westphalia, Berlin, and the Prussian Rhine Province.[28]Although the Swiss society of alienists had independent origins and a separate journal,[29] its deliberations were abstracted along with the regional German ones in the *AZP*. Berlin asylum doctors also had an independent organization, and it was pursuing its own plans for statistical reforms. A month before the Paris congress, Wilhelm Sander had reported to an assembly of Berlin alienists about ongoing deliberations with city magistrates and the Prussian ministry. It was useless, they agreed, merely to count patients in institutions, since so many of these come from outside Berlin, and so many Berliners are treated elsewhere. A proper census must begin with a full list of the names and residences of the insane. They also agreed on priorities. First on the list was heredity, whose importance "will scarcely be attacked from any side." Its investigation should extend beyond parents to grandparents and "collateral blood relatives" and should include neuroses, suicide, and the urge to drink.[30]

The Lunier commission published its report in 1869 in the *AMP* and sent out an abundance of offprints. The initial reaction seemed favorable. The Swiss, always liminal in relation to German developments, were again first to take up the French international plan. Wille managed within a month to convene a special meeting to debate the international suitability of the French proposals. He later explained how his countrymen had avoided any fundamental revision, seeking only to frame standards to suit "alienists of all nations" (*Irrenärzte aller Länder*). Their proposed modifications provided a starting point for discussions that year by the southwest German alienists, meeting in Heidelberg. In September 1870, at the all-German meeting in Innsbruck, the cosmopolitan spirit began to break down. Wille would subsequently point the finger of blame at alienist and statistician Wilhelm Tigges, who, choosing ease and convenience over the rigor of science, had demanded a lightening of the heavy burden of data.[31]

While doubts sprang up everywhere, the German response mattered most. The British did very little, while in the United States, a translation of the text, omitting the all-important tables, appeared forthwith in the *AJI* and then was ignored.[32] In the Bavarian State Archives, we find a circular letter in French, dated 31 March 1869, from the president of the Medical-Psychological Society of Paris to the Bavarian minister of justice. The letter and attached report found their way to the Ministry of the Interior, which had charge of these institutions, and then to Hagen in Erlangen for comment. Hagen meditated awhile then published a strong dissenting report (discussed below).[33] The response from Berlin was delayed to the point that Lunier wrote to a medical colleague, Eduard Croner, asking why the Prussian officials were so slow. Croner raised two concerns: a terrible excess of entries, many of them pointless or unclear, and the outdated technology of census lists. The modern way was to use counting slips (*Zählblättchen*).[34]

As Wille noted, Tigges had taken the lead in pressing to lighten the burden of data. German readers might not have expected this, since he had recently assembled the statistical part of the most copious asylum report ever seen, based on half a century of experience and 3,115 patients. Lunier's proposal, he calculated, would require every asylum to fill in 7,900 boxes every year. At the Innsbruck meeting, he proposed amendments to reduce the number to 2,900, and an alternative with only 888.[35] After some discussion, the delegates at this all-German congress chose a committee with three expert members—Tigges, Hagen, and Ludwig Snell (asylum director in Hildesheim)—to report on the French document. These men had recently proposed similar reductions to the psychiatric association of Rhine Province, in the context of intense debate with the Berlin alienists over census cards—not whether to adopt them, but how they could be designed to pack the most vital information into a limited space. Another regional meeting for Lower Saxony and Westphalia concluded that these questions should be saved for the next all-German meeting. The tide had clearly turned in favor of simplified data, to be recorded on census cards, even before July 1870.[36]

That was when war broke out. The standardization campaign now went quiet. In March 1871, Werner Nasse noted the regrettable interruption of progress toward uniform statistics, which, were it not for the war, should have led to a second congress, perhaps in Brussels.[37] In 1872, two years after the fact, the *AZP* published the report of a committee chaired by the aging Carl Flemming, with Heinrich Laehr, the power behind the throne, as first secretary. Flemming's committee had exchanged letters and then met for two days with the expert committee of Tigges, Hagen, and Snell in Kassel in the spring of 1870. Carefully and meticulously, Laehr explained, they worked through the printed document and at last reached agreement on a simplified version. It would revise the nomenclature of disease forms and reduce the tables to thirteen. They were just about to send these results to Lunier when packet service was interrupted by war. The last word from France had been a favorable reaction to Hagen's "excellent" (but critical) essay. Since then (here some delicately compounded negatives) "the question seems in Paris to be resting, to such an extent that we cannot presume that national ill humor, which regrettably has spread over the domain of science, must exclude us from carrying on with the deliberations." The Germans would abandon the international project and stand "on our own feet." The ambitions of the new Reich had opened up the irresistible prospect of integrating statistics of insanity with the census and its technologies. To this end, the leadership appointed a new commission representing all of Germany.[38] The Swiss now aligned themselves with the German effort and with census cards, based on an elaborate justification provided by Wille in a speech to the Swiss alienists meeting in Zurich.[39]

International standards, it appears, were never in the cards. Lunier's forms had to be long, since he lacked the authority to set aside French administrative requirements. There was a consensus on the importance of heredity but not on how best to record it. The technology of census cards held the prospect of tighter integration of diverse forms of data within Germany, which counted for more than integrated international statistics of madness.

But Should Asylum Data Be Standardized at All?

Hagen emerged from these discussions a radical dissenter from the cult of harmonized statistics and an advocate of institutional individuality. While offering the numbers from his asylum as a model for others, he doubted now that a shared statistical frame could ever justify merging their data. His critique did not depend on German exceptionalism but grew out of his engagement with Thurnam's reports. He was thoroughly familiar with the diverse asylum systems of Europe, initially as a result of an alienist education that began with a state-funded asylum tour through Germany, Belgium, England, and France.[40] His first publications were reviews of British reports on insanity for the *AZP*, among them Thurnam's *Observations and Essays on the Statistics of Insanity*, from which he learned the tricky mathematics of cure rates and death rates.[41] The reports of the Metropolitan Commissioners in Lunacy heightened his skepticism of bureaucratic numbers. In 1845, he concluded a review of one of them with the thought "that from tabular overviews of this kind, when unaccompanied by comprehensive notes, little is gained."[42]

He was far from putting blind confidence in medical judgment. In 1849, he again confronted a data-packed report of the Commissioners in Lunacy, this one concluding with an appendix on medical treatments. There, instead of forcing practitioners into a numerical grid, they relied on eminent physicians to describe their methods for different sorts of cases. These were juxtaposed in the report with mindless credulity—or was it devastating irony? Some say this, some say that, and the rest think otherwise. Mr. Prosser at Leicester advises for mania "the local abstraction of blood, counter-irritants, antimonials, hyoscyamus, enemata, purgative as well as anodyne, cold affusion, the application of ice, the warm bath, seclusion, a liberal but carefully regulated diet, with particular attention to the actions of the bowels," while Metcalfe and Simpson of York Lunatic Hospital prescribe "shaving the head, leeches to the temples, cold water or evaporating lotions to the scalp, active purgation, full doses of tartarized antimony, strict anti-phlogistic diet, and seclusion in a dark room."[43] Hagen concluded, as any reader must, that these remedies were baseless and ludicrously discordant. "In sum, the most various and contradictory methods and means are put into practice by the individual doctors." The public, which is not so stupid, must certainly ask how madhouse doctors can claim success when they put on display so much uncertainty and impotence.[44]

Hagen had come to mental illness from psychology and a kind of psychophysics. His own dream of a breakthrough in the understanding of insanity was bound up with measurement, but not that alone. He had a eureka moment when, after reading Adolf Zeising on the golden section of the ancient geometers, he detected this

FIGURE 7.4. Photograph of F. W. Hagen (1814–1888), who was an enthusiastic advocate of asylum statistics on the condition that they be gathered meticulously and under uniform conditions. He took this to imply that data from different institutions could not legitimately be merged. Image in German Wikipedia article on Hagen, from E. Lungershausen and R. Baer, eds., *Psychiatrie in Erlangen: Festschrift zur Eröffnung der Psychiatrischen Universitätsklinik Erlangen* (Erlangen: Fachbuch-Verlagegesellschaft, 1985), 16.

ratio or its square embodied in some basic measurements of skulls and brains. Zeising, who attracted a succession of admirers including Le Corbusier, associated this ratio with art and found a close approximation in the dimensions of pyramids. Hagen argued in an 1857 book on the topic that insanity might arise from unequal rates of atrophy in different regions of the brain during aging, disturbing the balance on which mental health depends. This was not simple ideology, but measurement-based numerology. His biographer called the book his *Schmerzenskind*, or "child of sorrow," on account of the trouble it caused him.[45]

He made his career in the new Bavarian state asylum system built up in the 1840s. He was an assistant physician in Erlangen from 1846 to 1849, then became director of a new institution at Irsee in western Bavaria, and finally in 1859 was called back to direct the institution in Erlangen, where he remained until 1886. In that year he was one of four expert physicians who pronounced Ludwig II of Bavaria incurably mentally disturbed. The reports to the ministry in the first decade of his career were called *Rechenschaftsberichte*, or "reports on the account," and were structured by a standard set of patient tables.[46] He performed the required calculations and focused his attention elsewhere.

At Irsee he published a three-year report in 1853, omitting, he stressed, all the technical, economic, and administrative details to focus on medical issues. Some readers, he remarked, might suppose that three years in a medium-sized institution implies too few patients to yield important results. "Das bestreite ich" (This I deny). The certainty of experience does not depend on the number of cases alone. Thousands of cases, if done superficially, will only mislead, while a few hundred, examined correctly, can be valid for all time. He attended closely to causes, especially to family disposition. The statutes of his institution required the "reporting observer" to specify the impulse that he judged to be causal, "whether predisposing (hereditary *Anlage*) or exciting." He insisted that identification of a disturbed or insane relative does not suffice to demonstrate causation, any more than would a similarity of hair color. It is necessary to collect numbers showing how many mentally ill parents suffer from the same condition as their children. Since only about one person in a thousand is insane or mentally defective, a showing that 134 in a thousand have an insane parent is conclusive. This reckoning, it may be noted, assumed that heredity was defined by similarity and was not polymorphic or degenerative.[47]

The study of heredity requires "exact and careful criticism" of every case, he said. The French too often shirk this responsibility, recording accidental precedents as if they were real causes. Such loose reasoning had raised masturbation into a major cause of insanity when it was more likely a result of organic disturbances. Hagen's case files are descriptive and hold back from diagnostic commitment. The evidence of heredity in Regina B., admitted in 1878, was her "hot-tempered father, who showed copious signs of mental illness," and a brother "who found himself in a state bordering on mental illness."[48] Hagen used cases in his printed reports to illustrate appropriate causal reasoning. He tended to favor hereditary or medical explanations over plausible alternatives. A farmer's son, engaged to be married, had an illicit affair with a servant girl and went mad after his intended called off the wedding. His illness might have been blamed on failure in love, but the young man died soon afterward of tuberculosis, and the autopsy revealed an irregular brain corresponding to his abnormally narrow skull. Hagen liked to check his tables against case-based causal reasoning. He valued the accumulations of experience, and even its statistical presentation, but he insisted that proper statistics depended on the solidity of the case analysis for each constituent unit. Heaped-up data, however copious, was unreliable.[49]

Hagen called himself a reluctant critic of standardization. Like everyone else, he explained, he had welcomed the Lunier Commission's declaration of the need for international uniformity, and merely called for some refinement and pruning. Initially, it was the sheer burden of work, the prospect of spending a quarter of every year preparing data, that most troubled him. Yet if every item had its logic, every deletion would be a choice for expediency over knowledge. If, however, these numbers were really so valuable, why would the prospect of so much data entry awaken "horror and aversion" rather than "joyfulness, delight, and hope"? Crucially, he noted, what took up so much time was not processing data, but preparing the materials. Almost every item involved ambiguity. Although the duration of illness prior to admission seemed as if it could be read off a calendar, it required a hard decision about the moment of onset. It was the same with assignments of outcome, and especially with determination of the form of illness and of its cause.

If uniformity were the goal, he argued, it could not be left to each asylum to analyze its own results. No, in that case the numbers would have to be sent to a General World-Central-Commission of Experts in Insanity Statistics (*Allgemeine Irrenstatistik-gebildete Welt-Central-Commission*), perhaps in Paris. This commission—"What a colossus!"—would face an extraordinary burden of work. Its members, knowing almost nothing of the constituent asylums, would inevitably turn up institutional differences in the forms of disease, mean lengths of stay, and rates of cure, and they would want to know why. State statistics, he generalized, proceed as if everything is clear, as if data can be worked up by composing machines. Science, by contrast, recognizes the need for expert decisions. The chief physician has to devote so much of his scarce time to preparing the data precisely because it is not automatic. Only in this way can he give assurance of its quality.[50]

What Kind of Statistics?

Hagen acknowledged the value of numbers and calculation, but not for all purposes. He described them as characteristic of state statistics, pertaining to political economy and public health, which can hardly proceed without cure and death rates. He thought them less suited to patient care, and for science they were most valuable in alliance with other approaches. The study of groups, he allowed, can sometimes provide new knowledge about individuals. Physiologists could never learn from a single lung cell that it produces carbon dioxide. Conversely, effective statistics depends on accurate diagnosis or classification, requiring close attention to individuals.

This back and forth between personal and collective, he observed, can be dizzying. In 1876, he called attention to a small book that relied on purely statistical reasoning to measure the effect of the events of 1870 and 1871 on insanity in France. That study, by Lunier, provoked him to reflect again on the value and disadvantages of standardization for understanding causes of madness. Lunier noted at the start a curious paradox presented by his problem. Although wartime trauma was identified by asylum doctors as the cause of insanity for hundreds of new patients, patient admissions were actually lower during the war and occupation than in the periods just before and after. Obstacles of travel during the fighting and other such inconveniences could not explain this dip in admissions. Hagen considered that Lunier's balance sheet gave too much weight to the statistics. He should have trusted the 375 capsule case narratives ("observations") that reflected his own and his colleagues' firsthand medical information. Lunier's mistake was to reduce these observations to illustrations of what could be known from the statistics. The clinical observations should have been allowed to stand on their own.[51]

The New German Standards

The collapse of Lunier's standardization campaign in Germany did not signify a triumph of individualized clinical expertise. Rather, it opened the door for standardization in one country. Already in 1871, Nasse redesigned the tables for Siegburg to conform to the Tigges revision, reducing their number to the ten he thought indispensable.[52] The final issue of the *AZP* for 1873 included a supplement (dated 1874), "Census Cards and Tables for the Statistics of Insane Asylums," a standard set of recommended tables, drawn up in 1872 by a commission composed of Hagen, Nasse, Roller, Sander, and Tigges and consecrated in 1873 by a congress of German asylum doctors in Wiesbaden. Insiders knew that the committee had been riven by conflict, but the members managed to agree on a frame for asylum statistics. They presented it as a consensus report, one that could unite German asylums from Baden and Bavaria to Prussia.[53]

This compression of the tables, with no retreat on standardization, was a triumph for Tigges and Nasse. On one topic only was the burden of recordkeeping to increase: hereditary causation. Tigges construed heredity as relevant to almost every question, and the new tables embodied this principle. Hagen, too, favored this move. Lunier's Tables 6 and 7 were for age at admission and age of onset of illness. Tigges wrote in his report: "The reviewer misses in these tables, as in many others, any allowance for heredity." He wanted, and he got, separate columns for age of onset in hereditary and nonhereditary cases, turning a tabular list into a basis for comparison. Lunier's Table 9 on causes inspired Tigges to remark that this was one of the most difficult issues of statistics, to the point that some speak of it as "terra incognita." The solution was more precise classification and sensible statistics. Since it was not possible to detect the causal force of heredity directly in individuals, it should be measured in terms of ratios of cases. Tigges challenged the distinction between predisposing and determining causes, arguing that heredity was sometimes sufficient by itself to produce insanity. As such, it deserved a more exacting differentiation than merely between direct and collateral relatives. The influence of parents, for example, is greater than that of grandparents or of aunts and uncles, and the tables should make this visible.[54]

They also looked for a relationship between the ancestral disorder and that of the descendants. Lunier's Table 9 for causes had already incorporated a second axis for disease form, and the focus on hereditary transmission in the proposed uniform German tables was still tighter. It split the table of causes into separate ones for each form of disease. The first of these, Table 5a, was for mental illness. It also pared away some of the causes that did not bear on heredity, while simultaneously expanding the table to specify degree of relatedness and whether on the paternal or maternal side, or both. Its columns were designed to bring out the role of specific conditions in family members. The first was mental illness, followed by nerve disease, urge to drink, suicide, "striking character and capacities" (*auffallende*

Tabelle Va.

Aufnahmen. Erblichkeit und Familienanlage a. bei einfacher Seelenstörung.

Nummer.	Grad der Verwandtschaft.	1. Geistes-krankheit.		2. Nerven-krankheit.		3. Trunk-sucht.		4. Selbstmord.		5. Auffallende Charactere u. Genies.		6. Vergehen.		7. Summa.		
		M.	Fr.	M.	Fr.	M.	Fr.	M	Fr.	M.	Fr.	M.	Fr.	M.	Fr.	Sa.
	A. Erste Aufnahmen.															
	I. Directe Erblichkeit, d. h. bei den Eltern fand sich:															
1.	Von Vater Seite															
2.	Von Mutter Seite															
3.	Von beider Eltern Seite															
	II. Familienanlage.															
	a. In aufsteigender Linie, d. h. bei Grosseltern, Geschwistern des Vaters oder der Mutter fand sich:															
4.	Von Vater Seite															
5.	Von Mutter Seite															
6.	Von beider Eltern Seite															
7.	b. In gleichstehender Linie, d. h bei Geschwistern fand sich															
8.	Unbekannt															
9.	III. Blutsverwandtschaft der Eltern															
10.	IV. Uneheliche Geburt															
11.	**B. Wiederholte Aufnahmen mit Erblichkeit**															
12.	Erblichkeit zweifelhaft															
13.	Nicht erblich															
14.	Summa															

FIGURE 7.5. Proposed table of hereditary causes, subdivided by relatives affected, against disease form, in an 1874 proposal for standardized German asylum statistics. Tables of causes of insanity in the German census reports had grown to monumental proportions by 1880. The biggest ones do not seem to have been much used, but the standardized categories were still in use for statistics of insane heredity in the 1930s. From the pamphlet *Zählkarten und Tabellen für die Statistik*, 14.

Charactere und Genies), and lawbreaking. The enshrinement of these categories by the census form ensured their place in psychiatric-eugenic discourse right into the Nazi era.[55] The basis for this and most other tables was a new admission form printed on the front and back of a census card. About a third of the card was used to record data on heredity, providing information for a table of degrees of relatedness. With so many numbers to record, efficiency was critical. Apart from the patient's name, birth place, residence, and date of birth, all information on the card was entered simply by underlining a printed entry. A shorter red card, filled out at the time of discharge, recorded outcomes, including, where appropriate, cause of death. Finally, for the sake of recordkeeping, the committee recommended a third form with just a few lines for each patient who remained at the beginning of the year.[56]

Unified statistics took time. For at least a decade after 1871, the Prussian statistical office, rather than the new imperial one, was the main site of standardizing efforts. Like its asylum system, Prussian statistics of insanity had developed later than in France and England. Damerow had criticized a count in 1852 for attending only to licensed institutions. Ernst Engel, arriving from Saxony in 1860, moved Prussian statistics very quickly into the first rank, and from the beginning, his field of vision extended to counts of insanity.[57] Standardization of asylum statistics was no isolated ambition but part of the planning for a new German census, which was to be revolutionized by card technologies. The Medical-Psychological Society of Berlin became involved in discussions of the new census in 1867, the very year of Lunier's initial proposal, and immediately fixed on heredity as an item deserving top priority.[58]

At first the cards were small, just 18 by 12 centimeters and 3 grams, the size of playing cards. The information on traditional enumeration lists, filled out by census takers on one line per household, was to be transferred to these cards, which would ease the sorting and tallying. But Engel was dissatisfied with the need to copy all this information from lists to cards. Stiffened and enlarged to 33 by 24 centimeters, the cards could be used for the initial data entry. Another innovation in Engel's census was to define the individual rather than the family as the statistical unit. Asylum patients, similarly, would now be recorded on one card per person, white for men and red for women. Christine von Oertzen describes how these cards were sent to data workers in households all over Berlin to be sorted into piles and counted off with hand movements like those of a dealer, then boxed up and returned for control. The process could be repeated, perhaps twice more, to record new combinations of categories. It was a brilliant paper technology, far better suited to the preparation of complex tables than the old census lists. In consultation with the society of alienists, the census office developed a special set of fully standardized cards for asylum patients. Already in 1873, according to a notice in the *AZP*, a printer in Siegburg was offering better ones at a lower price than in Berlin.[59]

The Prussian statistical officer Albert Guttstadt took a lead role in this work. He surveyed the field of censuses of insanity in 1874 in the journal of the Prussian Statistical Bureau. His paper described census taking in preunification Prussia and in other states then outlined the history of the Lunier initiative, how it had broken down, and what could take its place. The recommendations of the Association of German Alienists, he argued, were perfectly matched to the goals of the census office. Every white or green form (the revised colors) for male or female admissions should be followed when the time was right by a red one, indicating discharge. The filing system he envisioned was so perfect that there should be no need to file cards for continuing patients. The data work of asylum physicians was reduced to filling out the forms and assuring their accuracy. Local knowledge was otherwise irrelevant.

The system was German, not global, but Guttstadt's vision of impersonal, standardized data was uncompromising. "The material for the individual asylums must exclude the subjectivity of the individual reporter; the individual cases must be useable for general scientific questions without regard to the institutions concerned." These cards would be passed along to a central office (*Centralstelle*), which should retain possession of all documents. The tables achieve their value not in the particular institutions, he said, but only at the *Centralstelle*. It was not quite Hagen's General World-Central-Commission, since it applied for the moment only to Prussia, and soon afterward to Germany. A standardized array of globally ordered madness data would have to wait. And yet it appears, at the end of the day, that Lunier's ideal of asylum statistics emerged triumphant, while Hagen was left to tend his own garden.[60]

That garden, however, was well suited to the cultivation of *Anlagen*, an exemplary site of hereditary research. Hagen's meticulous care in assembling and ordering hereditary materials won him recognition as a leader of insanity statistics.[61] His faith in statistical homogeneity, fashioned within an institution rather than on a national or global scale, remained, in practice, fundamental to hereditary research. As census officials dreamed of universal statistical order, local authorities found ways of manipulating data that might enable them to measure the results of hereditary transmission and even to comprehend its mechanisms.

CHAPTER 8

German Doctors Organize Data to Turn the Tables on Degeneration, 1857–1879

Most often, the disease that is transmitted is transformed.

—Henri Legrand du Saulle (1873)

A statistical refutation of the medico-cosmic theory of *dégénérescence* seems as promising as smashing a shadow with a sledgehammer. However, a decade after Morel's articulation of this slippery, portentous doctrine, Wilhelm Tigges mobilized tabular data for just this purpose. Despite his statistics, a sense of intrinsic degeneration continued to develop as an international cultural movement widely appreciated by alienists, especially in France. Meanwhile, the data science of heredity gradually consolidated around methods for tracing the transmission of traits across generations. Insane heredity in France gradually separated from other parts of northern Europe and North America. Each version was, in a way, highly empirical, but the French theory of degeneration had almost no use for statistics.

In the new Germany, the great advantage of census cards was to ease the preparation of tables that could reveal causes by placing the relevant variables into a two-dimensional array. The new tabular system presumed that virtually every characteristic of asylum patients had a hereditary component. Neither Tigges, who spearheaded the effort to lighten the data demands on these institutions, nor F. W. Hagen, the most resolute skeptic of mass statistics and standardization, wanted to turn away from numbers. They struggled instead to sharpen the focus and in particular to advance from suggestive hereditary relationships to predictive calculations that could guide reproductive interventions. Such results should shed light on how the hereditary *Anlage* was transmitted, the role of dormancy, and what sorts of transformations were likely from one generation to the next.

The census reforms within newly unified Germany created the possibility for insanity statistics on a much larger scale. Already in the 1860s, by exploiting the growth of mental hospitals and extending reports over ever-longer periods, the tables were veering toward superabundance. Now, even without a census, uniform reporting from institutions all over Germany could turn figures in the thousands into hundreds of thousands. What was the value of so much data? Hagen held that the sacrifice of quality and of homogeneity must outweigh by far the advantages of copious records. While Tigges was more impressed by the potential advantages of large numbers, he too relied primarily on multiyear data from his own investigations. Massive new census reports on the insane tended to languish.

Institutional numbers, by contrast, had consequences. Three years before the Franco-Prussian War, just as the push for standardization was beginning to gain momentum, Tigges had ventured forth with tables from his institution to do battle with the theory of degeneration. As second physician at the Westphalian asylum in Marsberg, he prepared four hundred dense pages on its statistics, to appear beside the director's historical essay. Heinrich Laehr had the volume sent to all *AZP* subscribers as a special supplement for 1867. Was it Tigges's argument that impressed him, or the full half century of coverage, or the more than 3,000 patients, or the abundance of elaborate tables? Even true believers must have wilted in the face of so much statistical detail. Yet these tables were less conventional, and the diligence of the author still more remarkable, than at first appears. The labor of analysis was by far disproportionate to the number of pages, for Tigges almost always combined variables in order to clarify quantitative relationships. He folded results from other asylums into his tables as a basis for comparison, and he made use of new data technologies, as in his line graph of patients with and without hereditary antecedents according to age of onset.[1]

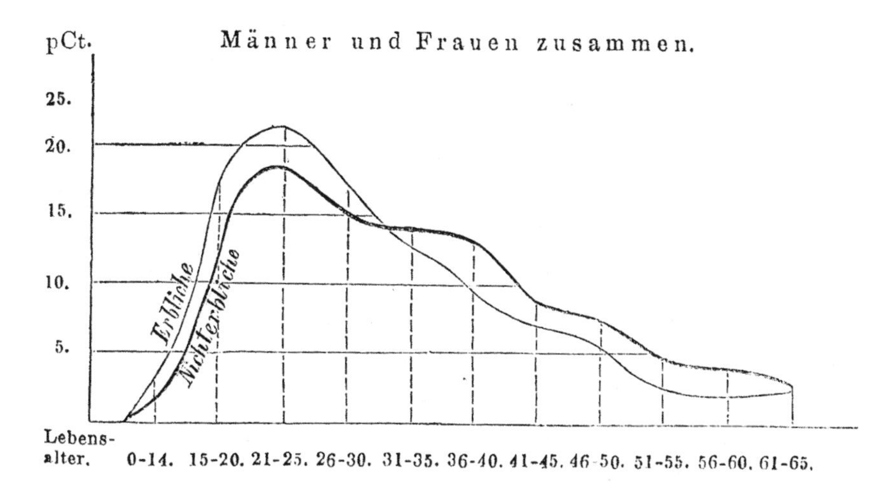

FIGURE 8.1. Line graph by Wilhelm Tigges comparing age distributions for first appearance of hereditary and nonhereditary insanity. A tendency toward earlier outbreak in hereditary cases could have been seen as evidence of family degeneration, but he drew no such conclusion. From Koster and Tigges, *Geschichte und Statistik*, 265.

Contesting Degeneration

The European preoccupation with degeneration from about 1860 to the First World War is well known to scholars.[2] It is easy, however, to misunderstand the logic of its relationship to the science of heredity. Biological degeneration was already widely discussed in the eighteenth century. Much of the brouhaha of the late nineteenth century was focused on a version articulated by Morel in books he published in 1857 and 1860. His version made a great impression on alienists, especially in France, and inspired European artists, philosophers, and novelists for at least half a century. Cultural historians have found it irresistible. It did not dominate the field, however, but was challenged by medical and scientific writers on heredity right from the start. While Tigges articulated his critique in statistical terms, using new tabular arrangements, he had definite ideas about how heredity works. While he allowed for mutability, behind it lay an *Anlage*, whose passage from generation to generation should be traceable.

Hereditary continuity was defined by similarities between ancestors and descendants.

For Morel and his followers, heredity was intrinsically a process of degenerative change within which nothing stood still. Pedigrees of degeneration—in contrast, for example, to Dahl's—were not a tool to track the recurrences of a taint or *Anlage* but, as in Émile Zola's novels, manifestations of a fateful narrative of decline. Hereditary insanity for Morel was a distinct form of illness with its own physical indications or stigmata, the "special physiognomy" of its victims.[3] It played out according to a pattern across several generations, from superficial symptoms appearing intermittently or late in life to disabling, irreversible ones that often were present at birth. It combined bodily and mental decay, each reflecting a progressively deeper penetration of the disease, culminating in cretinism or idiocy and in sterility, hence extinction. However, bodily conditions and delusional thoughts were not easily reconciled in a single typology. Like so many physicians, Morel wrote and thought primarily in cases, and his were uncontrollably various. Although he stuck by his grand explanatory categories, he did not paper over the exceptions. Tigges held that so many exceptions reduced Morel's argument to incoherence.[4]

Tigges and Morel shared a commitment to hereditary intervention but gave opposing rationales. Tigges thought in terms of the transmission of predispositions or *Anlagen*, whose presence should perhaps disqualify a victim of insanity from reproducing. Morel understood degeneration as irreducible process, a trajectory of decay that was set off by bad behavior of an ancestor, most often involving drink, and aggravated by ill-advised marriages. His advice on this topic, strangely, was utterly unoriginal. He copied it word for word from a treatise on inflammations by the pathologist and alienist Louis-Florentin Calmeil, the long-term assistant of the elder Royer-Collard at Charenton. Morel focused on marriage within families tainted by conditions such as epilepsy and idiocy, even calling for legislation to prevent them.[5] Traits running in families over several generations did not figure in his analysis. In 1867, Baillarger initiated a discussion at the Medical-Psychological Society concerning what he called the most important question of heredity. What proportion of children in families with a predisposition will be affected by this taint? Some physicians in attendance dismissed the question as too vague to be meaningful. Morel's answer pushed aside the language of predispositions and resemblances, focusing instead on the stigmata of heredity, which might be physical, moral, or intellectual.[6]

When, in 1876, Tigges articulated his understanding of hereditary mechanisms, he relied heavily on a German volume of commentary on Darwinian evolution. Its author, Oscar Schmidt, was a professor at the now-German university of Strassburg. Schmidt stressed that heredity by itself is neither a force for progress nor decay, but the ground of biological stability, passing on characters from parents to offspring. Biological change was the result of selection and adaptation. Tigges, welcoming this critique, accused Morel and his student Henri Legrand du Saulle of exaggerating the role of inheritance of acquired traits in a vain effort to establish a biological basis for degeneration. Heredity, he declared, can do no more than pass along *Anlagen* for the best-adapted traits to subsequent generations. His aims were entirely pragmatic, focusing on breeding as artificial selection with no mention of natural selection or of the origins of species. He thus reduced biological change to comparative statistics of survival and reproduction, on the assumption that like tends to reproduce like. Morel's mechanism of degeneration might be found in a few families, he conceded, but not as a rule. Brown-Séquard's recent experiments on guinea-pig neurology, showing medically induced epilepsy to be inherited in these creatures, must also be untypical. Darwinism, Tigges concluded, was more correct on these questions than were the theories of most alienists.[7]

These thoughts on heredity from 1876 appeared in a five-year report on the asylum Tigges then directed in Sachsenberg, an institution with a tradition of serious data collection going back to its first director, Carl Flemming.[8] Schmidt's Darwinian interpretations deepened his knowledge of heredity while supporting in the main what he already believed. His arguments against Morel's degeneration were, in the first instance, statistical and, in terms of his own intellectual development, pre-Darwinian. Already in 1867, he viewed Morel's "observations" as simultaneously slipshod and tendentious. They also were maddeningly heterogeneous, so much so that he found it hard to fix on a specific object of criticism. Why didn't Morel mobilize his own abundant statistics as evidence for or against degeneration? One thing, however, was clear: the theory involved some inherent process of decline from generation to generation. In a nutshell, patients with hereditary mental illness must, on average, descend from ancestors with less severe conditions. These included epilepsy, mild nervous hysteria, and psychically dubious (*psychisch zweifelhaft*) states. This last phrase referred to odd or delusional thinking that fell short of insanity. His proud achievement was to have collected and configured data to put Morel's doctrine to the test.

An inventory of relevant numbers appears on the four-page Table 6 of the Marsberg report, representing 532 patients whose insanity was known to be hereditary: 357 with at least one mentally ill parent, and 175 involving mental illness of a grandparent, aunt, or uncle. Tigges's criterion for classifying a patient as hereditary was the identification of a close relative showing any mental disease, including relatively mild nerve disorders that did not require asylum treatment. Just 32 of his hereditary patients had even a single relative from the parental or grandparental generation with epilepsy. The corresponding number for relatives with nervous diseases, including imbecility and hysteria, was 107, and for psychically dubious states, 44. The great majority of his hereditary patients had relatives with full-blooded mental illness. In sum, most asylum patients were no sicker than their progenitors. He thus was entitled to doubt whether the lash of degeneration "really influences the descendants so devastatingly" as Morel claimed.[9]

Table 10 (see fig. 8.2), similarly, was not calculated to add luster to Morel's theory. It involved elaborate processing of the basic patient data from Table 6 to reveal the relations of *Abstammung*, or descent, not to the diagnosed illness of the patients themselves, but rather to certain maladies detected in their brothers and sisters.

Tab. 10. (nach Tab. 6.)

Abstammung von	Vorkommen anderweitiger mit Geisteskrankheiten zusammenhängender Nervenleiden unter den Geschwistern der Aufgenommenen je nach der Abstammung. Descendenz.				
	Blödsinn. (*)	Epilepsie.	Psychisch zweifelhaft, mit Trunksucht und Selbstmord.	Sonstige Nervenleiden.	Summa.
Geisteskrankheit direct	1,67	1,02	1,84	0,82	5,3
- indirect	0,50	0,5	0,5	0,5	2
Epilepsie	—	7,3	—	—	7,3
Blödsinn	12,5	2,08	—	—	14,6
Hysterie etc.	1,63	1,63	0,81	3,25	7,3
Psychisch zweifelhaft	3,13	—	9,375	4,7	17,2
Sa.	1,87	1,18	1,67	1,18	5,9
Männer	1,93	1,4	2,28	0,53	6,1
Frauen	1,21	0,97	0,72	2,17	5,07
Angeboren	4,72	6,69	5,12	2,36	18,9
Sa.	2,4	2,3	2,4	1,4	8,5

*) Die Procentsätze sind nach der Gesammtzahl der Aufgenommenen und der Geschwister der betreffenden Kategorie berechnet.

FIGURE 8.2. Table in Koster and Tigges, *Geschichte und Statistik* (1867), 215, demonstrating the hereditary similarity of conditions suffered within families across generations. The main point was that hereditary patients with real mental illness only rarely came from families with less severe nerve conditions. Tigges offered the table as a refutation of Morel's theory of degeneration.

These siblings are described individually in the *Remarks* column of Table 6. Did these less unhealthy persons tend to have insane children? No, the rule once again was, like parent like child, even for mild mental conditions. The columns of Table 10 identify four such conditions: *Blödsinn*, or imbecility; epilepsy; "psychically dubious, including drunkenness and suicide"; and other nervous complaints. None of these nervous conditions were strongly linked by heredity with true mental illness. The largest percentages appeared at the intersection of row and column for mental weakness, or *Blödsinn*, (12.5); the intersection for psychically dubious states (9.375); and the intersection for inherited epilepsy (7.3).[10]

These percentage figures were complicated, reflecting the effort required to procure data on inheritance of nerve conditions that the asylum did not treat, such as dullness and epilepsy. With a little effort, we can infer from Table 6 the actual numbers corresponding to these percentages. Just six individuals made up the 12.5% of imbecile siblings who had imbecility in their ancestry. The corresponding number for the psychically dubious was also six; and for the epileptics, just three. Since Tigges knew and even used Poisson's formula for random error, he almost certainly realized that he was dealing with wide bands of uncertainty. Yet his calculation was not quite unprincipled. As he argued two years later in his first comment on the Lunier proposal: "The deeper we go into the study of causes, the more certain we become that consensus among statisticians and genuine advancement of knowledge will not be achieved by defining very wide groups, and still less as the groups become even larger, but instead through a precise apprehension of particular propensities and the most exact possible definition of particular groups."[11] The further advance of standardization should make it possible to achieve credible results even for highly specific conditions.

Troubling Imprecision

Although the high percentage of insanity cases involving hereditary causation was familiar to everyone, its measurement had never seemed satisfactory.[12] In 1873, Legrand du Saulle cast doubt on all prior results with that most basic form of meta-analysis, a tabular list. It was headed "Numerical proportions for heredity in insanity. Per 100" and included 50 published values from almost as many institutions. The numbers ranged from 4% to 90%, and while most were below 50%, there was no discernible pattern. Legrand du Saulle, a physician at Bicêtre, believed firmly in the force of heredity, and he felt bound to explain why so many of the numbers were so low. Large public asylums are incapable of proper investigation, and too many institutions do not look beyond direct ancestors. Adultery, often unknowable, reduces these measures, and many people are unable or unwilling to identify family afflictions.[13] There was also a more fundamental problem. "To understand hereditary transmission well, it is absolutely necessary to comprehend mental affections and major neuroses as varieties of a single species, taking this word 'species' in the sense of naturalists, that is, as a succession of organisms coming from similar parents and capable of reproducing together." A former intern at Saint-Yon under Morel, he credited his teacher with revivifying this exhausted tradition of research by turning attention from the ancestors to the descendants and showing that insanity is a highly polymorphic *dégénérescence*. Without this basic fact, alienists lost all hope of reliable statistics. Earlier classifications missed the key point that words like mania and melancholia merely describe symptomatic states. "Most often, *the disease that is transmitted is transformed*."[14]

In fact, Legrand du Saulle cared very little for statistics. Like Trélat, whom he admired, he multiplied instances of hereditary causation by always preferring it to rival explanations. The highest figure on his list, 90%, was based on an offhand remark in a book unencumbered by numerical data, *La psychologie morbide* by Jacques-Joseph Moreau de Tours, who simply had written: "As I understand it, and as I believe must generally be understood, heredity is the source of nine-tenths, perhaps, of mental diseases."[15] Following his teacher, Legrand du Saulle construed hereditary insanity as a distinct type of mental illness. His ally Gabriel Doutrebente, another veteran from Saint-Yon and the main source for his case material, had already drawn out the implications of inherited insanity for statistics. Their wild discordance, he explained, owed to a failure to include neuroses with mental illness in the ascending line as hereditary causes of mental disease in the descendants. A proper statistics of hereditary transmission would include epilepsy with insanity. Once we have understood degeneration, it becomes self-evident that heredity, though called a predisposing cause, is quite sufficient to produce insanity without the added stimulus of lost love or business failure. He, too, sounds rather like Trélat, though he allowed a modest secondary role for all the moral causes that showed up in asylum data.

Both physical and mental indications, from strange ideas to an ill-shaped head, enabled the physician to identify a hereditary case. Doutrebente wrote admiringly of Morel's insight in selecting Mme. Guérard for close family investigation. The master had discerned at a glance the hereditary nature of her delirium. The family, not the sick individual, was the proper unit of observation. Doutrebente composed his paper as a series of such observations, taking the form of tabular genealogies that captured Morel's theory in a visual display. They provided a template for degeneration leading to extinction (*néant*). Some families survived for five generations, some only for three, but the pattern was clear enough. He received a Prix Esquirol for the work, which also was reviewed enthusiastically in the English *JMS*. However, the notoriety of his charts, especially in Germany, owed much to Legrand du Saulle's use of two of them, redrawn, to illustrate his little book. Both of those families died out in exactly four generations.[16]

Although he provided some names, Doutrebente divulged nothing about his methods of gathering information. Opponents would claim that the theory depended on a shameless selection of families. That is probably too harsh. We may assume that he began with an institutionalized patient (here, Mme. Latouche) whose mother or father also manifested mental irregularities. In this chart, exceptionally, there is also a grandmother. Everything traces back, however, to the insane mother, located in the second generation, and the descent to extinction begins with her children. If the children of the third generation were healthy, the family would not be a candidate for inclusion. Yet the trajectory of degeneration was not quite a self-fulfilling prophecy, and in this uniquely large family, is not yet even conclusive. Several of the children here are in better shape than their diseased parent, and the progression through stages of degeneration flattens out or even turns back at some nodes. The table seems to leave open whether all grandchildren were old enough to warrant that the story has reached its end. Four grandchildren are entered on this table as intelligent, and one of these, in addition, as very young, and there are two very young children in the fifth generation. Yet a death drum of *néant*, the extinction of the line from failure to reproduce, distinguishes the last two generations from the third, and distinguishes, as well, these family trees of degeneration from Dahl's pedigrees and Tigges's multigenerational arrays.

OBSERVATION n° I.

Tableau généalogique de la famille de Mme LATOUCHE.

PREMIÈRE GÉNÉRATION.	DEUXIÈME GÉNÉRATION.	TROISIÈME GÉNÉRATION.	QUATRIÈME GÉNÉRATION.	CINQUIÈME GÉNÉRATION.
Aïeule maternelle intelligente, mais d'un caractère violent.	*Mère*, aliénée, folie circulaire.	1° *Fille*, nerveuse, hystérique.	*Enfant* sourd-muet, chétif, mort en bas âge.	Néant.
		2° *Fille*, très-intelligente.	2 enfants morts.	Néant.
			2 enfants vivants et intelligents.	Néant.
		3° *Garçon*, mort paralytique.	Néant.	Néant.
		4° *Garçon*, mort paralytique.	Néant.	Néant.
		5° *Fille* frappée de stérilité.	Néant.	Néant.
		6° *Fille* intelligente.	*Fils* dipsomane.	Néant.
		7° *Fille*, Mme **Latouche**.	a. *Fils* à tête faible.	*Fille* dégénérée, mal venue, morte en bas âge.
			b. *Fille* intelligente.	Deux enfants en bas âge.
			c. *Fils* intelligent très-jeune.	Néant.
			d. *Fils* mort à la suite d'une opération.	Néant.
		8° *Fille*, rachitique au dernier degré, morte très-jeune, qui n'aurait jamais pu marcher.	Néant.	Néant.

FIGURE 8.3. Gabriel Doutrebente's chart of degeneration and approaching extinction of the family of Madame Latouche, to be compared with the table by Tigges (fig. 8.2). The criteria of degeneration include earlier onset and greater severity of disease, implying that congenital mental weakness is more extreme than mental alienation (madness), which comes on later in life. The most decisive consequence of degeneration, however, is the failure to reproduce, which appears as a proliferation of descendants who died childless (*néant*). From Doutrebente, "Etude généalogique," 213.

Legrand du Saulle's book, which was taken very seriously, confronted German asylum doctors with a double challenge, first by alleging the incoherence of the customary measure of hereditary causation, and then by offering degeneration as the solution. For decades afterward, asylum statisticians cited this list of discrepant measures of "percent hereditary" with embarrassment. Carl Wilhelm Pelman, a Rhinelander by origin who was made head of the asylum of Stephansfeld after Alsace became German, reviewed the book for the *AZP*. He endorsed the French argument that hereditary insanity had been misunderstood, that it was really a distinctive form of mental illness with an unfavorable prognosis and could be recognized from certain stigmata. The book deserves many readers, he declared, and should be translated. Exploiting his binational situation, Pelman promptly commissioned an assistant to do the job.[17] At the 1874 congress of German asylum doctors, he extolled the tradition of administrative statistics at Stephansfeld and commended it as a model for unified Germany.[18]

Data for Hereditary Prognosis

Tables of the extinction of families fit well with the loss of hope that pervaded asylum systems by the 1870s. Although Tigges did not credit Morel or his students with any useful insights, he may have been moved by Morel's efforts to turn his attention to hereditary prediction, focusing, as Legrand du Saulle had put it, on the descendants instead of the ancestors. He probably noticed a paper in 1869 in the *AZP* by Richard von Krafft-Ebing, fresh from an apprenticeship with Christian Roller in Baden, lauding Morel's disease category of degenerative insanity as a new basis for family prognoses.[19] Asylum doctors often cited data on heredity to explain the irrepressible increase of insanity, an issue of great interest also to Tigges. The time had come to take on the task of quantitative prediction for potential offspring based on the characteristics of parents and other family members. Since very few children were already insane when information on their insane parent was being entered into an admission book, hereditary prognosis required a rethinking of numerical relationships and a long-term plan for collecting data.

In 1876, Tigges encountered a better model than French tables of degeneration in the form of a twenty-five-year report from Erlangen. This was F. W. Hagen's masterwork, a collective achievement of his staff, based, he explained, on the wisdom gained through long experience at his own asylum. Hagen had been present for sixteen of these years, including a few as assistant physician when it was new. As we have seen, he believed deeply in the accumulation of experience over time at an institution he knew intimately. The book brushed aside many bureaucratic categories to focus on what really mattered. Occupations, he argued, were so numerous that the numbers meant nothing without an appropriate way to group them. Insanity rates by geographical location were losing their value in an age of railroad travel. Marital state is certainly associated with mental illness, but how to separate cause from effect? He even dared to challenge the first article in the alienist creed: that incurability is the result of delayed admission. The correspondence between favorable outcomes and early treatment remained inconclusive, since these were often the least severe cases and might have cleared up on their own.[20]

Hagen described statistics as "the artificial creation of masses for the purpose of investigation." It is like looking backward through a microscope lens, enabling the observer to concentrate on an issue that matters and to abstract away the confusing details. He was, however, less willing even than Tigges to sacrifice the exactitude of each observation for the sake of a low coefficient of error. Problems of comparability, he said, arising from flawed data collection, create far more problems than mathematical shortcomings. He assigned paramount importance to systematic collection of the right kind of data. A tally of patients discharged in apparent good health provides no measure even of recoveries. It is necessary in addition to track their careers outside the institution.[21] Determining the effectiveness of treatment is more difficult still. It would require genuine controls to determine how many, if any, so-called cures were more than mere recoveries, the work of nature and not of medicine. He meditated on the possibility of a "parallel statistics" on the mentally ill outside institutions, perhaps through a historical comparison with earlier times when the prejudice against asylums was stronger. How else could alienists demonstrate that society is getting something for all it invests in asylum care? Meanwhile, he shunned aggressive therapies in favor of Bible-reading, instruction, singing, concerts, theater, and dance. He recalled a moving performance by his patients of Méhul's opera *Joseph*. Maybe that was recompense enough.[22]

Measures of heredity, to which Hagen assigned particular urgency, were as vulnerable to flawed reasoning and misleading data as cure rates. The declaration by a relative or health official of mental disturbance in a mother or uncle proved nothing, and even a systematic search for insane relatives was insufficient. Such inquiries merely provided data on ancestors or collateral relatives, where the *Anlage* had already worked its harm. The practical need was to anticipate the health of offspring based on disease characteristics of older relatives. This could serve as a guide to prophylaxis for those on the threshold of mental illness and, more crucially, measure "the intensity of the danger of taking ill for offspring of the mentally ill." Such knowledge had a clear bearing on advisability of marriage. A genealogy extending backward in time, however interesting, "lacks the force of demonstration by which to establish norms of behavior." To look forward required new techniques for compiling data on children of the insane. As usual, his investigations began with patients in the asylum. Since the children would be very young or even as yet unborn when their parent entered an asylum, their *Anlage* for mental illness might remain hidden for decades. The evidence would have to be statistical and sufficient to demonstrate collective tendencies, since cases can always be countered by other cases.[23]

Hagen's main contribution to the book he edited concerned the goals and methods of asylum statistics, including techniques for assembling data on discharged patients and their offspring. His institution, he explained, had from the start been assiduous in its recordkeeping, insisting always on firsthand information. The staff presented a detailed questionnaire (*Fragebogen*) to the physician arriving with each new patient then solicited private reports on these patients and their families. Hagen's rule was to exclude doubtful cases from the statistics. Having laid this groundwork, he entrusted the long chapter on heredity to second assistant physician Heinrich Ullrich. A few years later, when the *AZP* introduced semiannual reports on psychiatric literature, Ullrich was chosen as bibliographic specialist on statistics. His contribution to the Erlangen volume, which contained almost all its tables, conformed to the new Prussian standards. Since most of the data had been gathered years earlier, the analysis cannot have been straightforward. Ullrich recognized, however, that statistical harmonization in Germany was making it easier to achieve homogeneity. He also drew inspiration from the coding techniques, tables, and conclusions of Wilhelm Jung.[24]

Ullrich had definite ideas about studying the hereditary factor in insanity. The best way to understand its transmission would be to track the succession of organic structures, the diseased *Anlagen*, from generation to generation. Diagnosed conditions served as a proxy for these *Anlagen*. He found high rates of nervous conditions, including "peculiarities of character," the urge for drink, and eccentric behavior in many sane members of hereditarily burdened families, suggesting that the *Anlage* for disease was not quite invisible during dormant periods. His interpretation of minor neuropathies as signs of a latent mental illness was perpetuated in Mendelian interpretations of heterozygosity. Heredity appears in Ullrich's account as a wavering pattern of fluctuation, not as a fateful decline. The statistical tables from Erlangen had very different implications from Doutrebente's family diagrams. Ullrich did not claim to be able to track an individual *Anlage*, but he could measure its average force. Using a form of analysis that Hagen had been trying to transcend, he calculated that one of every 2.68 cases of mental illness was hereditary. This was a satisfying result, he thought, in good agreement with studies by Damerow, Jung, and Julius Rüppell, the best authorities, whose numbers he gave as 1:3.25, 1:2.95, and 1:2.7. There seemed to be hope of escaping the chaos of numbers described by Legrand du Saulle.[25]

Ullrich also gave an estimate of Hagen's most vital measure, the probability that children of parents with mental illness will themselves become insane. Like Tigges, Ullrich sometimes reported tiny differences as if they were meaningful, yet he recognized the inadequacy of his comparative figures. It would take decades to accumulate sufficient data on the second and third generations of hereditarily burdened families in order to test the theory of degeneration. On many points that mattered, the twenty-five-year Erlangen report had only qualitative results.[26] Hagen expressed greater satisfaction with his methods of investigation than with their conclusions. The data, alas, came in slowly, and he needed time. He dreamed of initiating a long-term study at a single location. We discover in his chapter a reprinted circular sent to physicians all over Bavaria asking those with information on discharged Erlangen patients to fill out and return a data form. In big cities like Munich, where many of these former patients would be unknown to the district physicians, he also sent inquiries to church offices. In either case, he checked and corrected all such information. He described this kind of work as more valuable than mathematical calculation of probable errors. In time, the data required should come in.[27]

Pedigree Shortcuts for Long Generations

The problem of hereditary insanity, whether explained by degeneration or not, appeared even more urgent to German alienists than to French ones. Morel wrote in an 1870 asylum report that, apart from general paralysis, the increase of the insane since 1838 was an illusion, the result of sick persons now having better options than prisons and poorhouses.[28] Others, to be sure, found Morel's theory more alarming than he did. Although Tigges's antidegenerationist arguments were endorsed by commentators in the *AZP* and in a leading medical reviewing journal, many German alienists, including Hagen, were less certain than Tigges of the incorrectness of Morel's theory.[29] And Tigges, like Hagen, was working urgently on alternative techniques to anticipate mental illness of the children based on the health status of their parents. He praised as exemplary Hagen's meticulous efforts to maintain the flow of data on former Erlangen patients and their families, yet even Hagen, after all these years, could identify only a fraction of inherited mental illness that eventually would erupt among the descendants of patients. Tigges did not want to wait. The most vital contribution of asylums to public health was at stake.

"From a certain number of mentally ill parents come a certain number of children, healthy and mentally ill," and we want to know what percentage will be sick. "The question is posed, whether there is a means and a way to press forward to the final goal of research," to calculate what proportion of mental illness in the general population is owed to ongoing transmission from mentally ill parents. These empirical investigations, like those of Hagen and of Doutrebente, began with current asylum patients. Hagen looked to earlier generations to determine if the insanity was hereditary, and to the children to determine how often the disease was inherited. Was there a better way than waiting for data on the children of his patients? One element of the solutions would be to compensate for incomplete family data by matching the children to adults with a similar background of family illness.[30]

Tigges had already adumbrated a method of generational comparison in his 1867 Marsberg report, as a correction to Morel's analysis of degeneration. Twelve years later, he laid out a specific proposal to orient research, not around the children of the insane, but the children of their parents. We might simply say, their siblings, but this glides by his innovative logic of investigation. The admission of a new patient should bring immediate attention to the parents, who assume a role in the study comparable to that of the patients themselves in prior studies, including Hagen's. The health of *their* parents and siblings, the grandparents, aunts, and uncles of the institutionalized patient, would give an indication of the presence of a family *Anlage* for mental or nervous illness. Did Tigges reason in a circle by using the condition of parents and their siblings to identify hereditary cases? His finding that parents with insane heredity had more children might reduce to the near-tautology that families with more members are more likely to have at least one diagnosed as insane. Tigges, in fact, recognized the problem without quite knowing how to handle it, except by searching for the right kind of data. Three decades later, Wilhelm Weinberg made these issues central to the quantitative study of psychiatric heredity.[31]

Tigges presented this research in 1879 as his contribution to a symposium on the redesign of—what else?—asylum census cards. The space for entering information on each patient was very limited, and much of it, he charged, was wasted. He proposed to clear away enough bureaucratic trivia so that each card could include an entire family tree, which could then be used to calculate the probability of mental illness in children of asylum patients and to compare that number with the probability for offspring of healthy parents. Already, he explained, he was correcting for age distribution and calculating this probability for parents at his institution. Tigges reckoned that a child of a patient was 162.9 times more likely to become mentally ill than a child of healthy parents. He rounded his multiplier to 160 then rounded it again to 150 for the summary of his Sachsenberg report in the *AZP*. Still, the hereditary reproduction of insanity appeared in his work as an immensely powerful force.[32]

Although Hagen, Ullrich, and Tigges were also curious about the mechanisms of transmission of mental illness, they focused on the empirical experience of transmission across generations. Even if it was not possible to know with certainty which families possessed a diseased *Anlage*, they did not doubt its reality. Their ambitions recall Dahl's: to track the transmission of the *Anlage* through the generations and to measure its contribution to the great social and medical problem of insanity in order, someday soon, to take action.

CHAPTER 9

Alienists Work to Systematize Haphazard Causal Data, 1854–1907

> What asylum statistics asserts about heredity is at bottom relatively worthless. From the considerable differences among institutional tables, we can much more readily draw conclusions about the observers than about the observed.
>
> —Paul Samt (1874)

> My examination reveals a hereditary taint on the side of an alcohol-dependent father who ended his life by suicide after a business collapse.
>
> —Thomas Mann, *Felix Krull*

Andrew Scull remarks that the teeming statistics of the asylum age "tell us more about the confiners than the confined." Asylum doctors themselves, as we have seen, were among the harshest critics of their own numbers. Still, it was comforting sometimes to experience a number as a simple fact, a moment of release from the burden of interpretation. Journal editors in Germany, who could be brutal in their attacks on naïve number-mongering, nevertheless used figures from asylum reports as filler. A report having come to them from Ohio, Venice, Petersburg, or Colditz might be summarized compactly as the numbers of patients admitted, cured, improved, unimproved, and dead, then reduced still further into a cure rate. Causes also were reduced to nuggets of data. A snappy paragraph in the *AZP* in 1850 passed on news that mental illness affected 70 of the 8,000 inhabitants of the Faroe Islands, a ratio of 1:110, and provided a short list of causes, with heredity in the lead at 22. Their source judged this to be an undercount.[1]

In insane asylums, as in so many institutions, official sanction gave numbers a validity that did not depend on independent truth. The pronouncement of a cure meant the patient would be discharged, and if a brother and sister were each diagnosed as mentally ill, this was relevant data for measures of the force of heredity, even if doubt remained. Asylums struggled at every level with data management, which became more complex because it was more consequential when a decision how to classify had immediate consequences for patients and their families. Once these numbers left the sites of decision, to be accumulated in books, files, or cabinets, they could be treated as neutral data. Techniques for merging and processing such data grew increasingly elaborate over time, especially after 1870. In German lands, they provided a basis for the study of heredity in medical research institutions, notably the new system of psychiatric clinics. In Britain and America, where non-medical statisticians and geneticists became the most prominent researchers on medical and psychological heredity, records of institutions made the work possible. In many cases, doctors had already created elaborate databases unbeknownst to the scientists, who welcomed the numbers as manna from heaven, the answer to their prayers.

The history of data and recordkeeping is a very human history, involving decisions with welcome or distressing consequences for people at moments of particular vulnerability. The data may subsequently be converted to infrastructure, the basis for bureaucratic and scientific work, again with hopeful or terrible implications. Accessibility and manipulability were greatly enhanced in the late nineteenth century by filing technologies and then by electronic ones. The basic structure remained intact throughout the following century.

Knowable Family Histories

Within Europe and North America, including institutions serving Europeans in the colonies, it was possible to impose a grid of patient characteristics, outcomes, and causes. Non-European populations did not fit well, even when the effort was made. The idea that insanity might be altogether different among these peoples, or even a disease of the civilized only, lingered on. The English in

India found enough lunacy there to justify a system of asylums, yet their cases diverged from the European norm. The leading cause, far outpacing all others, was ganja (cannabis). Heredity scarcely came into the picture, and the doctors never explained why. We may suppose that local informants did not volunteer such causes or tell stories of manic uncles and dotty grandmothers. British medical officers seem to have had little access to information on families or, indeed, on any etiological factors.[2] While an interest in ancestry and lineages is common among the peoples of the Earth, the ability and inclination to share knowledge of family illness is anything but routine. The first state asylums in California, as we have seen, also lacked information on heredity. It is no surprise that the segregated state asylum in Milledgeville, Georgia, in the final decade of the nineteenth century recorded fewer causes and much less hereditary illness for its black patients than for white ones.[3]

The recognition of insanity as a persistent public-health menace did not translate automatically into higher reported rates of hereditary causation. The rare historians who pay any heed to the ubiquitous tables in asylum reports might well anticipate, as I did, a progressive increase in the figures for heredity. If there is any such trend, it is weak and inconsistent, overwhelmed in every institution by the effects of changes of administration. At Hanwell near London, where the male and female divisions had different superintendents, the distributions of causes for men and for women diverged radically.[4] A few medical officers did not even include heredity as a cause, and the rest might or might not call specific attention to it. Hereditary influences on insanity had long been familiar to medical and lay reporters alike, to the point that case reports from early in the nineteenth century often mentioned casually when a patient was not hereditary. In 1885, in a fifty-year report from the asylum at Winnenthal in Württemberg, the assistant physician commented explicitly on shifts in recording practices. The effects of hereditary disposition, he remarked, have drawn much attention in recent years, yet these results were unavoidably inexact.[5]

By 1870, most asylum doctors were required to fill out a form containing specific questions about family background or hereditary causes. The rising medical hereditarianism of the late nineteenth century, reflecting new conceptions of professional responsibility and a growing sense of medical failure, was closely allied with new data practices, including case records as well as numbers. These were developed most impressively in Germany, where data on insane heredity grew to monumental proportions. Asylums and special schools were turned into sites for purposeful collection of hereditary data. Biometry and Mendelism, so often construed as the foundational methods of hereditary science, depended utterly on new techniques of data recording and filing.

English Cases

Assigning causation, if it went beyond merely recording a declaration by the patient or "friend," required some interpolation or untangling of the evidence. A young woman, Elizabeth T., admitted to Bethlem in 1854 was assigned "anxiety of mind" as cause, with a clear "no" for "relatives similarly affected." But after a report from a family member cast doubt on this, she admitted that her brother had died insane and that her own terrible anxiety arose from a fear that she would follow him. She was, perhaps, no victim of heredity but of hereditarian fears. Another new patient, Elizabeth Y., mentioned her niece, P, under "relatives similarly afflicted," and gave, as her own cause, "anxiety about an insane relative." That relative was P, who had been discharged from Bethlem just a few weeks earlier. Did P succumb to reciprocal anxieties, niece and aunt each dragging the other down by their bootstraps? But the problems of Elizabeth Y. went deeper. According to the medical certificate, she fancied that her food was poisoned, that she was threatened with violence by an unknown person or "stilettoing" by a neighborhood Italian, and that Papists wanted to confine her. She also charged her husband "with acts of adultery with a Servant Girl." For this charge there was no basis whatsoever, reported Dr. Albert Hall of Hounslow. How did he know? "This information I had from her husband." The minutes of the Bethlem subcommittee record that on 9 June 1854, this man, Thomas Y.—and not any conspiracy of Papists—had presented a petition for her confinement, supported by his own account of the causes and by two medical certificates. Did anyone imagine that he could be an interested party? A Bethlem physician subsequently interviewed Elizabeth Y. for a case history, but only after the legal process was complete.[6] Superintendent Charles Hood always insisted on a spirit of openness in his asylum, not for the sake of the patients but to promote statistical science. "The facts which occur, and the information which may be obtained in this Institution, ought not, I conceive, to be confined within its walls; it is the interest of science, and the benefit of humanity, that they should be made known." It was safer to publish statistical nuggets than to air family disagreements.[7]

The English system advanced toward nominal uniformity through parliamentary acts and resolutions of the Medico-Psychological Association, which might in turn be incorporated into rules sent out by Commissioners in Lunacy. Case books and admission registers in England had grown uniform by the 1870s to the point that they could be printed by medical stationers and sold from catalogues. At Hanwell, the case book devoted two folio pages to each patient. The first provided spaces to fill in basic information required at the time of admission and to affix the medical certificate. The bottom of the page had more little spaces for notes of the medical interview with a relative or acquaintance. The second page was set aside for case notes, which came to include more frequent and longer entries, in accordance with advancing legal demands. The doctors now included a case history for each admitted patient, which often, however, was nothing more than the items on the printed form strung together into a paragraph. It was like an Office of Medical-Legal Circumlocution. The law insisted on proof that admitted patients were genuinely ill and that the asylums were providing medical treatment. Medicine complied, legalistically, the letter followed and the spirit forgotten.[8]

It was hard to make these inscriptions consistent without a standard as to what made a case hereditary. In the Hanwell asylum books for 1873–1874, we discover that the friends of Emily G. rejected any role for heredity. When asked about relatives affected they mentioned two brothers and a son. By contrast, the sister of Sarah G. said her case was hereditary because of her mother's cousin. The daughter of Clara R. denied hereditary influence, despite a sister who was nervous and had severe headaches, while George M. considered the suicide by hanging of a maternal grandfather sufficient to mark Maria M.'s case as hereditary. Stephen R. was pronounced hereditary without any mention of relatives, and William S. because of a father and cousin who died of softening of the brain. Although William E.'s illness was not hereditary, it seemed relevant to the friend that his mother was paralyzed and his brother was "ultra religious, crying & praying a good deal." The friend of Christopher C. left open the question of cause: "Great aunt had a daughter who died insane & other relations from the stock, but whether the taint was through marriage is uncertain."[9] Two decades later, in 1893, the husband of Mary F. said her disease was not hereditary, despite a sister committed to an asylum, while the melancholic Ada B. required no more than a "light-headed" mother to earn an emphatic "Yes!" to heredity. The medical superintendent of the female department complained in 1876 of all the friends who held back information or who were too dull and stupid to provide "anything approaching a correct account of the antecedents of the patient." Despite these drawbacks, he claimed to have reliable histories for 25 of the 111 new admissions for the year, "and from these we gather that hereditary taint was the most frequent cause."[10]

At other times and places, for example, at Bethlem in 1894, the doctors did not even ask if the case was hereditary but made their own judgment based on what they could ascertain about relatives. The physicians were always alert to families with more than a single member in an institution. The more serious made efforts to correlate cases from multiple institutions. Around 1890, physicians at Hanwell were actively pursuing such data.[11] Occasionally, we find that a doctor has corrected the original record after learning of the confinement somewhere of a near relative. Emily S., admitted to St. Luke's Hospital in January of 1876, was entered as not hereditary, but the doctors reversed this a year later, with the comment: "A brother has lately been admitted into one of the Middlesex County Asylums. Some few months back a younger brother was drowned, this in all probability was accidental."[12]

The Futility of Standards

The Committee on Statistical Tables of the Medico-Psychological Association recommended in 1882 the registration of at least one cause in addition to heredity or previous attacks. This change, it stipulated, should not add to the burden of recording. Indeed it would not, if data entry was treated as an empty ritual. Practical administrators were more concerned to learn which among the new patients would be dangerous or suicidal.[13]

The English form of statistical standardization seemed often to reduce to specifying questions on mandatory forms while providing little or no guidance on the structure of appropriate answers. It reached a pinnacle of pointlessness in the consolidated reports of the Metropolitan Asylums Board from 1888 to 1913. These publications were credited to a statistical committee, which took a professional interest in the numbers and rearranged the forms often enough to obscure the results. Its responsibility extended to institutions for imbeciles, which had no tradition of registering causes, but not to insane asylums, which did. By far the most common cause given was old age, which could be asserted without external evidence. Congenital patients were assigned causes like teething and epilepsy. In 1897, any cause could be registered as either predisposing or exciting. By 1905, there was a third column, "predisposing or exciting," for cases in which these could not be distinguished. Although causes were to be classed either as moral or physical, one of the asylums, in Darenth, assigned only physical causes. While "Unknown" was the most common entry, sometimes overwhelmingly so, the list of causes kept expanding, requiring ever more folds in the tables provided to sum up years of results for multiple asylums. In the report for 1913, there are fifteen classes of etiological factors with up to ten subcategories. Hence most of the boxes, even on the summary table combining all institutions, were empty. Nonzero entries would spring up in some category for two or three years then melt back into the void, leaving a strong impression that such entries registered the crotchet of some physician and were nothing like neutral data.[14]

The London tables, already too large to be reproduced usefully in a book like this one, are scarcely in the same league as the monumental volumes of Prussian asylum statistics from the last quarter of the nineteenth century. The Germans made an excellent show of medical and bureaucratic efficiency in the registration of hereditary data. There were more than a million spaces to be filled in, most of which required sorting as well as tallying. In contrast to all the blank spaces in the London tables, virtually every space in the Prussian ones was filled in, year after year. Yet these incomparable tables seem to have held very little medical or scientific interest.

Anyone tempted to reduce such behaviors to cultural stereotypes should cast a glance over the registration of causes in New York at about the same time. In the report for 1890, every one of the 721 female patients admitted in the two years beginning October 1888 had an ascertained cause. For the male patients in the same period, 406 out of 732 were not ascertained. For 184 females and not a single male, the assigned cause was hereditary predisposition, while 234 female cases, but not one male, were attributed to moral causes. These differences seem to owe more to recording practices than to gender stereotypes. The leading causes in Kings County (Brooklyn) were alcoholism followed by heredity, while in Monroe County, which includes Rochester, both were zero.[15]

So many boxes filled, but to what end? In social and administrative statistics, as in science, completeness could be its own reward, reflecting on the moral character of the workers as well as on the efficiency of the organization. Also, statistical unification was an element of political unification. The Prussian tables reflected deliberations of asylum directors from all over Germany and had implications for the numerical coherence of the new German state.[16] Beyond these moral, practical, and symbolic reasons, there lingered a hope to understand causes, including the effects of an industrialized, urban civilization on susceptibility to insanity in its various forms. Alienists tried to clarify these questions using large-scale statistics.

The dream of international comparability was never abandoned. In November 1879, Ludwig Wille crossed town from the Basel asylum to a meeting of the Basel Economic-Statistical Society to reiterating his appeal to "alienists of all nations" for uniform asylum statistics.[17] Belgium revived the campaign for international statistics at a congress of "phréniatrie et psychopathologie," held in Antwerp in connection with the 1885 International Exposition. Among the leading themes of the congress were uniform statistics and the relations of insanity to criminality. Quetelet's nation was still admired for its censuses of insanity.[18] Ferdinand Lefebvre told of the successes of the Belgian system, and Hack Tuke of the English one. Albert Guttstadt sang the praises of the German model, and in his written report for the doctors back home, explained that the discussions in Antwerp were organized around his presentation. He reassured them that in view of the immense labor they had invested in the existing German statistical system, he had offered no concessions. Even so, he welcomed, in the name of greater uniformity, the appointment of a commission on which he would serve beside Tuke for England, Wille for Switzerland, and others from France, Austria, Russia, the Scandinavian lands, Holland, the United States, and Argentina. Yet none were to enter the promised land of perfect uniformity.[19]

Forms of Disease

Statistics largely defined the program of disease classification. Abnormalities readily recognized from their physical symptoms, such as cretinism, epilepsy, and, often, idiocy, were easily counted and inserted into tables. Insanity in its narrower sense, which usually presented no stigmata and rarely appeared before adulthood, was a much thornier problem. The ancient categories of mania, melancholia, and dementia, used by Esquirol, seemed too inflexible to mark off a disease so protean. Every attempt at a statistical classification ran up against this obstacle. Although asylum care was often based on symptoms, it was widely assumed that a proper diagnosis must provide useful guidance for treatment, and that illness form would correlate with cause. The tabular genre showing disease form on one axis and cause on the other goes back to Holst's census of Norway in 1828.

The *Diagnostic and Statistical Manual* of our own day, that inventory of treatable mental conditions, preserves in its name the ideal of statistical compilation. In the late nineteenth century, the push for an acceptable classification of mental illness was already driven by statistical ambitions to apply a consistent standard across institutions and in every nation. Inconsistencies in the assignment of categories seemed more harmful even than doubts regarding their validity. A new classification could be made real only by adopting new standards and new tabular forms.[20] And these data forms had legs. By 1892, the German ones were being reproduced in Tokyo.[21]

Since patients in the late nineteenth century were likely to be sorted into an institution for care or for cure based on an assessment of curability, classification was as much or more about prognosis as about treatment. A common nosology of the 1870s and 1880s distinguished between "primary" forms of mental illness, such as mania and melancholia, and an undivided secondary form (*secundäre Seelenstörung*). A case that had developed this far offered very little hope and might properly be consigned to a custodial institution. Heredity entered mainly into a different kind of prognosis, which took in the capacity of the patients to marry and have children and how many of their children would be mentally ill. Such considerations were often merged into discussions of disease classification and its implications. An essay on insanity in 1882 with some comparative international statistics integrated disease classification into a mélange of prognosis, heredity, and social policy. The categories are unreliable, the author conceded, yet we need them for planning.[22]

Emil Kraepelin's proposed divide between a manic-depressive form of insanity and dementia praecox was in part a response to debates on standardization, even as it drew from a wide field of social medicine. His data practices, too, were linked to statistics. He entered patient information on *Zählkarte*, cards similar in many ways to those used for patient registration and for the census. They provided a fixed array of spaces in which to enter designated bits of information regarding the patient and the course of the disease—a compact repository of cases that he could sort and rearrange in search of coherence. These have been called, in English, diagnostic cards, since that is how Kraepelin used them, with enumeration as only a secondary purpose. Such cards had a wide-ranging role in the information economy of the asylum and clinic. As Eric Engstrom points out, Kraepelin had used them ever since he was a medical assistant in Würzburg in the 1870s, and they reflected Bavarian recordkeeping from that era.[23]

For the sake of his database, Kraepelin favored a rapid turnover of patients. His was another instance of prognosis giving shape to diagnosis. Dementia praecox was, for him, quite literally a premature descent into the mental weakness of old age. It had a highly unfavorable prognosis, and he dispatched these patients as quickly as he could to a custodial hospital.[24] By 1900, he was putting more and more emphasis on social policy, including prohibitions on alcohol, the control of syphilis, and restrictions on reproduction, to combat mental illness. "Finally, the fact should not be underestimated that in the care of the insane in institutions we possess the only means by which it is possible to eliminate the most frequent cause of insanity, namely heredity."[25]

Records with No Purpose

Statistics of mental illness provoked some sharp controversies. Paul Samt, quoted in the epigraph, complained of the execrable quality of observations whenever they become too numerous. His real objection, however, was the misguided focus on human actions rather than brain tissues and nerve ganglia. "A mental illness would have a natural scientific explanation if its mechanics were decoded in the region of the unconscious, if we knew the laws by which concrete pathological forms develop in brain matter."[26] An 1866 *Lancet* review of asylum reports from England and Canada referred to statistics as so much "idle labor" on account of faulty categories. A decade later, a *Lancet* commission to examine English lunatic asylums wrote of "fruitless researches" leading to an accumulation of "false data." "A bog, over which myriad bewildering lights dance and delude the traveler, would be a faint picture of the perplexing and impracticable province that lies before the explorer who attempts to investigate the subject of lunacy by way of statistics." The problem here was the dishonest assignment of exciting causes in order to condemn or "to elicit sympathy for the afflicted."[27] In 1895, the Lunacy Commissioners in Scotland dismissed the annual tables as lacking even "the smallest statistical or scientific value." This concern was the erroneous merging of distinct populations.[28]

Readmissions and relapses provoked some of the hottest controversies in asylum statistics. The admission book, which at first appears boring and harmless, could be a cauldron of controversy. Every admission implied, eventually, a departure, whether by cure, death, or "elopement" (escape), whether at the prompting of the medical superintendent, the decision of a court, or the irresistible entreaties of family. A discharge closed the account. What, then, was to be done if the patient returned? It was the same patient, as the records showed, but it was a new case with a new numbered entry, portending a new result. The first outcome had been entered already into the official report. Should this now be altered? What if, instead, a patient were admitted to a different institution, perhaps a custodial asylum? This was not uncommon, but it complicated the metaphysics and introduced the practical problem of information sharing. Hagen's nightmare, a world central commission, might be the only solution.

Early nineteenth-century medical explanations offered a convenient evasion of this dilemma. Insanity involved a predisposing cause and a triggering factor. Even when asylum care restored health, the susceptibility remained. Little surprise if the discharged patient, returned to former circumstances, should again fall ill. On that hypothesis, a readmitted patient really was properly a new case. Alienists sometimes defended this view of things, though with hints of a bad conscience. At Siegburg, Maximilan Jacobi was challenged by a commission of investigation, appointed in July 1841 by the estates (*Landtag*) of the Prussian Rhine Province. They were by no means hostile to Herr Higher-Medical Privy Counselor Jacobi, but found him to be unreasonably insistent on using arbitrary criteria of curability to classify the insane as "decisively unsuited" or "only in the very slightest degree suited" to the institution. They also reminded him that most cures come from nature. Siegburg, they insisted, was not only a medical institution, but an arm of state policy, charged to care for its inmates and not only to deliver medical treatment. His claim of so many dubious cures was complicating state policy, and the commission could not ignore the 86 repeat patients on the books, who, in total, had been admitted 204 times. Proper accounts would reduce the 469 recoveries he claimed to 373. A lot of money could be saved by recognizing Siegburg as a charitable institution, rather than a medical one, and replacing the paid personnel with Sisters of Charity, as in Maréville, France. Jacobi, of course, defended his institution. The Sisters of Charity turned Maréville into a scene of horrors, he said. And insanity is like intermittent fever, striking people repeatedly when conditions are similar.[29]

Readmission as a statistical problem, though never forgotten, acquired new prominence when Pliny Earle raised it in his annual report for the Massachusetts asylum at Northampton in 1876. The *AJI* promptly reprinted the report, which was widely noticed, and later Earle published the original essay with seven supplements as a book, *The Curability of Insanity*. Asylum cures had been greatly exaggerated by the early directors of American and English asylums, Earle argued. He retraced some early claims of superlative results, including George Man Burrows with his 91% cures for new patients in London, the Hartford Retreat at 91.3%, Samuel Woodward's brief moment above 90% in Massachusetts, and finally John Galt's achievement in 1842 of a perfect 100% at the Eastern Asylum in Williamsburg, Virginia. Galt must have been a first-rate enterpriser. The key was to choose an advantageous formula. Cures should be divided by discharges, not by new admissions or patient population. It helped to exclude deaths, as Galt did. Perfection now followed as a mathematical necessity, provided the asylum was never compelled to release a patient while still insane. In fact, asylum directors in this age of wondrous numbers were not simply unscrupulous but had faith in their own magic. They received in addition a big assist from multiple cures of the same patient. The most extreme example that Earle could find involved a woman at the Bloomingdale Asylum in New York City, who had been admitted 59 times in 29 years, and discharged as recovered 46 times.[30]

The critique struck close to home. For 5 of those 29 years, Earle was the Bloomingdale director. When he arrived in 1844, he introduced a medical register with about 30 columns in which to enter information and rows wide enough for this enormous volume to take the place of a case book. One column, headed "No. of admission," was there specifically to keep track of readmissions. He subsequently tracked down the skeletons in these dusty closets in pursuit of a solution to a distressing problem. Why, decades later, were asylums unable to come close to matching the medical successes of their earliest predecessors? In 1844, he had written that for cases without constitutional eccentricity or weakness of intellect, a director should be able to cure 80%. He claimed to cure patients whose "particular propensities and hallucination" appeared hopeless to the nonalienist eye, such as this woman who arrived in 1846: "God forsaken her, torments of hell upon her! Would not eat because offend God. Smoked one pipe to please this Lord and one to please the Devil. Attempted suicide by drowning, hanging & cut throat."[31] It was an age of giants, but not on account of a few heroic cures. The percentages were the problem. By the 1850s, an 80% cure rate was an impossible dream. There was some consolation in discovering the tendentious accounting through which these men had achieved this irreproducible success. Thurnam, Earle now remarked, had had it right when he declared that for every ten patients, five will recover and five will die, while three of the five recovered will suffer additional attacks. Earle's explanation of the radical drop in cures was welcome news in London, where Hack Tuke of the founding family of the York Retreat invoked it in his 1881 presidential address to the Medico-Psychological Association.[32]

There were other repercussions. The superintendent at Worcester promptly recomputed its tables and found all its figures for recoveries to be in error. "It is a sad and almost cruel blow to the worth of the earlier tables of this Hospital."[33] A private institution in Vermont, subsisting mainly on state contracts, had to fend off critics who, ignoring a shift in the basis of calculation, claimed that the cure rate had fallen abruptly from 32% to 8%. Earle commented: "It is said 'figures do not lie,' but if manipulated they certainly can be made to misrepresent the truth." The *AJI*, in an unsigned review by its editor John Gray, gave credit for these pretended insights to an unnamed local party who had anticipated Earle by a dozen years. There in Utica, it barely mattered, since the rate of readmission there was very low.[34]

The Flexibility of Files

The British plan for rationalized recordkeeping at the end of the nineteenth century was to have everything in its place in a giant record book. German case records, by contrast, had long been kept in files. If the space set aside for a patient in a case book filled up, there was nothing to do but move to another blank page. The case file could simply grow fatter, almost without limit. Documents could also be separated and recombined. The shift to census cards or counting cards (*Zählkarte*) made for still greater flexibility. It was not yet the accountant's paradise, for the cards and files, unless copied, would have to reside in a single location. For the sake of research, Kraepelin fought to keep patient records at his Heidelberg clinic rather than give them up when the unfortunates who did not improve rapidly were moved to custodial institutions.[35] The ideal of a *Centralstelle*, or data center, privileged large-scale investigation and centralized management over local medical use. Despite his inexhaustible appetite for data, Kraepelin complained bitterly of all the meaningless numbers that the administration demanded from him. Yet the abundance of documents in patient files was too much to handle for a census bureau, which liked its information thin.[36]

A patient brought in to a Bavarian asylum in 1875, as elsewhere, would be promptly examined by a doctor. The history, supplied by an accompanying family member or health official, focused on the origins and progress of the illness and its causes, beginning with a family history. This case history was used to fill out the census card for admission, whose entries must, to a large degree, have structured the interview. The information was then transferred to a bound arrivals book.[37] The Bavarians kept detailed accounts of heredity. A patient admitted in January 1877 at the Munich asylum, but sent immediately to the nearby institution of Eglfing, was entered in the arrivals book for women as "direct" in the heredity column. The case history begins:

> Frl. Henriette W from Frankfurt am Main. Israelite confession, hereditary burden. The father suffered from periodic epileptic attacks and was in the asylum at Eichberg and Klingenmünster in 1861 and 1862, specifically in a condition of melancholia with suicide attempts. The mother was, according to the illness history a "nervous" lady.

Henriette W.'s mental illness began in puberty when she refused to eat, and so on. Each admission document assigned her a different number, and cross-referencing allowed movement back and forth among them.[38]

These careful registration procedures enhanced the ability of doctors and statisticians to keep track of readmissions and hence to make patients rather than cases the focus of attention. The files (*Akten*) provided a ready guide to the entries in admission books. Eglfing File 2413 is the paper incarnation of Ludwig S., who arrived on 1 May 1881 and was readmitted in August 1881, 1884, 1886, 1888, 1890, 1894, 1897, and 1912. Each admission corresponds to an entry number: 3522, 4477, and so on. The father, still living in 1881, had been mentally ill once, while the mother had died in 1865. The file indicates six siblings, two brothers and four sisters, of whom one sister had been mentally ill for ten years, and suggests that the insanity developed from insomnia. The first record in the entry books reduced the family relationship to "direct from Father and collateral a sister." The second entry mentions that the relatives were melancholy, while the third is a bare statistical entry: "direct and collateral."

The *Aufnahmszählkarte*, or statistical admission card, devoted a generous portion of its very limited space to information on heredity. Although some cards were filled out with handwritten particulars, most items could be completed simply by underlining or connecting the appropriate items with a curved line. The information required included the identification by relationship of abnormal or mentally ill family members, including, by 1891, children of the patient. The cards also specified the medical defects or conditions of specific interest, including blood relationship of parents. The two items following concerned criminality and disease form. The bits of added detail must have complicated this statistical registration. Conversely, the thin economy of statistics did not encourage rich detail, though it definitely enhanced data quality. Ludwig's page in the admission book recorded his transfer in 1904 to another Munich asylum, Gabersee, as well as his subsequent return to Eglfing. Until recently, it had required heroic efforts to keep track of patient transfers. Social and medical investigations of the insanity problem depended on this capacity to track patients and their relatives.[39]

Catching Up with the Germans

Evolving ideals of recordkeeping in Germany, somewhat paradoxically, were tending to put more responsibility in the hands of doctors in local institutions, for the sake of a form of research that had moved away from the annual report. Large-scale printed statistics, too, gradually lost their appeal. Heroic compilations like the Prussian folio volumes on causes of insanity moldered in libraries. Giant tables, assembled bureaucratically, served budgetary and demographic ends. Research on scientific questions seemed to require more supple tools of information: technologies of sorting and fil-

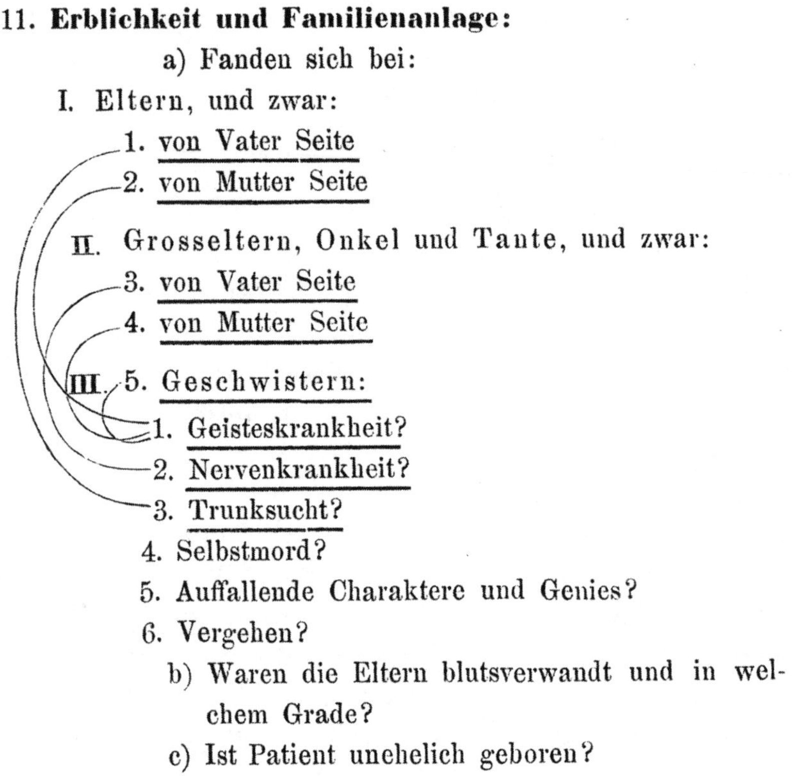

11. Erblichkeit und Familienanlage:

a) Fanden sich bei:

I. Eltern, und zwar:

1. von Vater Seite

2. von Mutter Seite

II. Grosseltern, Onkel und Tante, und zwar:

3. von Vater Seite

4. von Mutter Seite

III. 5. Geschwistern:

1. Geisteskrankheit?

2. Nervenkrankheit?

3. Trunksucht?

4. Selbstmord?

5. Auffallende Charaktere und Genies?

6. Vergehen?

b) Waren die Eltern blutsverwandt und in welchem Grade?

c) Ist Patient unehelich geboren?

FIGURE 9.1. Census card for drawing arcs to connect the relatives affected with the condition from which they suffered, part of the 1874 proposal for standard German asylum statistics included in a separate booklet at the end of the *AZP* volume for 1874, *Zählkarten und Tabellen für die Statistik der Irrenanstalten aufgestellt von dem Verein der deutschen Irrenärzte*. The German census in the 1870s developed brilliant technologies for recording patient data on cards and using these to construct elaborate tables.

ing. By no means did this imply the death of standardization. It required instead a reshaping of standards into uniform templates for personalized information.

In America, improved recordkeeping emerged in the 1890s as a key element in a new vision of the medically effective asylum. The annual report of the Taunton State Hospital for 1893 boasted of records made useful at last by prompt recording at the bedside. "A medical record, to be of any value, must be an accurate statement of the condition of the patient, and hence should be written by the physician himself who makes the examination, or at his dictation if written by another." The doctors were now required to examine each new patient within twenty-four hours of arrival and enter the results in the case book, then to repeat the process every week for the first three months. If he was very busy, he should put aside less important duties for the sake of complete records.[40] The inspiration came partly from Adolf Meyer, who arrived in America with a brilliant European pedigree that included education at the University of Zurich, postgraduate work in Paris, Edinburgh, and London, and additional work at Burghölzli in Zurich. He began in 1892 as pathologist at the new mental hospital of Kankakee, Illinois. In his first report to the governor, he explained that diseases run their course under one system in about the same way as another. "The main point is the conscience of all parties is soothed by the idea of superiority of therapeutic skill." Any method is good that does no harm. He worked there to institute a system of close patient observation and anthropometric measurement, supplemented by examination of relatives to detect signs of hereditary degeneration.[41]

In recruiting Meyer to the Worcester asylum in 1895, the directors authorized a research tour that began with several institutions in Italy, followed by six weeks with Kraepelin in Heidelberg, brief stays in other German institutions and in Zurich, and a visit to F. W. Mott's new laboratory in London. Meyer returned with a vision of teams of assistants trained in methods of case taking, including "concise and accurate recording at the bedside of the results of their observations." The old books, based on recollections of an assistant and recorded weeks after the occurrences noted, would now give way to a corps of interns accompanying the assistants on their rounds and setting down the results of batteries of tests that Meyer planned to introduce.[42] In 1901, he was hired as pathologist for the asylum system of the state of New York to set up a training school and work his magic of renewal, substituting for his predecessor's program of laboratory testing a new focus on clinical observation and information.[43]

In London, too, at the dawn of a new century, the winds of renewal were blowing. Keeping case books merely to satisfy a lunacy commission or defend against charges of "malpraxis" was one thing; "doing it from a scientific point of view for the advancement of the study of insanity is quite another thing." These heterogeneous case books contained much of potential clinical value, which might be extracted at the cost of enormous labor. "If, however, it were possible to frame some simple, uniform method of case-taking, and persuade the superintendents of all the asylums to use it, which possibly might be the most difficult of all tasks, then some definite and valuable statistical facts might be obtained." The author of this vision, Dr. A. H. Newth, envisioned "a simple method of case-taking," allowing the state of each patient at the time of admission, the progress of the case, the treatment adopted, and the result, to be taken in at a glance. For this, a uniform system would be required.[44]

Newth condemned the annual reports as costly and time-consuming, useful for nothing except "to light the fire." His ideal was an efficient system of cases. He especially admired the "check cards" that Dr. A. R. Urquhart culled for private use from the case books at James Murray's Royal Asylum in Perth, Scotland. While the Perth case books were "models of completeness," the cards provided a "bird's eye view" and allowed the rearrangement of patients for new inquiries. The bulky case books brought definite advantages. Urquhart compiled from them family trees for 331 of his patients. Newth, however, was coming to think that "the very completeness of detail in case-taking is a bar to the comparative study of cases." Keeping these books had also given him a case of "scrivener's palsy." He determined to emphasize the minimization of wasted effort, leaving it to Urquhart to balance economy with detail. So much accessible data was grist for the mill of biometricians David Heron and Karl Pearson, at University College, London, who analyzed his data and calculated hereditary correlations. Pedigrees like Urquhart's were no less suited to hereditary statistics in its Mendelian form.[45]

PART III

A Data Science of Human Heredity

It is the supreme law of Unreason.
—Francis Galton (1889) on the bell curve

The story thus far has focused on actors and institutions that are largely unknown to historians of genetics.[1] The main characters in the third section, though comparatively well known, appear altogether different against the background of a century of hereditary study anchored by asylum data. Mendelism, despite its key role in laboratory genetics, was more difficult to apply to animal breeding programs than agricultural geneticists wanted to believe, and the challenge of identifying genes for mental and behavioral traits of humans proved insurmountable. From about 1910, Charles B. Davenport and his allies sold many researchers in human sciences on the idea that insanity and feeblemindedness could be explained as single-factor Mendelian traits.[2] Others harshly criticized their data as well as their reasoning. In Germany and in Britain, leading researchers gradually put aside this mental Mendelism, pursuing in its place a strategy of empirical prediction. That meant studying the transmission not of genes for mind and behavior, which remained inaccessible, but of diagnosed conditions such as psychoses, nervous disorders, and low measured intelligence. The prospect of a match between genetic units and disease categories such as schizophrenia and manic-depressive illness was exhilarating to researchers. When this did not work out, even Davenport and his supporters reorganized their data to focus on looser categories such as general psychopathy or "neuropathic make-up."

The blurring of the diagnostic boundary between a healthy mind and a defective one was disappointing to doctors and psychologists.

Insanity and mental weakness, though unmistakable and disabling in some cases, had long been understood as grading into "normal" ability and behavior. A preponderance of expert opinion continued to identify heredity as the leading cause of mental defects. The diagnosed insane seemed now to proliferate uncontrollably, absorbing ever more resources in ever larger institutions. At the same time, a massive expansion of public schools from about 1870 turned "feeble-mindedness" into an urgent social and medical problem. "Mental weakness" in this form was no easier to isolate than mental illness. Psychologists developed an infrastructure of tests and classifications linking feeblemindedness to social problems, especially to crime. The enterprise of data gathering on these diverse conditions extended to schools, prisons, and other institutions. An ever-expanding range of experts was becoming involved in the battle against mental defect. As we have seen, eugenics, under different names, was an old story in alienist discussions. Now, it became a great public movement, nourished by data flowing in ever wider channels.

The intensified interactions between asylum medicine, university science, regulatory practices, and state commissions reflected something more than detached scientific ambitions. Psychiatry (as it was coming to be called) and science were brought together by a sense that the medical-social problems of hereditary mental defect now stood in the way of advancing civilization. A brave new world of rationalized armies, factories, and imperial colonies seemed to demand strong, efficient citizens, to be guided now by science. The ideal of a welfare state was taking form to assure that most citizens could be healthy and productive. Yet there seemed to be so many laggards and degenerates. Asylum patients, maintained at ruinous expense, many of them lingering for decades with no realistic prospect for recovery, stood as compelling symbols of a terrible failure. When children did not perform normally within the new public school systems, their weakness was now highly visible. An alliance of asylum doctors, psychologists, criminologists, educational specialists, and biologists searched for eugenic solutions. When Mendelian explanations based on pedigree research seemed to fall short, the experts sought out other ways to assemble and draw conclusions from all the numbers.

Alliances of science and medicine are often interpreted asymmetrically as the opening up of a hidebound medical profession to new knowledge. On the topic of medical heredity, the interactions were reciprocal. Between about 1890 and 1910, biologists and statisticians became familiar with the data work that had been going on for decades in mental institutions. In different countries, physicians and lay scientists worked out their own basis for collaboration. The British built up a field of quantitative population genetics to comprehend certain human traits. In Germany, psychiatric clinics were the main sites of this kind of study. When the results of Mendelian research proved disappointing, many returned to statistics of traits that appeared to be inherited. Gene talk, however, never went away.

Each of the first three chapters in this section focuses on a single country, first Britain, then the United States, and finally Germany. The familiar story of Galton's biometric methods giving way in the 1900s to bold Mendelian ambitions, followed a few years later by sharp criticisms and disappointed hopes, appears as the tip of an iceberg. These chapters emphasize instead the geneticists' reliance on data sources and data techniques, revealing close and indispensable connections with asylums, special schools, and prisons. The final chapter shows how the German program of research and intervention accommodated Nazi politics and exploited the immense data reserves of the Nazi era. Yet the German program of empirical prognosis, grounded in psychiatric, educational, and criminological data, was not uniquely Nazi. It formed the basis for American, British, and Scandinavian programs of genetic research and counseling right up to the Second World War and beyond.

CHAPTER 10

The Human Science of Heredity Takes On a British Crisis of Feeblemindedness, 1884–1910

> It will at first sight appear presumptuous that a layman should venture into a field which has been so much cultivated by the trained medical mind.
>
> —David Heron (1907)

Early in 1875, Francis Galton drew up an inquiry on the resemblance of twins to send around to a group of experts on "morbid heredity." On the evidence of this circular, he was inspired by a well-known book of 1859, *La psychologie morbide*, by Jacques-Joseph Moreau de Tours, the source of some remarkable examples of hereditary resemblance.

> He speaks "of two twin brothers who had been confined, on account of monomania, at Bicêtre. . . . Physically the two young men are so nearly alike that the one is easily mistaken for the other. Morally, their resemblance is no less complete, and is most remarkable in its details. Thus, their dominant ideas are absolutely the same. They both consider themselves subject to imaginary persecutions; the same enemies have sworn their destruction, and employ the same means to effect it. Both have hallucinations of hearing. They are both of them melancholy and morose; they never address a word to anybody, and will hardly answer the questions that others address to them. They always stay apart and never communicate with one another.
>
> "An extremely curious fact which has been frequently noted by superintendents of their section of the Hospital, and by myself, is this:—From time to time, at very irregular intervals of

> two, three, and many months, without appreciable cause, and by the purely spontaneous effect of their illness, a very marked change takes place in the condition of the two brothers. Both of them, at the same time, and often on the same day, rouse themselves from their habitual stupor and prostration; they make the same complaints, and they come of their own accord to the physician, with an urgent request to be liberated. I have seen this strange thing occur, even when they were some miles apart, the one being at Bicêtre and the other living at Sainte-Anne."[1]

Despite his strong faith in heredity, Galton was suspicious of such uncanny manifestations. His alienist commentators, similarly, did not go all the way with Moreau. Thomas Clouston, though claiming to have witnessed similar instances of family insanity, declared that Moreau had been carried away by his own story. "I confess that I should be much more apt to rely implicitly on a German or English account," for the Frenchman "cannot help making it as marvelous & dramatic as possible." It may also be noted that Moreau's titillating tale turned the patient's desire to escape his confinement into a proof of his madness. Galton charged Moreau with writing in too offhand a manner, with no evidence of "scrupulous exactness" in investigating the circumstances of the cases. A few months later, in a second phase of the inquiry, he drew on files of the London Orphan Asylum and related institutions to solicit information from the relatives of these children. The responses of parents, doctors, and guardians may be read among the Galton Papers at University College, London.[2]

Most of the replies were less elaborate than Galton seemed to wish. Henry Maudsley, the noted London alienist, was typical. He could not match Moreau's fine tales, he said, and described rather abstractly having "met with exactly the same sort of insanity in three brothers and also in two sisters." Clouston provided one of the better anecdotes, drawn from his experience at the Royal Edinburgh Asylum, which involved brothers growing apart as they reacted to each other and to their environment. Galton was seeking case stories to answer critics who claimed that his quantitative conclusions on "hereditary genius" were vitiated by a neglect of environment. In his published paper he offered these case reports as evidence that twins formed within a single ovum remain highly similar all their lives, even when separated, while twins from two eggs typically diverge, even when living in the same household.[3]

For well over a century, Galton has been credited—now mostly blamed—as a genetic pioneer and founder of eugenics. He always credited his cousin's theory of evolution by natural selection as the inspiration for his discovery of the power of heredity and of the urgent need for eugenic action. The inquiry on twins, however, illustrates Galton's reliance on other sources, in particular on traditions of asylum reporting. Alienists also provided ideological support and even scientific inspiration. Maudsley in 1867 surveyed an international collection of eminent alienists, including Moreau de Tours, concluding that the statistics pointed to heredity as the most important factor.[4] In 1864, in the second installment of a paper on "hereditary influence" for the *JMS*, Maudsley declared that mediocrity, laboring steadily, will never attain genius, which, instead, typically ran in families. Galton at that time was just coming to believe that a few are born great and that the rest must be content with modest achievements. The first installment of Maudsley's paper is still more striking. It seems to lay out in advance the challenge of Galton's first great project on human inheritance, *Hereditary Genius*. Maudsley argued in 1863 that "in almost every nation which possesses a history, families might be selected that have been remarkable for special characteristics." Two years later, Galton published the first fruit of his own investigation, which surveyed family achievements in a range of endeavors from rowing and wrestling to music and mathematics.[5]

Galton and Maudsley also seem to have shared a negative provocation for turning to heredity in the form of a much-discussed long footnote in Henry Thomas Buckle's 1857 bestseller, *The History of Civilization in England*. Buckle there dismissed claims for heredity as anecdotal and illogical. Many prominent alienists were provoked by Buckle's note, among them Maudsley, who wondered how

Buckle could fail to recognize that hereditary influence made civilization possible. To experts in mental medicine, such as New York asylum director John Gray, this rejection of "the whole theory of hereditary transmission" seemed perverse. The statistical evidence, he said, is overwhelming, as the "entire medical profession" now recognized. Galton's hereditarianism had a rich medical and alienist background, as he seems to have understood from the outset. He and Maudsley also agreed that madness was somehow allied with genius. Twenty-five years later, Galton published a communication on inherited lunacy in cats.[6]

By 1892, when he reprinted his book, public interest in his evolving scheme for the cultivation of hereditary talent was much intensified. Yet the primary focus of eugenic ambition remained the inheritance of mental and psychological weakness, not of brilliant achievement.[7] The terrible failure of asylum medicine to halt the growth of insanity inspired desperate calls to block somehow the terrible force of degeneration. Statistics of military recruits, factory workers, criminals, and schoolchildren showed that the problem was now quite general.[8] "Mental defect," which took in insanity and "mental deficiency" (or "feeblemindedness"), was the most worrying problem of all. These categories provided a focus for a wave of data-intensive medical-social investigations in prisons, special schools, and asylums.

Following Galton's lead, Karl Pearson developed a new mathematical basis for the statistics of human heredity. From the standpoint of data, the continuities with prior work are compelling. Pearson and his associates relied on numbers from asylums, special schools, prisons, and the various boards and commissions set up to investigate these populations. Professionals at such institutions were actively engaged in hereditary research long before they learned that statisticians and biologists shared their concerns. Their expertise was roughly on a par with that of the scientists. Mendelian geneticists, too, as they took up these pressing issues, depended on experts at mental institutions for their research tools and topics as well as their data. Many held that the processes identified in Mendel's pea breeding would demonstrate genetic causes of mental illness and even reconfigure the data on human variation, which had to be discontinuous. That issue, in particular, provoked fiery medical debates.

Galtonian Data of Mental Deficiency

Galton took great pride in his application of new mathematical tools. He had some predecessors in the study of inherited ability, he acknowledged, "but I may claim to be the first to treat the subject in a statistical manner, to arrive at numerical results, and to introduce the 'law of deviation from an average' into discussions on heredity." And yet, for all the fame of the bell curve, his "supreme law of Unreason," units of diagnosed pathology were the real building blocks of eugenics. The wave of pedigree studies began to accumulate for political, budgetary, and medical reasons, a decade before anyone cared about Mendel.[9]

Mental hospitals treating insanity housed mainly adults. Institutions for slow-learning children appeared later, when their needs attracted the attention of educational reformers. Special schools for such children were introduced in the same lands that had been first off the mark in developing asylum systems. As Mathew Thomson points out, the "problem of mental deficiency" arose in consequence of a massive expansion of public education. This movement was sealed by legislation in Britain in 1870, and at about the same time across Western Europe and North America.[10] The schools, as they enhanced opportunity, made new categories of problem children visible. Noel A. Humphreys, a vital statistician in the General Register Office, remarked that the replacement of "idiot" by the more inclusive label "feeble-minded" in the 1901 census had contributed to a large but spurious increase of mental weakness. For the first time, the census was tallying, as mentally defective, persons unknown to the Lunacy Commissioners because they had never yet been institutionalized. To statisticians, this was a new source of uncertainty in the numbers. To others, it encouraged a sense of teeming mental incapacity. The less disabled were soon taken to constitute a greater danger to society, since most were capable of forming families. They were not, like idiots, incapable of following a curriculum, but they could not keep up. Many were children of poor farmworkers or urban laborers, and many were diagnosed also with physical defects.[11]

British elites immediately recognized that a crisis like this one cried out for investigation by a distinguished committee, leading, in 1895, to the appearance of *Report on the Scientific Study of the Mental and Physical Conditions of Childhood, with Particular Reference to Children of Defective Constitution.*[12] From its forty-five members was chosen a statistical committee of five, including two fellows of the Royal College of Physicians and two fellows of the Royal Statistical Society. The thirteen-member executive committee also privileged expertise in medicine and practical statistics. It should be no surprise that the chairman bore the name of Galton. Having in his later years made a distinguished career in statistics of health and mental deficiency, he provided counsel to Florence Nightingale on the design of hospitals and served on the Statistical Committee that directed the Metropolitan Asylums Board. He also chaired the organizing committee for the Seventh International Congress of Hygiene and Demography, held in London in 1891, and was president of the British Association for the Advancement of Science (BAAS) in 1895. He had a leading role in the Sanitary Institute of Great Britain, for which he proposed the motto: "Prevention is better than cure."[13]

This eminent man of science was *Douglas* Galton, Francis Galton's cousin. Both were born in 1822. They were brought up separately, and not as an experiment. Whereas Francis lived on inherited money as a gentleman of science, taking up various research topics within a career that progressed from exploration, geography, and meteorology to heredity and eugenics, Douglas advanced through the Royal Engineers to the rank of captain and then was appointed to a series of high government posts linked mainly to sanitary engineering. These provided his entrée to the elite commissions on which he served in the final decades of his life. His high position in the BAAS is a bit ironical, given his cousin's attempt in 1877 to expel the statistical Section F as lacking

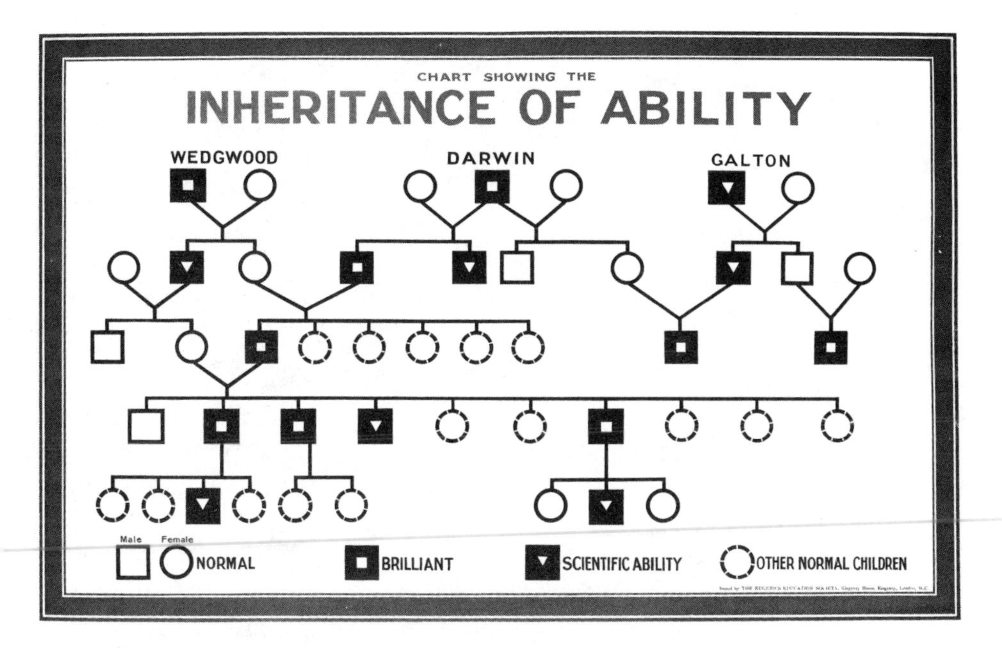

FIGURE 10.1. The iconic 1926 poster was made for the Eugenics Society (England), which liked to use the family distinction of the founders of eugenics to exemplify its validity. The three "brilliant" men in the third row are Charles Darwin, Francis Galton, and Douglas Galton. The Eugenics Society seems here to treat data work, unconnected with academic science, as sufficient evidence of scientific brilliance. © Museum of London.

the stuff of true science. The effort failed, and Douglas's career reveals what respect the premier institution of British science was prepared to grant to elite administrative inquiries, thereby placing data preparation almost on a par with mathematical and experimental science.[14]

If we open the cover and look inside the 1895 report on defective children, we see that the Galton cousins were brothers beneath the skin. The committee was obsessed with data, not only with its accumulation but also with modes of presentation devised to reveal important relationships. Their report referred to these as "co-relations."

Francis had published a paper called "Co-relations and Their Measurement" at the end of 1888. This was the moment when he first understood "co-relation" as a general measure of relations between two variables rather than something specifically biological,

FIGURE 10.2. Portrait of Francis Galton's cousin Douglas Galton (1822–1899), who, without ever mentioning eugenics, took a lead role in study of physical defects and mental weakness in children. He found that such children were often involved in crime, not because they were evil but from a heightened susceptibility, which, he concluded, tended to run in families. From Grace's Guide to British Industrial History (online), www.gracesguide.co.uk. In public domain.

Georges Cuvier's "correlation of parts." His method of drawing lines on graphs to estimate this quantity was supplanted in 1895 by Pearson's product-moment formula to calculate it. Douglas Galton's report paid no heed to these tools of a new statistical mathematics, relying instead on tabular reasoning. But he cannot be dismissed as plodding and unoriginal. His committee, which dealt with population counts rather than numerical measurements, employed advanced techniques of enumeration. Appointed in 1888, it began by drawing up "a list of all cases presenting each principal condition." In 1892, it shifted to more efficient "actuarial" methods involving cards. These call to mind German census techniques, but British life insurance offered a precedent of still longer standing.[15] "The Statistical Sub-Committee, taking the Register as the basis of this actuarial work, prepared a card . . . on which the list of defects is printed."[16]

Although teachers were not included on the committee, they were needed to assess the "mental state" of the children and to diagnose "mental dulness." Since the research did not intrude into homes, poor nutrition came down to indicators such as thinness and lack of strength, which, like the many forms of nervous and developmental defect, were revealed by a medical examination. For the sake of planning, the committee needed to identify and to number children in need of special schooling. Universal education was worth nothing to those placed in classes that went beyond their capacities. Although no one spoke of "stigmata," their heavy reliance on external "signs" or "indications" reminds us of the celebrated Italian criminologist and theorist of hereditary reversion, Cesare Lombroso. The committee however, preferred to rely on measures and statistics. "In the facts here given it is seen that defects in development and abnormal signs are largely co-related with mental dulness." According to the report, the teachers had been able to pick out almost all dull students from bodily features even before anyone asked. The idealization of normality, often identified with the normal curve, was never so extreme in Francis's scientific statistics as in Douglas's medical-administrative reports. It was all for the sake of an updated liberal cause, using knowledge "based upon

TABLE XVI. (*Cases seen* 1888-91).—*Co-relation of defects in Development with Nerve-signs: Low Nutrition and Mental Dulness.*

For Definition, and number of cases of Defects, see Nomenclature on page 72. For Groups of Cases see Catalogue on page 82.

Percentages are taken upon the number of children presenting the defects indicated in the first column.

Defects in Development. See Nomenclature, page 72.	With Group 29, Abnormal Nerve-signs, alone or in combination.				With Group 30, Low Nutrition, alone or in combination.				With Group 31, Mental Dulness, alone or in combination.			
	Number of cases.		Per cent. of cases.		Number of cases.		Per cent. of cases.		Number of cases.		Per cent. of cases.	
	Boys.	Girls.	Boys.	Girls.	Boys.	Girls.	Boys.	Girls.	Boys.	Girls.	Boys.	Girls.
All Development cases, Group 28	1,975	1,096	54·6	49·0	733	726	20·2	32·5	1,398	928	38·3	41·5
(1) Cranial defects	850	531	55·6	50·6	392	480	25·7	45·8	634	477	41·4	45·5
(3) Cranium small	177	372	54·1	50·4	151	399	46·1	54·1	165	353	50·4	47·8
(7) Cranium asymmetrical	40	7	47·6	43·7	18	3	21·4	18·7	35	6	41·6	37·5
(11) Ears defective	566	128	54·0	47·7	196	72	18·7	26·8	340	103	32·4	38·4
(12) Epicanthis	227	160	44·1	41·6	65	73	12·6	19·0	192	136	37·3	35·4
(13) Palate defective	441	262	55·4	49·9	173	155	21·7	29·5	324	232	40·7	44·2
(19) Nasal bones	131	95	54·3	44·3	16	19	6·6	8·8	87	77	36·1	36·0
(20) Growth small	119	110	56·9	52·6	88	101	42·1	48·3	78	79	37·3	37·8
(21) Other defects in development	555	355	61·1	55·0	189	172	20·8	26.5	383	263	42·0	40·5
(27) Features coarse	112	68	76·1	65·3	19	17	12·9	16·3	73	43	49·6	41·3
(34) Mouth small	16	10	59·2	58·8	8	2	29·6	11·7	8	10	29·6	58·8
(39) Palpebral fissures	61	57	62·2	68·6	22	16	22·4	19·2	41	39	41·8	47·0

FIGURE 10.3. Table of "Co-relations" of bodily defects against degrees of mental deficiency, prepared under a committee headed by Douglas Galton. He did not connect the work with his cousin Francis Galton's mathematical version of correlation, but he shared with his cousin the ambition to derive quantitative relationships from data. Table 16 from *Report on the Scientific Study of the Mental and Physical Conditions of Childhood*. . . .

scientific inquiry" to assign schools according to ability, "fitting each as a citizen to provide for himself," and to "render the population healthy, long-lived, and prosperous." The committee proposed policies to "lessen crime, pauperism, and social failure, by removing causes leading to degeneration among the population." The proposed interventions were educational rather than eugenic, keeping heredity in the background. But it loomed there, nevertheless. "The State becomes heavily burdened by the defectively made portion of the population."[17]

Contributing Data

Eugenics for Francis Galton depended on data, and he worked out a multitude of strategies to acquire it. For *Hereditary Genius*, he scoured biographical dictionaries and memoirs, making lists of members of families who had achieved distinction as judges, statesmen, commanders, literary men, men of science, poets, musicians, painters, divines, oarsmen, or wrestlers. Some of these he assembled into family trees as visual evidence of inherited ability. Such information, though available, had to be taken for what it was. He asked around to check identities and made modest additions and corrections.

At the 1884 International Health Exhibition, held in London and organized in part by cousin Douglas, his Anthropometric Laboratory was attractive enough to induce "no fewer than 9,337" visitors to pay 3 pence to be measured, while thousands more grew weary after waiting in line. He conceived his instruments, which gauged strength, speed, and accuracy of perception as well as bodily dimensions, as prototypes for what he hoped to perfect and systematize in schools. Already, he remarked, foreign governments had ordered some duplicates.[18]

He kept his laboratory going for a time after the health exhibition closed, but his ultimate goal, to investigate the inheritance of human abilities, required data arranged according to family. He appealed to a spirit of individual initiative for his *Record of Family Faculties*, also begun in 1884, relying on a printed booklet with instructions and forms for entering family data. He offered prizes for the most thorough pedigrees.[19] This liberal-voluntarist mode of eugenic data gathering survived and flourished well into the twentieth century. It was allied to an ideal of raising the standard of human reproduction through appeals to rational self-interest and the public good. Individualism, however, had its limits. Even Galton wanted to unleash the pressure of public opinion on the undeserving and unfit, and stubborn irresponsibility seemed to invite sterner measures, backed up by legitimate expertise.

Solid data did not magically appear when eager volunteers received instructions and standardized forms. The next step was inspection and curation by scientific or medical experts. If, as it seemed, the insane and feebleminded came predominantly from poor and uneducated stocks, they could not well be relied on to provide scientific data on the taints that weighted down their family lines. Here, inquiry into the principles of heredity meant expert identification of superior and defective lines. Even enlightened public opinion was not sufficient for the serious business of reversing hereditary decay. When Pearson took up these questions, he drew up new circulars to incorporate modern ideas of hereditary transmission and techniques of multiple correlation. This was work for mathematicians.[20]

Galton's *English Men of Science*, like *Hereditary Genius*, employed multiple sources of information to identify high achievers and to retrieve family relationships among notable men. Some scholars have wondered at the looseness of his reasoning, but they fail perhaps to credit his resourcefulness in devising statistical tools to draw conclusions on heredity from data that happened to be available. Compilers of reference works, for example, may have had an idea of distinguished family lines, but not of subjecting them to statistical analysis. By the 1890s, Galton was prepared to endow new institutions for this purpose, and Pearson, to whom he looked as a successor, was already proving to be an exemplary institution-builder at University College, London. Pearson's passion by this time was evolution, especially human evolution and eugenics. He is known mainly for new mathematical formulations and for eugenic ideologies, but he devoted as much energy and enterprise to constructing appropriate data reserves—and in a great variety of fields—as he did to new tools of analysis. This alliance of mathematics and data depended on copious calculations performed by colleagues and assistants, many of them women.[21]

One example, very much in the Galtonian tradition, involves a study published in 1906 by Edgar Schuster and Ethel Elderton, respectively Galton Fellow and Galton Scholar in National Eugenics. They mined three reference books of Oxford alumni to compare the level of honors attained by fathers and sons over much of the nineteenth century then calculated correlations to measure the effect of biological inheritance. Schuster pushed the work one step further, drawing on Crockford's *Clerical Dictionary* and Forster's *Men at the Bar* to assess the relations between quality of degree and professional success.[22] In 1907, *The Oxford and Cambridge Review* printed a proposal for "A Bureau of Biometry" at the great universities to investigate the relationship between physical and mental ability and the importance of both, along with family background, for success after graduation. They presented the work as a model for inquiries in state elementary schools.[23]

One of Schuster's first papers, published in *Biometrika* in 1906, was a study of hereditary deafness based on a compilation by the American doctor Edward Fay, *Marriages of the Deaf in America*. With an endowment from Alexander Graham Bell, Fay gathered data on 4,471 marriages of the deaf in America, including on deafness in the siblings and other relatives of each partner and on causes. Often the cause was illness, and Fay held that the role of heredity was greatly exaggerated by common opinion. He prepared tables showing the numbers and percentage of deaf children for marriages of various descriptions: for example, one partner hearing, the other deaf, and the hearing partner having deaf relatives; one partner "congenitally deaf," the other "adventitiously deaf"; or one partner with deaf relatives and no information on the other. Ignoring data that "cannot be regarded as a true fair sample," Schuster calculated correlations based on fourfold tables of deaf and hearing fathers (or mothers) against deaf and hearing children.

His work had none of the fine detail of Fay's teeming tables, he acknowledged, and could only be approximate. Since, however, the study of rare conditions like deafness depends on a large-scale survey, which is outside the capacity of any private individual, he was grateful even for flawed data. He hoped the British state might proceed with an anthropometric survey recently recommended by the Interdepartmental Committee on Physical Deterioration.[24]

The Central Metric Office and the Fairly Impartial School Teacher

Much, perhaps most, human data for eugenic and anthropometric use was accumulated without active consent by social and medical institutions. Schools, prisons, and asylums were particular favorites. Such studies often presumed that diverse human defects were linked and then proceeded to stitch them together with threads of data. In 1901, for example, Pearson and Alice Lee gained access to skull measurements taken by the Cambridge Anthropometric Committee. A mathematical friend from his own days at King's College, W. H. Macaulay, interceded with the university registrar to extract data on the degrees ultimately taken by Cambridge students. They concluded, based on low calculated correlations, that skull size and shape had little or no relationship to "intellectual power."[25]

Prison systems were among the richest sites of data on bodily measurements. Most of the measurements were generated within a system of criminal identification based on the match of certain physical dimensions that had been introduced about 1879 by the French criminologist Alphonse Bertillon. Francis Galton and Pearson each had a role in deliberations by the Metropolitan Police in London about data suited to this purpose. In 1901, the first volume of their new journal, *Biometrika*, included an article on criminal anthropometry by W. R. Macdonell, relying on data from what he called the Central Metric Office of New Scotland Yard. These measurements were supplied by John G. Garson, a Scottish physician and anthropometric expert in the Royal Anthropological Institute, who, during the previous decade, had consulted for the Home

Office on implementation of the Bertillon system of measurement and classification. Macdonell fit Garson's measurements to one of Pearson's asymmetrical frequency curves to show that a sample of 3,000 criminal measurements were of a single type. They were categorically different, however, from the measures of 1,000 Cambridge undergraduates. He did not conclude that criminality was caused by these differences, but only that the criminal population had its origins in "a different section of the community."[26]

A few years later, Pearson backed up his own doubts on the significance of skull size and shape using head measurements of 5,000 school children, which he correlated with assessments of intelligence by their teachers.[27] By placing notices in the *Journal of Education*, *School World*, and the *Schoolmaster*, he recruited masters and mistresses of more than 200 schools to fill out detailed schedules of measurements on their pupils. He defined the problem in terms of sibling relationships rather than of parents and children so that he could rely on the judgment of the "fairly impartial school teacher." The alternative would be unacceptable. "Even if relatives and friends could be trusted to be impartial, the discovery of the preparation of schedules by the subjects of observation might have ruptured the peace of households and broken down life-long friendships." He interpreted the results as confirming the inheritance of mental ability, but not any causal significance of brain size.[28]

Pearson's team relied primarily on medical and educational institutions for their data of human heredity. When he began looking to asylums and special schools, he quickly discovered that their physicians and administrators were already deeply engaged in the analysis of hereditary data. These men, the heirs of Black, Thurnam, and Hood, were now more numerous and professionally diverse. He respected their expertise, accepting without question what they diagnosed as illness or defective intelligence. He also took on faith their data, and even their tabular statistics, provided they did not neglect correlated variables or mix up cause and effect. Of course the numbers had to be large enough to avoid assigning significance to meaningless fluctuations. The design of population surveys and the estimation of statistical error was his business. This led to a few bitter disputes, but doctors and psychologists most often deferred to his authority on statistical questions and often sought out Pearson's assistance.

Institutional Data on Mental Defect

The British asylum population almost doubled from 1890 to 1910, as the last hopes of relieving this epidemic through treatment melted away.[29] The problem of mentally deficient children added a new dimension to the problem. The movement for special schools had been inspired by the French physician and reformer Edouard Séguin, who, in the 1840s, set up such a school and then wrote a book on "moral treatment," meaning education, of "idiots." After 1848 he moved to Ohio and then to New York, extending his ideals to another continent and language. Most such institutions were residential, including the pioneering Royal Earlswood Asylum for Idiots and Imbeciles in Surrey, England, which opened in 1848. It modeled its patient records on those of insane asylums. For causes, their sources mentioned fright of mother, injuries at birth, childhood diseases, and convulsions or fevers during teething. By 1859, there was a specific question about family members with cerebral disorders, and from the 1880s, one about whether the disease was hereditary. When answered in the affirmative, these forms asked for information on one or more specific relatives and the condition(s) from which they suffered.[30]

Up to the end of the century, most of these schools were supported by charity rather than by state funds. The largest and most prominent such institution, the Royal Albert Asylum in Lancaster, became noted for its records on patient heredity. In 1892, its medical superintendent, George Edward Shuttleworth, prepared data on 1,200 patients, which, in combination with 1,180 patient forms from Darenth in the London system of hospitals, supplied the statistical data for an article on idiocy in the monumental *Dictionary of Psychological Medicine*.[31] In 1912, soon after Pearson inherited responsibility for Galton's Eugenics Laboratory, the Royal Albert Asylum sent him 2,900 information cards on its patients, the results of a long-term investigation of inheritance of feeblemindedness. With this treasure chest of hereditary information, supplemented by 1,200 slips from the Edinburgh Charity Organisation Society, he could delegate eugenics researcher Amy Barrington to construct pedigrees of family illness. Here, as so often, biometric investigation of hereditary defect was intertwined with researches performed by physician-statisticians in asylums or on the commissions that supervised them. Similar records were mobilized for Mendelian research.

The biometricians' favorite sources for the statistics of insanity and mental deficiency were the commissioners and asylum superintendents of Scotland, who, like Norwegian ones, relied heavily on population surveys and censuses. David Heron, of Galton's Eugenics Laboratory, complained in 1907 that annual asylum reports were useless for statistical study of hereditary transmission. The percentage of patients with an insane relative meant nothing, he said, without a suitable control showing the probability of encountering mental illness in families of the healthy. Heron favored the hypothesis, convenient for medically untrained statisticians, that mental defect arose from a constitutional susceptibility, or "diathesis." Such an instability of "degenerate" stocks could be studied and tallied without requiring a precise diagnosis. Even so, the work of compiling rigorous pedigrees was painfully tedious. He longed for a "General Register of the Insane," to be preserved in the office of the Lunacy Commissioners, and an index number identifying every insane person.[32]

Heron's dreams were fulfilled thanks to some Scottish officials whom we met in chapter 9. "At this stage, Dr. John Macpherson of the Scottish Lunacy Commission came to our aid and kindly furnished the Laboratory with a progressive history of 1319 insane patients who were admitted for the first time to Scottish Asylums in 1868." Next, A. R. Urquhart arrived with 331 family trees based on patients at James Murray's Royal Asylum in Perth. W. S. Gossett ("Student"), on sabbatical in Pearson's laboratory from the Guinness Brewery, used these records to estimate the percentage of Scots who had ever been in an asylum. Heron then calculated a coefficient of inheritance for insanity between 0.52 and 0.63. This number was very acceptable to the biometric faithful, just what the statistician ordered. Urquhart, proud to be associated with a scientific project, included measures of correlation in his Morison Lectures at the Royal College of Physicians in Edinburgh in 1907. Although his own calculated figure for inheritance of insanity, at 45%, was disappointingly low, he contrived in his final period of observation to raise it to 48%. "The important biometric system advocated and instituted by Professor Karl Pearson will greatly enlarge our knowledge and correct our prepossessions if the desirable data are forthcoming."[33]

Pearson's exchanges with the Scottish alienists were initiated mainly from their side. The scope of these interactions is dizzying. Macpherson, an Edinburgh physician, wrote Pearson in January 1904 asking permission to reproduce some tables and curves from his philosophical book, *The Grammar of Science*, and from his Huxley Lecture at the Royal Anthropological Institute, "Laws of Inheritance in Man." Macpherson was preparing some statistical lectures of his own on human variation in relation to insanity. By 1907, he and Pearson were discussing standard forms for recording its hereditary transmission. Macpherson explained that he was about to issue a thousand blank schedules to be filled out by school teachers. He included two old papers by Sir Arthur Mitchell, well known for his statistical work as commissioner in lunacy, to familiarize Pearson with Scottish censuses of insanity.[34] Another key figure here was J. F. Tocher, who paid a visit in 1904 to Pearson's laboratory and who secured permission from Macpherson, Mitchell, and John Sibbald (the third Lunacy Commissioner) to publish the Scottish lunacy measurements in *Biometrika*.

Tocher subsequently organized the schoolteachers of Scotland to carry out a "pigmentation survey" of school children, an indicator of "racial" variation, which came to be published in *Biometrika* in 1908. In 1910, appealing to the memory of Sir John Sinclair and the great Scottish statistical tradition, he sounded the tocsin in the *Eugenics Review* for a "national eugenic survey" of children throughout the United Kingdom to assist in the "grading of stock" in a racially heterogeneous population. Tocher used the language of race in reference to modest distinctions of skull shape, hair color, and pigmentation. A full eugenic assessment, however, would require data on other variables and for whole families. He would have liked to survey the entire adult population for the information it would provide on longevity and fertility. Alas, this was out of reach.[35]

In 1910, Heron published a statistical review of the Pioneer School Survey recently released by the London County Council. More insistently even than Tocher, he pressed for studies based on routine medical inspections of school children. That meant putting data into the form required by modern statistics, and he invoked the Edinburgh Charity Organization as a model. There were many differentiating factors, he stressed, that needed to be sorted out, beginning with the "local races" formed by varying proportions of Irish, Jewish, Scandinavian, Anglo-Saxon, and other elements. He firmly rejected any unique standard of normality defined by an average over all parts of the country, since each local race had a distinct character. "It is idle, for example, to compare Lancashire and Devonshire children, or either, with a most misleading 'British Association Standard' and to attribute the results to an influence of factories or of the rural environment." Heron praised Tocher's Glasgow surveys as exemplary, though he regretted their lack of access to "the pauper, the mentally defective, and the criminal."[36]

Degenerates and Degeneration

The frequent appearance of pauperism, mental defect, criminality, and bodily abnormalities together in a single pedigree chart was common knowledge in the new eugenic era. In a 1910 paper, Pearson and Elderton discerned in the statistics a hereditary link of epilepsy to mental defect, deaf-mutism, and dwarfism. "Superficially and for the time being only we may possibly look upon it as the inheritance of some defect in a general development-controlling determinant."[37] Such biological language, unusual for Pearson, supported the statistician's preferred focus on general disposition or diathesis over specific disease forms. It derived from research for the *Treasury of Human Inheritance*, an encyclopedic data project of the Eugenics Laboratory. Galton and then Pearson recruited several doctors early in the project to assemble pedigrees of diseases and deformities. One of them was Harold Rischbieth, a Cambridge-educated surgeon and MD, who drew from a large bibliography of case literature in several languages for his material on hare lip and cleft palate. Deformities of architectural form or size, he explained, may be attributed "to a defective developmental determinant in either gamete which leads to an arrest in the development of the zygote."[38]

Pearson and Elderton used Heron's term "general degeneracy" to mean "the correlated appearance of improbable mental and physical defects in a group of blood relations." This formulation, too, links up with the *Treasury*. Walter Jobson Horne, also a surgeon and physician out of Cambridge and a specialist in laryngology, wrote up a brief entry, illustrated by pedigrees, on inherited deaf-mutism. Two of his tables, redrawn to conform to the precise specifications of the Galton Laboratory, came from Ludvig Dahl's 1859 book, including Table 3, the descendants of Ejvind. The conditions Dahl included—deaf-mutism, insanity, idiocy, and epilepsy—were precisely the ones mentioned by Pearson and Elderton. Dahl's tables provided the most compelling illustration Horne could find of multiple defects in a single lineage. German researchers quickly noticed Dahl's tables in this English compilation.[39]

Pearson gave the impression that many or most wrongheaded ideas owed to a failure of physicians, biologists, politicians, and popularizers to give proper heed to statistics. Often, however, the opponents who plagued him were similarly committed to numerical evidence. Almost every topic he took up was already a site of energetic data collection and, often, of distinctive tools of analysis. He more often fixed on enemies for using numbers incorrectly rather than for neglecting to use them. He even waged battles against his own students, dedicated quantifiers like Udny Yule and Major Greenwood at home or Charles Davenport and Raymond Pearl abroad, when they strayed.[40]

His debates on degeneration were also of this kind. Many of his opponents, heirs to the great tradition of asylum statistics, were now seeking ways to move beyond bureaucratic data-mongering to real statistical science. Physicians who attributed British decline to alcoholic decay or to "causes inherent in the germ plasm" and who set about measuring these effects were among the most influential commentators on insanity and mental weakness of their day.[41] European temperance movements were cradles of eugenics, and many who endorsed the noninheritance of acquired characters made an exception for alcohol, which might act directly on the germ plasm. These were matters on which doctors were ill-equipped to judge, Elderton wrote. "The medical man as a rule has no opportunity of dealing with a random sample of the general population; it is the social worker who goes into the homes who alone can appreciate the extent of the drinking habit, and record the economic conditions of the working population."[42]

In 1912, Pearson crossed swords with Dr. Frederick W. Mott, a pathologist for the London County asylums, over the doctrine of "ante-dating," a variety of degeneration whose evidence was strictly statistical. The offspring of insane parents, he found, tend to have their first attack at a younger age than their parents. He credited the finding to Maudsley and cited a handbook of life insurance as well as one of Pearson's papers on tuberculosis.[43] His best evidence consisted of 3,000 data cards documenting the insanity of 750 closely related persons. Pearson and Heron dismissed his result as a statistical artifact. A simple model, they explained, shows that offspring will exhibit a lower mean age of onset than parents even with no impulse from inherited toxins. Insane persons who die young bear no children, and it is hard to get data on ancestors of those who became insane late in life.[44]

Born Criminals and the Inheritance of Mental Characters

Pearson, a master of craniometry, was contemptuous of Lombroso's fixation on head shape and other stigmata of hereditary regression. "Whole schools of criminology have arisen based solely on such assertions," he said, despite a complete failure to demonstrate any association of crime with physical and mental characters. In reality, "nobody knows whether crime is associated with general degeneracy, whether it is a manifestation of certain hereditary qualities, or whether it is a product of environment or tradition." We expect Pearson to come down on the side of heredity, especially, as here, when speaking under a title like "Nature and Nurture: The Problem of the Future." Indeed he did, but he was always skeptical of claims for inheritance of particular human actions or behaviors. In place of hereditary tuberculosis, he spoke of diathesis; and of inherited talents and vulnerabilities rather than a specific tendency to literature, mathematics, or crime.[45]

On the question of criminality, as usual, he went straight to the source for his statistics, and as usual, his contacts had definite ideas already about how to read the numbers. In this case the key figure is Horatio Bryan Donkin, Pearson's friend from their days in the protofeminist Men and Women's Club in the late 1880s, an Oxford-educated physician who had treated Karl Marx as well as the club member and author Olive Schreiner.[46] Donkin acquired the title Sir Bryan after leaving private practice to become medical director on the Prisons Commission. In 1903, he introduced a paper in *Biometrika*, an analysis of anthropometric measurements on 130 criminals by the medical officers of the English Convict Prisons. The Pearson archive holds a carbon copy of an eleven-page paper, "Report upon the Aims, Methods, Progress, and Results of a Statistical Investigation now being conducted for the Prison Commissioners at the Biometric Laboratory, University College."[47]

That report, written for prison authorities, begins: "Modern statistical discoveries and the recent applications of disinterested and exact methods of enquiry" have shown how much of the current sociological understanding of crime reduces to conventions and assumptions, without evidence. It fingers Lombroso as the inspiration for biased studies that have "found the evidence they sought" and that lead to a denial of criminal responsibility. Donkin put in motion a campaign to acquire the needed evidence and to break the hold of superstition, to be "piloted" by prison inspector Dr. Herbert Smalley. The report goes on to outline a regime of measurements and observations to be entered on a blank form that also appears in the files. This biometric initiative culminated in 1913 with a book by Charles Goring, *The English Criminal: A Statistical Study*. The preface explains how Goring was put onto statistics in 1901 by Dr. Griffiths, who had been responsible for the measurements in the 1903 *Biometrika* article. Griffiths at first had in mind a Lombrosian investigation linking bodily abnormalities of criminals to their offenses. Soon the work came to the attention of Donkin, who referred Goring to W.F.R. Weldon, the Linacre Professor of Zoology at Oxford and for a decade Pearson's closest scientific ally. Weldon, perhaps as anticipated, sent him on to Pearson's laboratories, where, from 1911 to 1913, he wrote his book. His conclusions seem to be overdetermined. He dismissed the project of reading mental and moral tendencies from bodily traits as misguided. Head size and shape must in no way be interpreted as stigmata of criminal tendencies. Instead, heredity and environment conspired to fashion these men as weak in body and especially in mind, hence vulnerable to bad influences. While Goring thanked Pearson for indispensable contributions to every dimension of the work, the initiative for this statistical-eugenic project came from within the prison system, perhaps first of all from a push for criminal identification at the "Central Metric Office."[48]

The story goes on. In October 1910, Donkin delivered the Harveian Lecture, "On the Inheritance of Acquired Characters," to the Royal College of Physicians in London. The title implied no plea for soft biological heredity, which, he explained, science had properly rejected. He was thinking instead of cultural or moral inheritance. While informed students of crime now reject Lombroso's and Max Nordau's idea of the criminal as a hereditary degenerate, he declared, it had seized the public mind. A similar scientific fallacy had arisen closer to home: the claim of a "prominent writer on biology" that there had been no proper genetic knowledge until Mendel demonstrated unit segregation of hereditary factors. But that writer, revealed in a footnote as the pioneering geneticist William Bateson, was applying this insight falsely to criminals, treating them as sharply distinct from normals.

ON

INHERITANCE OF MENTAL CHARACTERS

THE HARVEIAN ORATION FOR 1910

DELIVERED BEFORE THE ROYAL COLLEGE OF PHYSICIANS OF LONDON ON OCTOBER 18TH

BY

H. B. DONKIN, M.D.OXON., F.R.C.P.

MEDICAL ADVISER TO THE PRISON COMMISSIONERS FOR ENGLAND AND WALES; MEMBER OF THE PRISONS BOARD; CONSULTING PHYSICIAN TO WESTMINSTER HOSPITAL AND THE EAST LONDON HOSPITAL FOR CHILDREN, ETC.

"Let parents choose betimes the vocations and courses they mean their children should take; for then they are most flexible: and let them not too much apply themselves to the disposition of their children as thinking they will take best to that which they have most mind to."

Francis Bacon.

London:

ADLARD AND SON, BARTHOLOMEW PRESS

BARTHOLOMEW CLOSE, E.C.

1910

FIGURE 10.4. The title page of Horatio Bryan Donkin's 1910 Harveian Lecture on inheritance of mental characters. From work on the mental deficiencies of children, Donkin (1845–1927) moved to prison medicine and to statistics of causes of criminality. He was subsequently involved with commissions charged to investigate the seeming epidemic of feeblemindedness. He took an interest in hereditary explanations but remained skeptical, especially of the attribution of mental weakness to a single Mendelian factor.

Donkin quoted a claim from the concluding section on eugenics in *Mendel's Principles of Heredity* (1909) "that in the light of such knowledge public opinion will welcome measures likely to do more for the extinction of the criminal and degenerate than has been accomplished by ages of penal enactment." Bateson went on there to assert "that in the extreme cases, unfitness is comparatively definite in its genetic causation, and can, not unfrequently, be recognized as due to the presence of a simple genetic factor." He then called for "new conceptions of justice" based on genetics, with its power to anticipate. The Mendelian factor for criminality betokened criminal responsibility in advance of any crime.[49]

Bateson here treated "criminality" as a discrete trait, analogous to round versus wrinkled peas in Mendel's garden. The social Mendelians, as we may call them, supposed that many human defects, including criminality, were recessive, meaning that the hereditary criminal (a "homozygous recessive") must carry two hereditary factors (we call them *genes*) for the recessive trait, criminality. A heterozygous individual, with one normal factor and one criminal one, would he healthy, but if crossed with another such individual, would produce, on average, one criminal offspring for every three normal ones. The cross of a heterozygote with a double recessive (whether pea or criminal), yields on average one normal for each recessive. These ratios, 3:1 and 1:1, were understood from the beginning as necessary and sufficient evidence for a simple Mendelian trait. It was, of course, easier to identify a wrinkled pea than a hereditary criminal. Bateson and the social Mendelians, while recognizing their reliance on the analogy with plant breeding, worked to bolster their conclusions with swelling rivers of data.

To Donkin, a physician and prison reformer, this was a clear instance of the biologist's indifference to facts of observation. The dangers of misunderstood heredity had been made plain to him, he said, during his service on the Royal Commission on the Control of the Feeble-Minded, where the evidence was ruined by "confusion of thought" and "inaccurate language." Donkin praised Galton's biometry and eugenics while dismissing as "meaningless" the opposition of nature to nurture. Nature, he said, has rendered man responsive to nurture.[50] He recalled practicing in a children's hospital, before he took up prison work, and being moved by the image of so many mental defectives, "unable to shift for themselves and very likely to take to a life of misery to themselves and multiform evil to others." It was reason enough to segregate these children in special schools, as indeed would be ordered in 1913 by the Mental Deficiency Act.[51]

Donkin's lecture was in part a commentary on the royal commission he mentioned, which had issued its report in 1908. This consisted of six volumes of evidence from the UK, a seventh volume of observations on American institutions, and a book of findings and recommendations. On some key points, the report was inconclusive. The statistics could not establish how many mentally defective children had a defective parent. The witnesses did not even agree on how to understand or apply the adjective "inherited," although a majority identified *heredity* as the most important cause of mental defect. The commission, while complaining of loose definitions of heredity, endorsed the claim by a series of eminent asylum doctors of its "almost overwhelming probability," from a "biological standpoint." It now seemed urgent to determine what the correct biological standpoint might be.[52]

The next year, in 1909, the Royal Society of Medicine took up the charge to provide scientific clarity on these issues, holding a series of discussions devoted to inheritance of mental weakness. Its president declared, on opening the meeting, "There is no class of disease in which the conviction that they are hereditary is more firmly fixed both among the lay public and the profession than many forms of diseases of the nervous system."[53] But what were the mechanisms of human heredity? Both Bateson and Pearson were invited to give evidence. The gestures of deference to their hosts by these proud men suggest how unusual it was to invite mere scientists into the sacred halls of medicine.

The proceedings, seemingly as anticipated, turned into a debate between medical biometry and medical Mendelism. Sir William R. Gowers, a London neurologist, introduced the second day with rejoicing that the mists concealing Mendel's discoveries were at last being swept away. But these theories, he continued, are not much use to medicine. "The human race is not open to Mendel's essential methods, and its mere complexity of development involves innumerable differences from lower forms of life." Gowers, having devoted much of his career to statistics, preferred to count patients rather than genetic factors. He proudly presented data from his practice showing 47% hereditary causation for epilepsy and insanity, much higher than in hospital data. Since even his numbers were incomplete, the true value must be at least 50%. This was just as Pearson wanted it.[54]

Gowers was selected to challenge Bateson, who responded that he could easily demonstrate, sometimes quite precisely, the applicability of Mendel's work to diseases and congenital deformities in man. He began with a lecture, using Mendel's results on peas to explain segregation and dominance and to introduce Mendelian ratios. Next he gave examples involving the color and other conditions of eyes, based on new research by Edward Nettleship and Charles Chamberlain Hurst. He listed some other Mendelian conditions, including brachydactyly (abnormally short fingers and toes) and "hereditary chorea." Some of his ratios matched well with Mendel's numbers, while others diverged, on account of imperfect records, he said. Tuberculosis involved an infective organism, so he could not track it fully, and the same might hold for cancer. Insanity depended too often on environmental influences to present clean ratios.[55]

The third session, one week later, featured Pearson. Charles Mercier, a prolific author on statistics of insanity and mental defect, introduced it with a declaration that Mendel's law, being a law of probability, required data on a large scale and could be true only in the long run. Sadly, the statistics of the Lunacy Commissioners were pointless, "a gigantic waste of time and labour" and "of no value for scientific purposes." Dr. Arthur Latham elaborated that official statistics on inheritance of tuberculosis were "based largely on old wives' recollections, on uncompleted family records," and so on. Pearson, he added, in relying on these numbers, could not avoid certain basic errors. When Pearson rose to speak, he endorsed every effort to improve the quality of statistics, including Mercier's call for proper controls. He insisted above all on the need for a storehouse of excellent pedigrees and described the dozens of visits and letters required to check and complete even one. Pedigrees from the north of England and Scotland, he continued, are far superior to anything from a London hospital. People in the north have a stronger memory, and the "pride of family" is far more intense.[56]

Pearson cast doubt on Bateson's charts. "My own standpoint is that there is no definite proof of Mendelism applying to any living form at present; the proof has got to be given yet." He challenged Bateson specifically on albinism, which, he argued, does not divide clearly between presence and absence. Dr. Vilhelm Magnus of Norway had sent to him and to Bateson the record of a woman who had albino children with two unrelated men. This, Pearson declared, was so improbable as to exclude any possibility of albinism as a Mendelian recessive, moving Bateson, in an impromptu discussion, to concede that albinism was "an example of a character that did not follow the rules."[57]

Two weeks later, in the final session, Dr. George Percival Mudge, of the London Hospital Medical College, took up the biometric challenge. He gave the impression that many or even most human conditions reduced to Mendelian traits, but cautioned that one had to be wary of superficial resemblances. On this score, Pearson had repeatedly failed, particularly on albinism, a genuine Mendelian trait. What Pearson called partial albinism was always the result of some other cause or disturbance, such as arrested development. Mudge set biometry and Mendelism in radical opposition, contrasting the biometric pursuit of masses of data with the Mendelian's focus on "individual cases, carefully studied and rigorously analysed."[58]

This last claim is hard to square with Mudge's own practices. He had recourse again and again to Mendelian ratios as proof of determination by a single factor, and his experimental papers on inheritance of rat color were packed with numbers. Bateson understood very well that Mendelism was about quantities, which he deployed with some skill, even as he consigned biometry to the other side of an impassable barrier. "Actuarial" methods, he once wrote, are appropriate only to a science in its infancy. "In nearly every case to which the method of accurate experimental breeding has been applied, it has been possible to show that the phenomena of heredity follow precise laws of remarkable simplicity, which the grosser statistical methods had necessarily failed to reveal." Bateson and Mudge put their faith in the simplicity that would emerge in biology and medicine when things were properly sorted out. Mudge detected Mendelian segregation even in the skin colors resulting from racial mixing of Europeans with American Indians. Here there was no dominance but intermediate color for heterozygotes, while homozygotes must resemble one or the other racial ancestor.[59]

Pearson, while conceding that Mendelism might sometimes hold, objected that often, and certainly in the case of interracial skin color, the claimed ratios depended on observational practices that neglected variability. Perhaps unconsciously, Mendelians exploited variation to get the numbers they wanted. This was how Weldon and Pearson had interpreted the unreasonably precise ratios from Mendel's original paper after Bateson made it famous. Pearson accused Mudge of a more egregious manipulation. When his own study yielded data showing a ratio of 1 to 4, Mudge "hunted up some other experiments, which were made twenty-three years ago on the same subject, in which the ratio given was 1 in 2," and added them together to get the desired 1 to 3. "A glorious proof of Mendelism," Pearson sneered. Indeed, Mudge had supplemented his own data with some from a 1906 paper by zoologist Leonard Doncaster, who had in turn incorporated results from 1885 by a German doctor, Hugo Crampe.[60]

Pearson, though coming off rather well in this exchange, seems to have been troubled by some specific criticisms he had made. Shortly afterward, he inserted a paper on race and skin color into *Biometrika*, introducing it with the declaration that those who still doubt some claims made for Mendelism are yet "ready to emphasise the paramount service of Mendel in drawing attention to the great factor of segregation in many inheritance problems." Some traits at least were sufficiently distinct to be sorted and counted. But skin color in racial mixing, which for centuries had exemplified blended inheritance, was certainly not. He chose it to ridicule "theorists" such as Mudge.[61]

Pearson's preferred response to critiques by Bateson and Mudge did not focus on Mendelian ratios, nor even on the inheritance of discrete traits, but on the status of biometry as a master discipline. As he insisted to another medical audience at the end of 1908, biometry is no "ism." It "neither pledged itself to 'mutations' nor to continuous variation; it was solely an attempt to apply exact methods to vital statistics of any kind." Mendelian claims should be demonstrated by the scrupulous application of biometric methods, which require immense patience. "Ten or twelve years of collecting evidence was required, and at the end of that time they would know to what extent Mendelism did or did not apply." Sadly, he concluded, Bateson and his allies were not willing to wait so long before jumping to conclusions.[62]

Neither, of course, was Pearson. Both sides looked to hereditary data for a resolution of their conflict and as the foundation for new advances. These data would come from collaborations with doctors, psychologists, teachers, and criminologists in institutions that held hundreds or thousands of individuals. In neither version, Mendelian or biometric, could the new science of human heredity break with tradition of institutional data work on populations of special concern to the medical-social state. Biometricians and geneticists alike depended on stockpiles of data from asylums, schools, prisons, and population surveys and on assessments by special commissions linked to law, medicine, engineering, eugenics, and poor relief. The new genetics was but one element in the mix, and not even the most promising one.

CHAPTER 11

Genetic Ratios and Medical Numbers Give Rise to Big Data Ambitions in America, 1902–1920

> Take the single matter of feeble mindedness. I presume there are over a score of institutions for such people in the country and in many of them as I know by correspondence, not only records of heredity but detailed studies of families [have] been made.
>
> —Charles B. Davenport (18 Mar. 1909)

> It would seem quite as proper and important that each person of the family in which we are interested should have a numerical name as that each plant should be so designated in the plant breeding nursery. A name with eleven letters is not over long.
>
> —Willet M. Hays (1912)

Although Mendelian genetics had its origins in experimental breeding, it was applied almost immediately to human abnormalities. While "blended" characters, as Francis Galton called them, seemed to resist Mendelian analysis, the new geneticists looked for discrete, or "segregated," factors underlying continuous variability. Congenital and lingering conditions such as tuberculosis as well as mental deficiency and insanity had long been supposed hereditary. Even a known infectious agent left room for an inherited diathesis (a nonspecific factor) that could determine susceptibility to disease. Some conditions, however, could not be neatly classified, with insanity and feeblemindedness in the first rank of diagnostic ambiguity. Mendelian medical heredity, in contrast to the more generalized "biometric" or biostatistical variety, was scarcely thinkable without neat categories. The suppleness of biometry, in which Pearson took such pride, appeared to Bateson as its Achilles heel, encouraging its application to sloppy data to reach meaningless conclusions.[1]

Pearson countered that Mendelism, too, depended on numerical data, adding that Mendelians often tortured their data to provide spurious confirmation of what they supposed to be already known.

Bateson's plant experiments, like Mendel's, were data intensive, and chromosome mapping in Thomas Hunt Morgan's fly lab at Columbia University was an exercise in applied probability. The most ambitious eugenic project in America was undertaken in the name of Mendelism by an American biologist who followed and then fell out with Pearson. Yet Charles B. Davenport, commander in chief of eugenic (or social) Mendelism, lived his life for data. Although his research was funded at a princely level, he recognized that his laboratory could never come close to matching the financial resources of state institutions. He soon discovered that the doctors and psychologists in these institutions had already built up impressive expertise on methods of researching heredity. He allied with them to train and send out eugenics fieldworkers, who constructed pedigrees of patients from dozens of institutions. Davenport turned up the classic Mendelian proportions, 3:1 and 1:1, for almost every condition, however ill-defined, that mattered to eugenics. Although these claims were subjected to withering criticism almost from the beginning, his project appeared for some time as a remarkable success story. Afterward it became one of the best-known tales of eugenic and genetic hubris.

Animal Models and Human Heredity

When Raymond Pearl, another biometric pioneer, arrived in London for an apprenticeship with Pearson, he could scarcely believe what cold, cramped quarters confined the professor and his assistants, or how little time remained after teaching in which to carry out such prodigious research. "The great biometric laboratory of University College is all comprised of one room with two windows, the size of the room being just that of Room 3 at Ann Arbor."[2] Pearl, like Davenport, was among those captains of science made possible by the plutocratic wealth of the gilded age. Already, in America, dollars made the visionary. Davenport, in an autobiographical reflection, stressed the strong mathematical education he had received at Brooklyn Polytechnic. He began teaching biology from a statistical perspective in the early 1890s, and his book on statistics and biological variation introduced his colleagues to Pearson's methods.[3] By 1902, when he met Galton, Weldon, and Pearson in England, he was esteemed, not least by the Carnegie Institution, as one of America's leading scientists. From Paris, a few months later, he wrote to Willet Martin Hays of the Agricultural Experiment Station in Minnesota to propose that the newly created American Breeders' Association (ABA) "urge upon the trustees of the Carnegie fund the establishment of a station for studying heredity." By early 1904, the station was almost ready, with Davenport, the well-paid director, hatching schemes for chicken breeding.[4]

These birds were a secondary topic in his first letter to Bateson, whom he invited to serve as a correspondent of the new laboratory. Bateson, accepting, endorsed the merits of chickens for his experiments then reminded Davenport that postage to Europe was 5 cents per 1/2 ounce. While English scientists minded their pennies, the Carnegie solicited their expert advice on grants to American scientists they thought undistinguished, promising resources beyond their dreams. In a 1907 report to the Carnegie Institution on American research, Pearson complained of men who gather data indiscriminately then wonder what to do with it. "The first point in any work in this field is to have a real biological, anthropological, or sociological problem which *needs* solution." Three years later, he removed Davenport from the board of editors of *Biometrika*. The great statistician was simply hostile to Mendelism, Davenport complained, while Pearson charged Davenport with careless and superficial work that threatened to discredit biometry.[5] An American textbook by the geneticist William Castle sized up the situation in 1916, declaring Pearson's data to be sound though his analysis was corrupted by hostility to Mendelism, while American eugenic data was unreliable on account of fieldworkers trained to assume that inheritance reduced to the presence or absence of a Mendelian factor. Davenport claimed in 1926 that "the development of Mendelism has led to the general introduction of mathematics into genetics." Pearl, a few years earlier, called one of Davenport's books a "bad tabulation of bad statistics."[6]

Davenport never thought of limiting his genetics to farm animals; he had been keen on eugenics at least since his first discussions with Galton and Pearson. His correspondence with Alexander Graham Bell demonstrates the interpenetration of agricultural and eugenic breeding. Bell, a learned advocate for the deaf, was also a noted breeder of sheep at his estate on Cape Breton Island, Nova Scotia. Davenport initiated the correspondence in 1904 as he was setting up his laboratory, asking for four-nipple sheep to use in hybridization experiments that, he hoped, would demonstrate inheritance of this "sport," or anomaly, "in accordance with Mendel's laws."[7] Bell responded that his sheep had had multiple nipples for generations and so were not true sports, but he was happy to assist the research. An ensuing request for six-nipple sheep flummoxed him, since, though reluctant to donate an animal so rare and valuable, he considered himself a benefactor of research rather than a profiteer. In May 1906, he sent Davenport the breeding results of sheep born that year, and in July he arranged for Davenport to receive the recently issued census report on the blind and deaf of the United States, which had been prepared under his direction. The chapter on marriages of the deaf and their progeny was especially worthy of attention, he said. "It has generally been assumed that the laws of heredity that are known to apply to animals also apply to man but I do not know of any large collection of statistics that demonstrate the proposition with the exception of the Census returns relating to the Deaf." He volunteered to ask the Census Office about their unprinted tables on marriages of the blind. "It has just occurred to me that this might be a matter of interest to the Carnegie Institution, Station for Experimental Evolution, and that you might perhaps like to obtain copies of these Tables."[8]

Davenport certainly did want those tables, but he was now focusing on "inheritance of color" in offspring of whites and blacks of the human variety and was planning a trip to Jamaica. Bell told him of the Maroons of Jamaica, descendants of slave refugees now living across the bay from Halifax, Nova Scotia. Potentially of still greater value were the census records, including manuscript returns going back to 1810, stored in the rooms of the Volta Bureau, which he had established to study deafness. Bell, who had used them to investigate the ancestry of deaf-mutes, was sure that Davenport could extract similar results for inheritance of color.[9] Records like these were now Davenport's eugenic sustenance. Bell inquired a month later if he had ever investigated "Deafness in White Cats that have blue eyes," which Darwin had mentioned. Early in 1907, they discussed whether American eugenics should continue to be organized within the American Breeders' Association. Davenport said yes, on the grounds that the laws of heredity were indifferent to species. Mendel's principles could not have been worked out in full from studies of man alone.[10]

Hopelessly Vicious Protoplasm

In these years, the preeminent journal of genetics in America, and perhaps the world, was the *American Naturalist.* Davenport was an active member of its sponsoring society, the American Society of Naturalists. Its business brought him into contact with men like anthropologist Franz Boas and psychologist Edward Thorndike.[11] It was, however, the ABA that, in 1906, formed a committee on eugenics as one of the "general subjects" in a miscellany that included Animal Hybridizing, Breeding for Dairy Production, Cooperative Work in Plant Breeding, Prize Competitions, and Theoretical Research in Heredity.[12] Hays, now Theodore Roosevelt's assistant secretary of agriculture, was equally devoted to eugenics. In 1910, when the society upgraded its newsletter to the *American Breeders Magazine: A Journal of Genetics and Eugenics*, Davenport held forth there in praise of eugenics. Man's nature "follows the laws of the rest of the organic world," and the affiliation with breeders would provide "dignity and safety" against quacks and popularizers. "Our greatest danger is from some impetuous temperament who, planting a banner of Eugenics, rallies a volunteer army of Utopians, freelovers, and muddy thinkers to start a holy war for the new religion."[13] Soon, he raised eugenics into a section of the society while letting the other sections wither. He envisioned a swarm of committees and subcommittees devoted to feeblemindedness, insanity, deafness, eye defects, and so on.[14] The work required oceans of data and would not be cheap. Free-lovers, indeed, were not welcome, but the religious fervor he swore to lock out had already established a stronghold within. Ten millions to "redeem mankind from vice, imbecility and suffering," Davenport proclaimed, does far more good than ten millions for charity. There loomed, after all, a satanic enemy. "Society must protect itself; as it claims the right to deprive the murderer of his life so also it may annihilate the hideous serpent of hopelessly vicious protoplasm."[15]

In March 1907, when Hays inquired what tasks this new eugenics committee might pursue, Davenport was ready. The problem, he began, is so large that twenty men could work at it full time. Their work must be statistical. Thinking, no doubt, of Galton's book *English Men of Science*, he proposed to study the ancestries of James McKeen Cattell's "1000 first men of science in America." He next mentioned the inheritance of physical characters in man and then hereditary mechanisms, including prepotency, hybridization, transmission of acquired characters, and the possible effect of maternal impressions on the fetus.[16] "Supply blanks to lying-in hospitals," he inserted here, in unwitting obeisance to asylum tradition of hereditary investigation based on filled-out forms. Also on the list were "insanity in various forms, criminal tendencies, various idiosyncrasies."[17] Early in 1909, the committee worked up a proposal to add eugenic questions to the US census. The suggestions were too late to be incorporated into the 1910 census, as they promptly learned from the House and Senate committee chairmen.[18]

Undaunted by this failure, the ABA formed a committee in 1912 to explore the possibilities of securing data through the Census Bureau, the Bureau of Health, "and other societies and institutions." Davenport sketched out his hopes in a letter to Elmer Ernest Southard, a pathologist for the Massachusetts asylum system: "In brief, what is needed is a qualitative census. I suppose the original purpose of the census was military and a simpleton will stop a bullet as well as a genius but now-a-days it is the fashion to take stock of natural resources and it would seem that the State should take stock of its germ plasms and their product."[19] An undated document from about the same time outlines a project for cooperation with the U.S. Bureau of Industry and Immigration to map the geography of flawed heredity. "Are there foci of criminality and defectiveness in Europe which are supplying an exceptionally large proportion of the immigrants who become charges on the State? If so, locate centers and find their characteristic defects." At other moments he emphasized non-European races such as American Indians alongside these immigrants from less desirable parts of Europe.[20] But government statisticians did not cooperate. In 1917, he transmitted a modest resolution to the director of the census, Samuel Rogers, proposing that in 1920 and thereafter the census enumerators should record, for every male, the father's name. It sounds easy, Rogers replied, but the United States has 100,000,000 inhabitants, "and the task of collecting, compiling, and tabulating these data is an enormous one."[21]

Davenport's most ambitious data scheme did not depend on legislation or federal agencies. He would line up experts to mine the records of institutions for persons with distinctive hereditary traits. Mainly, this meant socially undesirable characters for which states were increasingly taking responsibility. Each type of institution had its experts, who, he soon discovered, were just as interested in heredity as he was. This realization opened a new world of opportunity to him. He articulated his updated dream to David Starr Jordan, the biologist-president of Stanford University and Davenport's choice as chairman of the ABA Eugenics Committee. They should create subcommittees to investigate "feeble-mindedness, pauperism, psychiatry, the deaf, dumb and blind, cripples, criminals." Happily, they would not need to start from scratch. "There is a large number of institutions devoted to aberrant individuals belonging to each of these classes and some of them have departments of research with directors interested in the subject of heredity. Take the single matter of feeble mindedness."[22]

A folder in the archives names six possible members of the subcommittee on heredity of the feebleminded, four for insanity, five for epilepsy, two for deaf-mutism, and four for criminality. He could only think of one candidate, W.E.B. DuBois, for the subcommittee on "heredity in Negro-white crosses," and one also for the heredity of eye defects. Six more bodily conditions, such as cancer and physical strength, remained, for the moment, "unorganized," as did his envisioned subcommittee on "laziness and physical basis of poverty."[23] The focus on persons "under the control of the state" was not by choice, Davenport explained some years later in a letter to Bell. Originally, his office had intended to study gifted persons as well, but it is "*very much harder*" to secure their cooperation. He now recognized the advantages of this unintended focus: these traits had already been "very carefully analyzed" and "minutely classified," and the fact of state control made the individuals and their families much more accessible to research. Finally, state institutions covered half the costs and conferred a measure of public recognition on the work of eugenics.[24]

Data for Mendel

Davenport's ideal subcommittee included a man of consequence as a figurehead chairman and an industrious secretary to organize the labor. He installed himself as secretary of the whole Committee on Eugenics and set to work on its first and most urgent need, data. Happily, he now realized, the data were all around.

> They lie hidden in records of our numerous charity organizations, our 42 institutions for the feeble-minded, our 115 schools and homes for the deaf and blind, our 350 hospitals for the insane, our 1,200 refuge homes, our 1,300 prisons, our 1,500 hospitals and our 3,500 almshouses. Our great insurance companies and our college gymnasiums have tens of thousands of records of the characters of human blood lines. These records should be studied, their hereditary data sifted out and properly recorded on cards and the cards sent to a central bureau for study in order that data should be placed in their proper relations in the great strains of human protoplasm that are coursing through the country.

With these reservoirs of information, he could pinpoint the "lines which supply our families of great men" as well as the insane and feeble-minded, blind and deaf, prisoners, criminals, and paupers. The same data would advance understanding of genetic mechanisms, the "method of heredity of human characteristics."[25]

The Eugenics Record Office (ERO), which he founded in 1910 next door to the Station for Experimental Evolution, became a storehouse of hereditary data. The records of so many schools, prisons, and asylums provided a million points from which to launch investigations. A series of donations by Mary Williamson Harriman, heiress to a railroad fortune, and others provided funds to train and to employ, over the next fifteen years, 257 eugenics fieldworkers. He always looked to the institutions where they labored to pick up as much as possible of the cost. The work drew on interpersonal skills that he regarded as distinctively feminine: a degree of tact in dealing with defective persons and the grace to coax their relatives into discussions of conditions that ran in the family. He took immense pride in the system of fieldworkers, which, he thought, raised his data to an incomparably higher level than the familiar institutional tables.

He was not shy, however, about using pedigrees from earlier researches, going back to Dahl.[26] He also followed Galton's example, encouraging eugenic-minded citizens to assemble records of as many family members as they could and to send them to the ERO. The office printed *The Family-History Book*, incorporating forms and schedules supplied by physicians and institution directors, to guide these volunteers, and *The Trait Book*, laying out standards.[27] The family relationships could be reduced to a pedigree of ancestors and other relatives and annotated with markers of notable talents or, more often, psychic and bodily defects, based on descriptive histories of each individual. All these individual and family histories were recorded on cards and filed in drawers designed to preserve them indefinitely. The enthusiasm of the Honorable James Wilson, US secretary of agriculture, in his presidential address to the ninth annual meeting of the ABA, was so fierce as to sound like mockery. But these men were crippled by irony deficiency. "You have developed in your eugenics section a great experiment station and institution of research . . . containing fire-proof vaults. . . . Your one or two dozen scientific eugenists who are devoting their time to assembling the genetic data of thousands of families in those fire-proof vaults are making records of the very souls of our people, of the very life essence of our racial blood."[28] The files were designed for research and as a resource for eugenic counseling. Davenport routinely assumed such counseling would be grounded in Mendelism.[29]

Most of his subcommittees never got off the ground. Bell accepted the invitation to chair a subcommittee on deafness but favored a more positive eugenic approach. He had his own institutions for recording families of the deaf, and in place of Mendelian predictions of deafness, he called for the exploitation of family histories gathered by life insurance companies to anticipate longevity. Bell's cousin, Elias J. Marsh, had studied this question in his capacity as the medical director of the Mutual Life Insurance Company of New York.[30] Davenport's man to chair the subcommittee on criminality was Professor R. G. Henderson of the University of Chicago. He found time to complain "that the name of Heredity of Criminality does not quite express what is desired," but left the work to the secretary, Max G. Schlapp of Cornell Medical School, who agreed on this point and wondered if a "Committee on Criminality" or on "Causes of Criminality" might be "more to the point." Davenport countered that the sociological aspects of criminality had been much studied already and that the charge of this committee was to take on "the hereditary basis which makes criminality possible . . . , the sensitive protoplasm rather than the stimulus." And then the correspondence petered out.[31]

Only two subcommittees, those concerned with insanity and feeblemindedness, were able to generate any momentum. Not by accident, these were conditions already endowed with abundant family data and with allied techniques of analysis. These committees, each with Davenport's forceful encouragement, concluded that the data on their defect confirmed its status as a Mendelian recessive

FIGURE 11.1. Cottage No 4—Group III, one of the not-yet-massive residences that housed, in total, a population of patients approaching 4,000. From *Seventeenth Annual Report of the State Hospital at Kings Park to the State Hospital Commission for the Year Ending September 30, 1912*, plate between pp. 22 and 23.

trait. In each case, the institutional expert—Aaron J. Rosanoff for insanity and Henry Herbert Goddard for feeblemindedness—had professional reasons to hesitate at the leap of evidential faith required. In each case, eugenics fieldworkers had an important role—Gertrude J. Cannon and Florence I. Orr for insanity, Elizabeth Kite for feeblemindedness. The stories of these subcommittees, which interpenetrate, reveal to what extent hereditary investigations were driven by the historical trajectories of data and expertise at these medical-social institutions. Davenport, instantiating science, coordinated the work, and his Mendelian language of unitary factors was vital for the public relations.

The Subcommittee on Insanity

As in Europe, distress at the unrelenting expansion of mental illness was stimulating bold scientific initiatives in the United States, led by Massachusetts and New York. These ambitions derived partly from sources within mental health systems while reflecting a general spirit of Progressive reform. In 1896, after six years of planning, the New York Commission in Lunacy opened its Pathological Institute in New York City. The founding director, Ira Van Gieson, included in his first report a grand historical survey of treatment of the insane, the dark prelude to a luminous future. Commencing with a "Period of Revenge," it had advanced through periods of "Indifference" and "Humanitarian and Empirical treatment," and now stood at the threshold of "Scientific Study, Rational Treatment and Preventive Medicine." Already, the prognosis was favorable for patients without hereditary brain defects. Such defects must have been numerous, since the report for the state system showed a recovery rate under 5% (951 recoveries out of 20,843).[32] The New York asylums, including the first in Utica, had begun their advance into the last, glorious epoch by acquiring a freezing microtome for slicing brain samples along with a card catalogue "so devised that each specimen receives its special number with a history of the case and a record of each successive step in the examination." It is characteristic of this new scientific age that laboratory specimens rather than patients were the basic units of these information technologies. Since the causes of mental disease are material ones, wrote the Utica superintendent, the filing system must naturally adapt.[33]

Van Gieson was partial to neurological explanations. A dislocation of the arms of a nerve cell leads to discordance of the "spheres of higher consciousness." Over a few generations, the defect would worsen and become irremediable. Psychiatry, the reigning science of mental illness, was now "flapping about in the doldrums. . . . It has shut itself up within the asylum walls, discouraged original work and thought and met deservedly the fate of China and ancient Egypt. As a science psychiatry, at present, is dead, and a mummy may be its symbol." The renewal so desperately needed would arise from the integration of other sciences, which were making brilliant progress. There was a new psychology of mental disease, which examined the nervous systems of cockroaches and conducted experiments on sane humans. The Institute's psychologist could be found "at the laboratory table, gathering facts, using instruments of precision, conducting experiments." Cellular biology was laying a new foundation for heredity based on the examination of male and female components, "intimately wrought together and distributed in equal amounts in the process of cell division." Insanity, arising from defects of this germ plasm, was also a problem for anthropology, whose assigned role was to identify "the initial and intermediate stages in the course of degeneracy." The anthropologist depended on precision instruments to measure skulls and test sensory acuity and on statistics to compare the number of cases of insanity without hereditary predisposition to those caused or complicated by heredity. So much paper data required an archivist, and Van Gieson even named her: Miss Marie Onuf. In a reflexive move, he announced the founding of a new journal, the *Archives of Neurology and Psychopathology*, and republished his long essay in its first volume as "The Correlation of Sciences in the Investigation of Nervous and Mental Diseases."[34]

Van Gieson's radical attempt to fashion an institute for basic research within the state asylum system rang alarm bells in the legislature, forcing his resignation. He was replaced by Adolf Meyer, whose ambitions favored clinical medicine over nerve cells. "A great part of psychiatry is administrative knowledge," he wrote in 1904 in his first full report. "Psychiatry must be met with the same clear determination as a business proposition, not as a field of vague aspirations and ambitions." As in Massachusetts, his goal in New York was to supply "full statistical particulars" so as to make the "statistical evidence" reliable. He complained that the data on the 72,228 admissions in 1901–1902 was marred by various little contradictions. The figures should be checked rigorously and inscribed onto cards. "In this way you obtain a cross-index of your material according to causes."[35]

Meyer was cautious about hereditary explanations, especially after he encountered a Swiss study by Jenny Koller that seemed to challenge it (see chapter 12).[36] Davenport nevertheless invited him in November 1909 to become president of the subcommittee on insanity. Possibly he was unaware of this skepticism, or perhaps it mattered less to him than Meyer's devotion to data. It helped that the men agreed on appointing Southard, no less a data enthusiast, as secretary. Southard was not in the least daunted by Davenport's ambitious paper forms. Quite the contrary; he already had acquired his own supply and was trying to stimulate interest among Harvard medical students. He had begun preparation of an inventory of families with two or more patients in Massachusetts institutions. Southard endorsed Davenport's claim that research could be a profitable investment for the state by helping "to dry up the springs which feed the flood of protoplasm too weak to withstand the strain of untoward conditions."[37]

In March 1910, Davenport asked his advice on several appointments, including Rosanoff for the subcommittee on inheritance of epilepsy. This suggestion met with silence, and in late October he elaborated that Rosanoff, though a Jew, was "a very nice sort of Jew." Southard replied immediately that his hesitation had nothing to do with religion. The problem was Rosanoff's tendency to make strong claims on insufficient data. "I think the writers on insanity are altogether premature with their statements that such and such data 'indicate Mendelian inheritance,' and so the actual status of matters is that many writers have not the slightest idea of the differences which exist in types of insanity." Without rigorous diagnostic categories, Rosanoff could only make blind leaps to Mendelian conclusions.[38] It is a strange comment to make to the high priest of presumptive Mendelism. Perhaps he thought Davenport, a biologist, was being led astray by flawed psychiatry.

Rosanoff's early career exemplifies the bold ambitions of asylum medicine in the new century. About 1904 he was appointed junior assistant physician at the King Park State Hospital on Long Island, just twelve miles east of Cold Spring Harbor. The new medical superintendent, William Austin Macy, argued in the annual report that a good recovery rate was scarcely possible given the character of the city and its insane. The hospital was huge, approaching 4,000 patients when, in 1908, Rosanoff and Macy prepared a paper on institutional scale for the annual report. Even a thousand patients is too many for the superintendent to hope to know them individually, they declared, and once this threshold is passed, the state could just as well seize the advantages of a division of labor. A large institution permits administrative efficiencies and high-quality research. It can afford a well-paid pathologist and a well-equipped lab with an attendant for purely technical work. Not least among the advantages for research is the greater variety of "clinical and pathological material," that is, living patients and dead ones. To support this point, they supplied a seventy-page case list with all kinds of patient data extending to paternal, sibling, and maternal heredity, the raw material for statistical analysis of the causes and conditions of illness. The next year's report explained cold-bloodedly that they were proceeding with a "special study of certain organic cases. . . . Fortunately, many of these cases reach autopsy . . ."[39]

On 14 October 1910, Rosanoff joined a group of charity professionals at the Skillman (New Jersey) Village for Epileptics in a discussion of methods of standardization. He presented there a paper, coauthored with Gertrude Cannon, who had just resigned from her post as eugenics fieldworker. According to a report, they showed "that (excluding certain types) two insane parents will have only insane offspring and that normal parents, both of whom belong to insane strains, will, in the long run, have one quarter of their offspring defective." This assertion of Mendelian ratios must have provoked Southard's acid remark, just two weeks later, about unsupported Mendelian assertions. Davenport, who had taught Rosanoff his genetics, defended the study to Southard, mainly on the basis of the excellent fieldwork. Cannon's research, he said, was comparable in quality to the expert pedigrees gathered by Elizabeth Kite for Henry Goddard, institutional psychologist at Vineland, New Jersey, who also attended the Skillman meeting. The excellent charts of the fieldworkers, based on research in homes and communities, were in sharp contrast to slipshod institutional pedigrees, Davenport said. Southard, however, was no more impressed by Goddard's data than by Rosanoff's. Kite's pedigrees, he objected, showed even more feeblemindedness in offspring of one feeble-minded and one normal parent than the 50% allowed by Mendel's rules.[40]

FIGURE 11.2. Undated photograph of Aaron J. Rosanoff (1878–1943), who spent the first part of his career at the huge Kings Parks Asylum on Long Island. He already was interested in hereditary explanations when he began collaborating with Charles B. Davenport of the nearby Eugenics Record Office, who taught him some techniques of data analysis and convinced him that mental disease required a Mendelian explanation. Portrait courtesy of the National Library of Denmark: http://www.kb.dk/images/billed/2010/okt/billeder/object145923/da/.

We cannot know what exactly Rosanoff said at the meeting on 14 October. However, ten days earlier he had presented the same work to the New York Neurological Society, and that paper, coauthored by Cannon, came out the following May in the *Journal of Nervous and Mental Disease*, edited by the New York neurologist Smith Ely Jelliffe. The printed version, billed as a "preliminary study," spoke of Mendelian inheritance—not of insanity, but of "neuropathic make-up," scarcely a recognized term in the medical lexicon. Davenport used a similar term a month later when he described the paper to Southard as revealing "that, contrary to all our anticipations, a neuropathic condition is inherited as tho it were due to the absence of a single unit character and that this neuropathic taint may show itself indifferently in any one of a number of kinds of insanity: now in maniac depressive insanity, now in dementia praecox; now in senile dementia, now in paranoia and so on." He added that if both parents are insane, all the children will show some type of insanity.[41]

Nobody, then or now, could imagine that Mendelian proportions were contrary to Davenport's anticipations. It sounds disingenuous at first. He was surprised not by the Mendelian ratio, but by its grab-bag object, "neuropathic make-up." The key to the mystery is preserved in Eugenics Record Office Bulletin No. 3, billed as a reprint of the Cannon-Rosanoff paper. It is dated 3 May 1911, implying simultaneous publication with the "preliminary study" in Jelliffe's neurological journal, and the texts are identical. The ERO reprint, however, includes at the end something extra: the proceedings of the session of New York neurologists, consisting primarily of a comment by Davenport himself, who had been present as Rosanoff's guest. He was unreservedly positive, yet on the crucial question of disease specificity, his recorded spoken words contradict the printed paper. The rules of inheritance, he had said, are always the same, whether it be dementia praecox, manic-depressive insanity, senile dementia, epilepsy, or feeblemindedness. It followed that "two feeble-minded parents could have only feebleminded offspring, and that two epileptic parents could have only feeble-minded or epileptic children."[42]

Davenport's intervention makes no sense unless Rosanoff's original presentation was about inheritance of specific, recognized disease conditions rather than an all-embracing neuropathic make-up. Sometime within about two months of the neurological meeting on 4 October, and probably before the charity workshop at Skillman on 14 October, the authors abandoned their claim for inheritance of specific mental disease conditions. The revision was not one that Davenport could have welcomed. He had until then been proud to distinguish himself from the London biometricians by his reliance on precise diagnoses, made possible by the trained eugenics fieldworkers. Jelliffe, a neurologist, firmly rejected this catch-all category. Although he had been absent from the October neurological meeting, by Christmas he had read Rosanoff's paper. It figured prominently in a long letter to Davenport, his old high-school classmate, dated 26 December, nominally a response to Davenport's just-published essay on eugenics in *Popular Science Monthly*, yet specifically relevant to Rosanoff's paper. Davenport's reliance on the concept of "insanity," Jelliffe wrote, was a "great mistake," one that "goes back to the teaching of Morel in the early 50's." Insanity is so heterogeneous, with such diverse causes, that it is more properly understood as a legal category rather than a medical one. Having read Jenny Koller, Jelliffe was now skeptical of claims for the overbearing power of heredity. Mainly he insisted on the absolute need for hereditary researchers to employ a proper disease classification, such as Kraepelin's, rather than imagining that a mere legal category could be biologically heritable. The Cannon-Rosanoff paper, a conscientious effort with deep flaws, must have misled him.[43]

Davenport wanted to agree on disease specificity. His comments to the neurologists referred specifically to inheritance of dementia praecox and manic-depressive illness, whose absence from the Cannon-Rosanoff paper he had explained as a reluctant capitulation to unanswerable evidence. He replied immediately that their reliance on so vague a category was "in opposition to all my prepossessions. . . . This conclusion I must say has astounded me in my attempt to interpret it." He obviously was no mere dupe of Rosanoff's erroneous reasoning. Davenport now proposed a distinction between eugenic realities and medical ones. For other purposes it might be appropriate to insist on distinct disease forms, but practical eugenics was something else, "and if it appears empirically that for the latter purposes the whole classification reared with so much care is useless and can be lumped, then I for one do not see why it should not be lumped in formulating eugenical advice."[44] Jelliffe, unconvinced, inserted a sharp editorial critique of all-encompassing disease categories in the January 1911 issue of his journal. The "great bane of psychiatry" in regard to factors of heredity, he declared, "has been the hopeless confusion of statistical studies for lack of fundamental nosological conceptions."[45]

It remains unclear just what led Davenport and Rosanoff to give up disease specificity in the inheritance of mental illness. Clearly, Davenport was the driving force for universal Mendelism. He never asked *if* Mendelian ratios applied, always *how*. All the evidence suggests that his role in the formulation of "neuropathic make-up" was fundamental. The authors thanked him in the paper for "guidance, advice, and assistance." Rosanoff cannot have been any happier than Davenport with this result. As a physician, he looked on accurate diagnosis as a professional responsibility, and many American psychiatrists shared Jelliffe's conviction that Kraepelin's nosology was at last making it possible. In 1912, Rosanoff tried out a modified genetic ontology, a "hypothetical germ-plasmic determiner for complete mental development" made up of discrete units, such as for epilepsy or manic-depressive insanity, with a hierarchy of dominance relations.[46] Flummoxed, however, by the problem of distinguishing dominant from recessive characters, he and Orr finally took refuge in the supposition that degree of dominance could vary. None of this mattered for their published conclusions. The final version of their study was decked out with many more tables and an intricate coding of defects, yet they scarcely altered a word of their conclusions, which seem to have owed more to Davenport's Mendelian theories than to painstaking pedigrees. Rosanoff, putting no faith in cures, described prevention as the task of

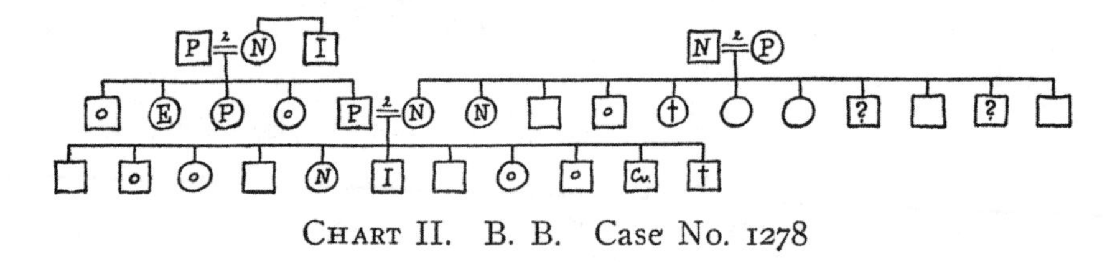

FIGURE 11.3. Pedigree tables assembled by Aaron J. Rosanoff and his first eugenics fieldworker, Gertrude Cannon. The tree begins with two sets of grandparents, each coded as type 2 marriages, meaning that (as they inferred) each individual appeared normal but carried one recessive gene for neuropathic defect. The marriage on the next line is also of this type. *I* in the table, means insanity and *N* is for "feeble-mindedness, hysteria, or other pronounced neuropathic manifestation." The authors inferred the presence of invisible (recessive) factors for defect based on observed conditions of their descendants. Psychiatric Mendelism required that one in four offspring of marriages like these should, on average, show neuropathic defect. From Cannon and Rosanoff, "Preliminary Report," 276.

eugenics, and treatment as "palliative measures to combat antisocial manifestations."[47]

Anticipations of a match between Mendelian genetics and specific disease categories recurred like a refrain in debates about psychiatric heredity but went nowhere.[48] Most eugenic pedigree charts displayed myriad defects of mind and behavior whose tangled webs could never be unraveled. Rosanoff and Davenport were unable in the end to go beyond diathesis to diagnostic precision. Jelliffe, unconvinced by his old classmate's Mendelian discoveries, declared his preference for the hereditary researches of English biometricians. He also praised Jenny Koller, Otto Diem (who built on her research), and Wilhelm Tigges. Even so, he did not disdain to publish, a few months later, Davenport's paper (with David Weeks of the Skillman Institution) on the Mendelian inheritance of epilepsy.[49]

The Feebleminded Subcommittee

Davenport's choice as chairman of the subcommittee on feeblemindedness, Dr. Arthur C. Rogers of the School for the Feeble Minded in Faribault, Minnesota, was no figurehead. He remarked in a report to the ABA on the wealth of data being gathered by

American institutions for the feebleminded and cited researches going back to 1880. Three of these institutions—in Faribault; Vineland, New Jersey; and Lincoln, Illinois—deserved special attention for having established departments of research. Vineland was where Henry Goddard, secretary of this subcommittee, worked as institutional psychologist. He attributed its abundant patient histories and pedigrees to the "full, free, and hearty" cooperation of the parents of these children as well as "adroit questioning and cross-reference" on the part of the professional staff. The trust of the families was crucial to the great goal of data accumulation. "The field worker comes to them as the superintendent's personal representative with a letter from him recommending her and urging the parents, for the sake of the child, to tell all they possibly can, and to send her on to other relatives or to any one who may be able to give the information, which may be used to help their child, or some one's child." Here and elsewhere, interviewers relied heavily on the pretext that eugenic information would benefit individual patients. The subcommittee commenced its investigations at these exemplary institutions, aided by eugenics fieldworkers. Davenport and his associates paddled forward in an already-flowing stream.[50]

In these years, the science of feeblemindedness was brimming with new or resuscitated ideas. Davenport even thought of employing a eugenics fieldworker trained in psychoanalysis to work with Vineland patients. By examining the children with their siblings, parents, and aunts and uncles, a psychologist might be able to determine "how far the defects of the morons are specific and are inherited as units." Superintendent E. R. Johnstone, like Davenport, had hopes of identifying particular families, often of immigrants, who had brought feeblemindedness into a region.[51] Goddard, similarly, required no persuasion to take up research on patient heredity. On the contrary, he was already a master, and his first letter came as a revelation. Davenport initially wrote to Johnstone on 9 March 1909 to ask about "data concerning feeblemindedness." Goddard, delegated to reply, explained that while the Vineland admissions forms had always requested information on heredity, the responses were meager and unreliable. He was just then engaged in drawing up new blanks to be sent out after admission, not just to the family but to other informed parties, such as their physician. "This is yielding fruit slowly," he continued, and as evidence he passed on an early result for a newly admitted child, born in an almshouse to an unmarried, feebleminded mother with four feebleminded children. She had since been married three times and had given birth to six more feebleminded children. The Department of Research at Vineland was seeking funds to hire "a man and probably also a woman" and put them on the road to seek out information on families of its nearly four hundred inmates.[52]

Davenport, who was then designing his own blank forms, was stunned. He replied immediately. "I can hardly express my enthusiasm over these blanks," he declared, "and my enthusiasm that you are planning, I trust, extensive work in the pedigree of feeble minded children." He was especially taken by the idea of putting a man and woman on the road. Would Goddard join the ABA so that Davenport could appoint him to the subcommittee on feebleminded children? Within weeks, Davenport was making arrangements to tour the Vineland facility. He owed his proudest research innovation to professionals at a home for feebleminded children.[53]

Meanwhile, he had written exultantly to David Starr Jordan of his delight to have found this unexpected ally. He marveled at the "detailed family studies" as well as data collection going on in a score of institutions for the feebleminded. Goddard was suitably deferential, and Davenport's own plan for form-filling was not working out. It also appeared that his great hopes for the Census Bureau to circulate forms in its limited enumeration areas would be frustrated. He was still awaiting the "printed blanks from our dilatory printer." Needing information on "perhaps 1,000 large families," he inquired if Jordan might persuade some enterprising students at Stanford University to gather up such data. Jordan said he could, and even promised to take this investigation to other universities west of the Rocky Mountains.[54] At the end of March, Davenport bundled up a thousand printed forms for dispatch to Palo Alto. He would gladly send more, he said, and then he explained how these industrious students could extend the entries to a still "longer heredity" than the form called for. But fate favored fewer shorter family forms. "I have distributed most of the blanks which you sent me," Jordan wrote in November, "but the students who take them are staggered by the large amount of writing and by their not having thought much about matters of heredity." Jordan concluded cheerily that the forms would at least stimulate thought.[55]

Davenport preferred data. He complained to Meyer in May 1909 that "I have met with practically no response with the full family blanks because they deal with so many matters difficult of access and not sufficiently pointed for the physician." He did not scale down his ambitions. A few months later he was discussing with economist Irving Fisher, another leader of American eugenics, the design of Hollerith cards for eugenic data.[56] Harry H. Laughlin, at this time a school superintendent in Kirksville, Missouri, earned his position as eugenics deputy at the ERO in part by diligently registering complexion, form of hair, and eye color as Mendelian family traits. Laughlin also contributed to the solution of "the mulatto problem" by acquiring an expensive microscope. "Can we consider a blend as the Mendelian inheritance of minute units?" he asked in March 1908, referring now to shorthorn cattle. The answer was yes.[57]

Davenport also advocated Mendelian inheritance of mental ability to Goddard, whose endorsement of these doctrines is a little surprising. As Leila Zenderland points out, Goddard made his reputation as a pioneer of Binet testing, which treated the distribution of intelligence as a continuum. Children who would never surpass a mental age of four he classified as idiots, while an age of 5 to 8 meant imbecility. Goddard's new category, the "moron," was for those between 9 and 12, while (in the United States) feeblemindedness took in all three classes. On what basis could he draw a sharp hereditary line between mental ages of 12 and 13? His defense was strikingly hesitant. "The writer confesses to being one of those psychologists who find it hard to accept the idea that the intelligence even acts like a unit character. But there seems to be no way to escape the conclusions from these figures."[58] Rogers, in a coauthored book of 1919 on the feebleminded "in the Vale of Siddem," dissented from Goddard's conclusions on just these grounds. Intelligence was too complicated to be reduced to a Mendelian law. Rogers and the fieldworker Maud Merrill quoted Lewis Terman, the new guru of IQ testing, for the doctrine that low intelligence could be defined only relative to the demands placed on the individual.[59]

The pose of reluctant submission to facts is one we encounter often. "Any theories or hypotheses that have been presented have been merely those that were suggested by the data themselves," Goddard intoned. Yet his hesitation was real, even if his data were very far from raw. He began his work with a list of causes assigned by parents and physicians, a long-familiar genre. In just 173 of his 327 cases had they assigned any cause, and only 30 of these involved heredity. He reasoned that other causes, such as neuropathy, pointed to heredity, while certain diseases, such as measles, could occasion feeblemindedness only in connection with hereditary weakness. With adjustments like these, he reached 164 hereditary cases plus 34 "probable." His field workers then sought out feebleminded relatives of these children. On the hypothesis of a recessive Mendelian unit, he had to infer from the traits of relatives the presence or absence of a single factor for feeblemindedness in each normal sibling. This often amounted to arguing in a circle. When he had no information at all, he assumed arbitrarily that exactly half bore a defective factor. He knew from Wilhelm Weinberg's critique of one of Davenport's papers that a recessive factor will remain invisible unless at least one offspring inherits it from both parents, and he made the appropriate correction. There were six feebleminded children born in families with two feebleminded parents, clearly violating his Mendelian law unless he supposed that the husband was not the father. And so he did. Goddard did not conceal these inferences, revisions, and corrections to the data on which his result depended. At the end he added up all the numbers and compared them with his prediction. The agreement was excellent: 704 normals to 352 feebleminded expected, 708 to 348 observed. "Such results are difficult to account for on any other basis."[60]

And so, despite painful doubts, he stuck by his own makeshift numbers. Davenport's unflagging Mendelism counted for much here. Goddard, who seems unaware of the biometric critics of monogenic causation, was sincere in his credulity. This is especially evident from the Kallikak episode, his most treasured "natural experiment," which made him famous, and then infamous. That story involved a soldier from the era of the American Revolution, Martin Kallikak (Greek for "goodbad"), who initiated one line of defective descendants by a feebleminded girl and another of excellent ones with a sound moral wife. Zenderland shows how eugenics fieldworker Elizabeth Kite, an obliging investigator and gifted writer, convinced Goddard, Davenport, and legions of readers, most of them primed to be deceived, that she could reliably detect feeblemindedness from a glance or an anecdote.[61]

The Kallikak Family presented the Mendelian unit character for feeblemindedness as an assumption rather than a result. Goddard anticipated its demonstration in a big scientific book that was soon to follow. His faith was sustained specifically by the success of Mendelian psychiatry, especially by Rosanoff and Orr's final report in 1913. They, too, acknowledged the strangeness of their Mendelian result given the motley defects they had merged into neuropathic make-up: "imbecility, epilepsy, deteriorating psychoses, periodic psychoses, paranoic conditions, involutional psychoses, the slighter psychopathic states, and certain eccentricities." There must also be specific "unit determiners," such as for feeblemindedness, as well as this encompassing one. In the end, Rosanoff and Goddard alike looked out to the match of their numerical results with theoretical expectation and concluded that it was too good to be denied. Both overcame grave professional doubts to preserve a faith in unit Mendelian factors. It was a relief to reach conclusions in accord with the latest discoveries of science.[62]

Mendelism Was Biometry for Other Ends

In March 1910, still feeling the sting of Pearson's rejection, Davenport wrote to Frederick Adams Woods, a prospective member of the Eugenics Committee, criticizing the biometricians for a lack of "biological, let alone medical training." "Does it not seem to you as a medical man a bad procedure to lump a lot of insane individuals of different generations to get at the hereditary index of insanity? Does it not seem more hopeful to study some well defined form of insanity such as melancholia and work it out carefully by first hand studies on a few families? Do you think it would be a fair statement that mass statistics are good when nothing better is available?"[63] Before the year was over, he went over to the other side, inviting Jelliffe's scorn. It was difficult to maintain disease specificity of inheritance when almost every pedigree chart, starting with Dahl's, was crowded with the heterogeneous symbols of a multiplicity of defects. His reliance now on lumping to get his Mendelian proportions, however, was also, in a way, a concession to statistical values, the privileging of numerical data. Just as Pearson was satisfied with parent-child or sibling data on mental traits when he achieved correlation measures of about 0.55, Davenport scoured his data for the all-holy three to one, or sometimes one to one. It was all the evidence he needed. He criticized Goddard's masterwork for faltering at this conclusion.[64]

Even before 1900, agricultural breeders had used hybridization techniques like Mendel's to develop new varieties. By 1900, they were working on an industrial scale.[65] In the 1910s, experimental animals with short life cycles, most notably the fruit flies cultivated in T. H. Morgan's fly lab, yielded Mendelian variants in comparable abundance. Somatic medicine, meanwhile, was beginning to identify rare genetic conditions such as chorea. Even before George Huntington began recording the inheritance of this disease, and one year after Dahl's book, the Norwegian Johan Christian Lund had described this "inherited disease" and compiled a family table.[66] If a human condition so common and so costly as insanity or mental deficiency had yielded readily to rules of taxonomy and statistics, we might now have Thurnam's, Dahl's, or Jung's laws of heredity instead of Mendel's.

With its paper forms and files, the ERO put forth a legion of Mendelian unit traits. Davenport, who contributed a botanical classification and then a eugenic one for Melvil Dewey's decimalized library system, used a card catalogue for triple-filing of individual records by name, disease form, and place.[67] In letter-sized cabinets, he arranged and filed individuals according to distinctive behaviors, talents, and disease forms. He gathered records of insanity and feeblemindedness from institutions using established research techniques that began with a patient or inmate, the propositus, and moved outward, weaving a web of parents, grandparents, siblings, and other relatives (see chapter 12). Since many social problems seemed to arise from lack of self-restraint, he worked up a series of studies on the "feebly-inhibited." One, *Hereditary Nomadism*, for example, was about persons who gave in to the temptation to drift, run away, or emigrate. His cards revealed that such behaviors ran in families and could be grouped to generate the ratios he wanted. The wonderful miscellany of behaviors represented in these files was made possible by eugenic-minded volunteers who sent in printed forms such as the "Single Trait Sheet" that he distributed to one and all. It began with a blank line on which to enter a trait, then a line for the propositus, and then a set of ordered spaces on which to inscribe names and recount behaviors in affected relatives.[68]

The proof of the Mendelian factor was a numerical ratio. There was no other. Pearson's group joked that Mendelians were simply incompetent biometricians. It has been shown over and over that Pearson's categorical rejection of Mendelism is a myth. He even gave mathematical arguments for its theoretical compatibility with the biometricians' law of ancestral inheritance. To be sure, he opposed the assumption of complete dominance, and he mobilized data to challenge various particular claims for Mendelian inheritance.[69] There was plenty of low-lying fruit. About 1911, for example, Radcliffe Nathan Salman extended his expertise on potato breeding to the "racial" heredity of Jewishness. Jewish distinctiveness had remained mostly constant since the fifth century, he explained, and consisted mainly of facial traits that were recessive to gentile ones. Pearson's student Harold J. Laski responded in *Biometrika*, comparing the work to Davenport's claims about

Mendelian heredity of the mulatto and Mudge's rat pedigrees, concluding: "It is not, indeed, too much to say that the endeavor to make man a complex of sharply-defined unit characters has failed, and failed completely."[70] Aleš Hrdlička, a member of the ABA Committee on Eugenics and expert on interbreeding between whites and Indians, was similarly dismayed by Davenport's explanation of skin color. He wrote: "You blame the anthropologist for making so little use of 'the new points of view'; is the cause of this not to be found in the fact, perhaps, that . . . he finds so little that agrees with his more general observations and helps him to understand the vital phenomena with which he is confronted?"[71]

The biometricians were especially harsh on Mendelian explanations of moral and mental characteristics. Provoked by praise for Davenport, Rosanoff, and Goddard ("the American researchers") during the International Eugenics Congress in London in the summer of 1912, David Heron of the Eugenics Laboratory initiated a series of attacks. He focused on bad data practices such as inconsistent categorization of a single individual in different pedigree charts. He found much circular reasoning, when, for example, the parents were symptom free but had one or more mentally defective children, and the theory was made to work by assuming the presence of a recessive factor. These were not mere errors of execution. They were codified in the manuals that instructed fieldworkers to search for defects in ever more remote relatives when the condition of a descendant required a recessive factor, and otherwise to stop. Heron made clear that the scientists, not the workers, were to blame.[72]

The second installment of this critique was by Pearson and Gustav Jaederholm, a Swedish school expert who was working in Pearson's laboratory. Jaederholm described the factors considered by nurses, medical officers, and teachers in deciding to place a student in a special school. They included "capacity and willingness to learn, power of self-control, habits of cleanliness, moral order, power or desire of attention to instruction, fits, possibly epilepsy, and a vast variety of semi-physical deficiencies." Heron called attention to descriptions used by Rosanoff to justify categorization as neuropathic, including "high strung, cries easily," and "very queer, lives alone, boards out cats."[73] Were these suitable criteria for diagnosing a biological anomaly? Pearson's justification for rejecting the Mendelian factor for feeblemindedness did not depend on mathematics. Segregation between types, what he had called Mendel's most valuable contribution, was nowhere apparent for mental weakness. School performance showed no sharp breaks, only continuous curves. Jaederholm came over from Sweden bearing data, the most welcome of gifts. Pearson calculated that almost half of the Swedish children who were three or more years behind normal could be explained as the tail of a Gaussian bell curve.[74]

Heron's diatribe, picked up by the *New York Times* in November 1913, became a terrible embarrassment to Davenport and his allies. Rosanoff, claiming his right to an immediate reply in the *AJI*, invoked medical "judgment and experience" against this mere layman, who "assumes an attitude, in relation to a psychiatric issue, which is in opposition to a view universally held by psychiatrists." He had, he said, merely extended well-known genetic principles to a new case.[75]

The story is not of isolated failings by a few bad scientists. David Barker has shown that in Britain as well as America, the very biologists given credit by historians for discrediting eugenic dogmas endorsed "naïve" one-factor explanations of epilepsy, insanity, and feeblemindedness. Bateson, one of the first, celebrated the achievements of the "American students" in 1913. In 1921, he anticipated progress "of a magnitude that statesmen perhaps have never conceived" once jurisprudence learned to reason with the Mendelian factor for criminality.[76] A paper by Hamish Spencer and Diane Paul shows that the critiques by Heron and Pearson include all the main points made by the new human geneticists of the 1930s.[77] R. A. Fisher was, like Pearson, a far more active eugenicist than Bateson and almost incomparably expert on the mathematics of heredity. Yet he was an early convert to Davenport's single-gene explanations. As late as in 1924, he referred to "strong evidence of a mendelian factor" and still stronger evidence of clear differentiation between the normal and the feebleminded. On this basis he anticipated that strict eugenic policies could reduce mental defect far more quickly than many scientists had supposed.[78] Barker, exaggerating, comments that the flaws in his reasoning should have been apparent to an undergraduate geneticist. Fisher had by then replaced Pearson as the world's leading statistician.

The biometric critique of Mendelian psychiatry was well known to human geneticists in the 1920s and 1930s. Lionel Penrose, a forceful opponent of eugenic policies with no reservations about Mendelism, cited Heron and Pearson and set out from their lines of argument when he challenged single-gene explanations in his 1933 book on mental defect. His own conclusions on its inheritance relied primarily on statistical analysis of phenotypic (rather than genetic) data from medical-social institutions. Insanity and mental deficiency remained the most urgent topics of human genetics through the 1930s.[79] The reshaping of psychiatric heredity by Mendelian genetics was dubious and fleeting. It is absurd to describe biometricians as standing in the way of a promising Mendelian program.

CHAPTER 12

German Doctors Link Genetics to Rigorous Disease Categories Then Settle for Statistics, 1895–1920

> The customary statistics of heredity is a narcotic, providing a deceptive gratification of the need for causality.
>
> —Julius Wagner-Jauregg (1906)

> Wir leben in einer Zeit der Renaissance der Erblichkeitsforschung. (We are living in a Renaissance of hereditary research.)
>
> —Ernst Wittermann (1913)

Until about 1908, German psychiatry knew "Mendel" exclusively as Emanuel Mendel, a Jewish asylum director and specialist on general paralysis whose data on inheritance supported the growing suspicion that this dread disease began as syphilis. This Mendel made his career in Berlin, taking a particular interest in reproduction of the mentally ill. He also was concerned about heredity and advised against marriage for anyone whose insane parent had diseased relatives or had been sick at the time of conception. If a woman developed a psychosis during pregnancy, an artificial abortion should be considered. Gregor Mendel's research on transmission of traits in hybridized peas, though initially published in 1866, impressed no one until 1900, when finally it was noticed and taken up by botanists and breeders. In mental medicine, it was primarily the "American researchers," notably Davenport, Rosanoff, and Goddard, who, about 1908, convinced European doctors that mental defect in some form might reduce to one or more Mendelian traits.[1]

Ernst Rüdin, then a privatdocent in Munich, commented in 1911 that it had sparked by then a decade of original work on plant and animal hybrids but almost nothing as yet in psychiatry. The time was now ripe, he declared, and he was right.[2] For example, Emil Oberholzer, trained by Eugen Bleuler at Burghölzli in Zurich, came to the 1913 annual meeting in Vienna of the German Association of Naturalists and Physicians armed with numbers to demonstrate that inheritance of schizophrenia revealed the same proportions as Mendel had recorded with his peas. It must, he concluded, be governed by a recessive Mendelian factor. For tainted but healthy (hence, heterozygous) parents, he explained, there is "never more than a single sick child" out of three to four offspring; "only one or two sick out of 5, 6 and 7 children; for 8 children, always exactly 2; a third schizophrenic child arrives only with still more of children." These figures amounted to empirical confirmation of misunderstood probability relations, requiring very good luck for the researcher or ample flexibility in the diagnosis. The discussants in his session did not notice. The next year, in a version presented to the Association of Swiss Alienists, he drew out the reassuring implication that if one among the four children of healthy parents develops schizophrenia, the other three must certainly be spared.[3]

The leading German authority on medical genetics was the Stuttgart physician Wilhelm Weinberg. Unlike Davenport, he understood from the outset that genetics did not require what he called "Mendel numbers" at every turn. He conceived psychiatric genetics as a long-term project, and he made his mark largely with the tools he devised to deal with imperfect knowledge of the genetic constitution of a patient's parents. Rüdin, who took charge of hereditary research at Kraepelin's psychiatric clinic in Munich, knew little of statistics until he met Weinberg. In 1911, as an organizer of the International Hygiene Exhibition in Dresden, he asked Davenport for natural objects that "demonstrate the important rules of inheritance," Mendel's rules. If, he continued, Heron was correct to say that segregation into distinct types had never been demonstrated for a psychiatric condition, this must be because none had been properly investigated.[4] Mendel's theory was altogether more satisfying than Galton's, he continued, since it applied to distinct, individual traits, leading to "rules of hereditary mechanism," and even "biological laws of heredity." Biometry, by contrast, applied merely "to some cases (Bateson also asserts here, only to accidents)"—in German: "für manche Fälle (Bateson behauptet auch hier nur für Zufälle)." *Fälle* means "cases," while *Zufall* is "chance" and *Zufälle* are "chance events," making a little pun of cases and chance. The joke disappears in English, leaving only clichés about the failings of statistics. Rüdin piled them on: "mere statistical formulations that are valid only for masses . . . , at best a mean proportion."[5]

By 1914, Rüdin had changed his tune. Chance, in the form of war, delayed publication of his massively researched book on inheritance of insanity until 1916. He argued there for the need to calculate probabilities of mental disease in children based on the traits of their relatives. He had hoped to confirm a Mendelian mechanism, but his data were nowhere close to the theorized proportions.[6] Mendelian genetics now seemed increasingly to offer no way forward for research on inheritance of mental illness. Rüdin did not abandon hope. Despite the persistent failure to achieve a quantitative match with Mendelian prediction, his group, and Weinberg in particular, could not forsake the haunting yet elusive simplicity of genes for mental defect, which must be lurking somewhere, wherever people were strange, sick, unproductive, or immoral. In practice, however, the group refocused its efforts on new methods of statistical prediction based on measurable or diagnosable traits. The work of "empirical hereditary prognosis" gained international recognition as a practical technique of psychiatric genetics, one that depended little on an identifiable *Erbanlage*. It was, as before, a data project, relying on records from asylums, special schools, and an ever wider range of medical and social institutions.

A New Survey Technique and Its Consequences

The age-old measure of hereditary causation—the percentage of patients with a mentally ill relative—was mostly discredited by the 1890s. The scandal of its uncontrolled variability, made visible in Legrand du Saulle's list, was one reason. Another was uncertainty as to how widely the net should be cast in search of defective kin. Already in 1862, Heinrich Neumann's report on an asylum near

Breslau in Silesia declared that since one person in a hundred is deranged, the claim for heredity required an appreciably higher proportion for relatives of the insane. In an American census report in 1895, John S. Billings called for controls: "As we have no data with regard to the number and classification of insane relatives for persons not insane it is impossible to determine the amount of influence exercised by heredity in the production of insanity."[7]

That very year, Jenny Koller published the first such data in a dissertation carried out under Auguste Forel, asylum director at Burghölzli (and a noted ant biologist).[8] He associated degeneration with drink, whose effects on the Swiss population, he claimed, were revealed by its insanity numbers, the highest in the world. In fact, his own instructions for the 1888 census may have had a greater role than alcohol in raising the measured rate of insanity in Zurich above that in other lands. They stipulated that registration of insanity must not be limited to severe cases but should include the feebleminded of every age. Census takers were also provided a list of discharged asylum patients to ensure that none were overlooked. The statistical authorities feared, however, that registration of heredity would be counterproductive. "Among specialists it will perhaps appear strange," the report explained, but the elusiveness of data must lead to incomplete results, and an undercount could not be permitted.[9]

Koller, who had also studied at the Charité hospital in Paris and worked as substitute second physician at the Swiss custodial asylum in Rheinau, was up to date on questions of heredity. She endorsed August Weismann's evidence that the germ plasm was unaffected by acquired traits. If a man became mentally ill after fathering ten children, this would not alter the probability for subsequent births. However, the proportion insane might rise due to alcoholic poisoning of the germ.[10] Weismann's work also convinced her that the statistical evidence for the greater hereditary receptivity of women must be flawed. A decade later, her ally Otto Diem identified this inequality as "a pure statistical artifact" arising from the mother's more complete knowledge of her family and greater willingness to share it. Koller thought the hereditary transmission

FIGURE 12.1. Portrait of Jenny Koller (Thomann) (1866–1949) from the mid-1890s. Her doctoral thesis, published in 1895, was the first to go beyond measuring "percent hereditary" by determining the same number for a comparison group. I thank Heidi Tewarson, Koller's granddaughter, for authorizing the use of this photograph.

of the urge for drink could not depend on sex, but that in women it often remained latent. The removal of impediments that destroy body and brain, including syphilis and alcohol, would encourage the action of regenerative factors. Like her teacher, she blamed alcohol for the uniquely high level of hereditary taint, or burden (*erbliche Belastung*), in the canton of Zurich.[11]

The financial report at Burghölzli did not extend to disease forms and outcomes until the 1870s. Forel, arriving as director in 1879, introduced a *Formular* modeled on German census cards. Immediately, heredity began to appear as a cause of overwhelming importance, among the highest percentage figures ever recorded. Forel was proud of his numbers: "We have endeavored to make these inquiries as exact and reliable as possible and have not taken into account any doubtful submissions." He succeeded in tracking down family information for 180 of 222 patients and found a hereditary burden in 153 of them, or 85%. His report categorized these cases as direct and indirect, and then as hereditary on father's side, mother's side, and among siblings, finding a near equality between paternal and maternal influences.[12]

Although the Zurich archives do not allow access to filled-out forms on patient heredity, Carl Jung's notes on Sabina Spielrein have been published in a modern biography of this patient-turned-psychoanalyst. Her peculiarities inspired Jung to initiate a fateful correspondence with Freud, for she was a decidedly interesting case. Jung reported suspicions that she was plotting to seduce him. He wrote and rewrote the peculiarities of the family, as she narrated them, onto the form for hereditary burden. Her father was overworked and neurasthenic; her mother, a dentist, hysterical and nervous. The eldest brother had hysterical crying fits, the second was hot-tempered and subject to tics, and the youngest, severely hysterical, found happiness in pain. Jung's description dates from 1905, after Bleuler took over from Forel and redefined anamnesis in terms mainly of mental processes. He also did away with the patient tables.[13] Case notes like these show how statistics were shaped by the attitude of the physician.

Hereditäre Belastung.

N°

☛ Glieder, über welche keine oder ungenügende Auskunt erhältlich ist, sind mit ?, solche, über die nichts Belastendes zu erfahren ist, mit 0 zu bezeichnen.

Name :

Vater :

Vater des Vaters :

Mutter des Vaters :

Geschwister des Vaters, (Onkel, Tante), Anzahl und Krankheiten derselben :

Mutter :

Vater der Mutter :

Mutter der Mutter :

Geschwister des Mutter : (Onkel, Tante), Anzahl und Krankheiten derselben :

Geschwister :

Kinder :

Zahl, Alter Krankheiten derselben :

Nebenlinien :

☛ Organ. & funktionelle Neurosen, Psychosen, auffällige Charaktere, Trunksucht, Selbstmord, Verbrechen. Blutsverwandtschaft. Unehel. Geburt, Tuberkulose, Diabetes.

FIGURE 12.2. Mockup of patient form for entering information on hereditary burden of relatives, Burghölzli asylum (1892). It has separate entries for the father and mother and for all grandparents, for aunts and uncles on both sides, for siblings and children of the patient, and for side branches. It specifies the relevant categories of hereditary burden, most of them familiar from German sources: functional neuroses and psychoses, striking character, alcohol dependence, suicide, and crime, and requests information on blood relationship of parents, birth outside of marriage, and tuberculosis and diabetes. Courtesy of Staatsarchiv des Kantons Zürich, which reconstructed for me a blank *Formular*.

From the structure of Koller's paper and her comments on the work, it appears that Forel had suggested she examine the key role of alcohol in the propagation of defective inheritance. He also supplied her with Burghölzli patient records from 1881 to 1892, which composed her database of hereditary causation among the mentally ill. Koller used Hagen's methods for tracking hereditary transmission from parents to offspring. Starting with a middle generation, she looked for indications of heredity in the ascendant line then gathered information on diseases of the children. Her counts of hereditary transmission from ancestors with psychical or alcoholic conditions were in line with expectations, but they were very low for "organic" ones such as apoplexy and nerve disease. Forel had already noticed the discrepancy and wondered if family organic disorders might be associated with reduced mental illness in the offspring. He thought of testing this hypothesis by gathering data on ancestors of healthy people to compare with the ancestors of the mentally ill. After making a start by investigating families of 110 personnel and surgical patients from Burghölzli and Rheinau, he turned the project over to Koller. She extended the asylum research and added information from personal acquaintances. In total, she was able to get family data on 370 healthy persons to compare with the relatives of mental patients. The controls, she acknowledged, were imperfect, since they were on average ten years younger.[14]

Applying the usual standard, she found that a full 59% of the healthy were hereditarily burdened. This number was not much lower than the hereditary figure for the mentally ill, and a proper statistical match, she said, would further narrow the divide. Her tables showed, however, that hereditary influences on the healthy were more frequently indirect, involving nerve disease, drink, and migraines rather than true mental illness. While emphasizing her empirical results over explanatory claims, she stated clearly that if 59% of the healthy were hereditarily burdened, heredity could not be destiny. The effect of factors such as nerve disease or senile dementia must be slight. She did not dismiss heredity, then, but focused it. Most of the hereditary burden of the mentally ill was direct, coming often from parents, and involved real mental illness.[15]

A. Unter den 218 erblich belasteten Gesunden sind belastet durch:		pCt.
1. Geisteskrankheit	57	26,1
2. Nervenkrankheit	38	17,4
3. Potatio	55	25,2
4. Apoplexie	36	16,5
5. Dementia senilis	10	4,6
6. Auffallende Charaktere	16	7,3
7. Selbstmord	6	2,8

B. Unter den 284 erblich belasteten Geisteskranken sind belastet durch:		pCt.
1. Geisteskrankheit	113	39,8
2. Nervenkrankheit	29	10,2
3. Potatio	63	22,2
4. Apoplexie	16	5,6
5. Dementia senilis . . .	8	2,8
6. Auffallende Charaktere	51	18,0
7. Selbstmord	4	1,4

FIGURE 12.3. One of Jenny Koller's key tables, emphasizing that while the "hereditary burden" (*erbliche Belastung*) of the mentally ill was only modestly higher than that for the healthy, much of the excess involved relatives with serious mental illness rather than minor nervous disorders. From Koller, "Beitrag," 280.

Unter den 218 erblich belasteten Gesunden sind belastet durch:

Vererbungsfactoren	Directe Erblichkeit	pCt.	Indirecte Erblichkeit	pCt.	Collaterale Erblichkeit	pCt.
1. Geisteskranheit . . .	22	10,1	26	11,9	9	4,1
2. Nervenkrankheit . .	24	11	10	4,6	4	1,8
3. Potatio	31	14,2	23	10,5	1	0,5
4. Apoplexie	17	7,8	18	8,2	1	0,5
5. Dementia senilis . .	1	0,5	8	3,6	1	0,5
6. Auffallende Charaktere	7	3,2	7	3,2	2	0,9
7. Selbstmord	2	0,9	4	1,9	—	—
Summe der Vererbungsfactoren = 218	104	47,7	96	44,0	18	8,3

(Die zweite Tabelle siehe umseitig.)

FIGURE 12.4. A more detailed table, illustrating the tendency of mentally ill asylum patients to show direct rather than indirect hereditary influence, often from close relatives such as parents, and to have relatives with real mental illness rather than nervous diseases and the like. A companion table, omitted here, showed the hereditary burden of the healthy to be often indirect and to involve relatives with conditions other than mental illness. From Koller, "Beitrag," 281.

These results spoke to an ongoing debate between polymorphic heredity, in which offspring often depart from the parental form, and similar (*gleichartig*) heredity, implying transmission to children of the traits of the ancestors. French degenerationists backed a polymorphic view, as did Freud, who in 1896 published a paper in French comparing heredity to a multiplier in an electric circuit, which magnifies the deviation of the needle but cannot determine its direction. "Assuredly," he continued, our opinion on this issue must be settled by "an impartial statistical examination."[16] Koller's statistics supporting similar heredity over general psychopathic inheritance made a strong impression.[17]

When, in 1902, Julius Wagner-Jauregg chose a theme to illustrate the importance of new scientific understandings of hereditary burden in his inaugural lecture at the second psychiatric clinic in Vienna, Gregor Mendel was not yet in the picture. In 1906, in a new inaugural lecture for the first Viennese psychiatric clinic, it was the same. The breakthrough that most impressed was based on Koller's statistics, supplemented in the second lecture by Diem's continuation with improved and expanded data.[18]

Diem, who also had studied at Burghölzli, took up practice not far from Zurich, in Herisau. While endorsing the need for large numbers, he selected his patients and comparison groups to maximize comparability. From Wilhelm Jung and Tigges he had learned to define statistical groups as precisely as possible, the key, he held, to a proper calculation of chances. He had no truck with Morel's "unified psychopathic disposition."[19] His ally Wagner-Jauregg declared that Morel's "law of transformation" violated the very concept of law. Koller and Diem had at last shown how to distinguish real effects of heredity from illusory ones. This new knowledge took shape on tabular forms bearing the standard German census categories. Their tables showed a weak link between mental illness and striking character, and none at all with nerve disease, apoplexy, or suicide. The grossest measure of heredity, the percent hereditary, scarcely distinguished the mentally ill from the healthy in Diem's tables. Wagner-Jauregg even denied, for a brief moment, the need to discourage marriages of

	Hereditäre Belastung in Prozenten von	
	370 Geistes-gesunden	370 Geistes-kranken
I	59	77
II	28	57
III	25	53
IV	4·5	21

FIGURE 12.5. Table constructed by Wagner-Jauregg using Jenny Koller's data, designed to show that a modest disproportion between the "hereditary burden" of the mentally ill and that of the healthy increased greatly when he focused on hereditary influence of parents and grandparents rather than collateral relatives (line II) or on relatives with genuine mental illness rather than nerve disorders (line III), and especially in combination (line IV). From Wagner-Jauregg, "Erbliche Belastung," 1157.

the mentally ill, arguing, "The customary statistics of heredity is a narcotic, furnishing a deceptive satisfaction of the need for causality." Koller had spoken of forces of regeneration, while Diem announced that the sword of Damocles no longer hangs over the head of anyone with a psychical anomaly in their ancestry. It is sufficient, he concluded, to lead a sensible life, avoiding stimulants and alcohol.[20]

This moment of indifference to hereditary forces was fleeting. Diem explained that he had begun as a supporter of conventional hereditary doctrines and then, as his numbers multiplied, lost faith in the practical value of his work. When, at the end, he examined the data more closely, he recognized a strong hereditary influence of parents on their offspring. Rüdin, in a critical comment on his friend Diem's article, argued in the other direction: the hospital patients who made up most of the control group, though free of mental illness, were yet hereditarily burdened in a way that might account for the defects of their relatives. Wagner-Jauregg, however, insisted on hereditary specificity, illustrated by a little table of four lines abstracted from Koller's data. Line I showed 59 and 77, the percentage of individuals from her comparison and patient groups with any defective relatives. Line II applied to those whose hereditary burden derived from their parents, and Line III, to all those with mental illness in the family. These showed a doubling of mental illness in relation to the comparison group. The ratio in Line IV, for those having a mother or father with true mental illness, was almost 5:1.[21]

Family Research and Similar Heredity

The theory of hereditary degeneration was embodied in tables of family decline, novellas awaiting their novelist. Robert Sommer, author of a noted text on family research and heredity, remarked that the reciprocal influences of scientific and artistic work had never appeared so compellingly as in the theory of degeneration, while an unnamed British reviewer of Zola's *Le Docteur Pascal* in the *JMS* took the scientific content seriously enough to quibble about its imaginative dimension, which veered too far from strict accuracy.[22] There were of course German advocates of Morel's theories, including Griesinger and Krafft-Ebing. The Leipzig neurologist Paul Julius Möbius proposed an ontology of *Urschleim*, or protoplasm, the primordial source of nervous and mental diseases, which oozed into novels by Theodor Fontane. Kraepelin, at first a supporter of Morel, moved toward an emphasis on hereditary similarity after 1900, and many supposed that his disease classification required it. Among them was Paul Albrecht, a military physician at Treptow in Pomerania, who surveyed the arguments in 1912 and presented family statistics as allied to similar heredity.[23] By the turn of the century, many German psychiatrists were rejecting medical narratives of family degeneration as flawed reasoning from improper evidence. Clinical experience, wrote the Munich doctor Karl Grassmann, could not support French neurologist Joseph Dejerine's claim that inherited psychosis might manifest in the next generation as nerve illness. Despite a few validated exceptions, he concluded, modern research reveals similar inheritance to be more typical than degenerative. Dejerine described *dégénérescence* as a distinctive achievement of the "physicians of our country," and Germans were more and more inclined to agree.[24]

Degeneration assumed many forms and had many sources, including Esquirol's notion of a disease of civilization. Kraepelin's essay on the topic in 1903 was in this tradition. He backed up the theory with personal observations of low insanity levels in Java and Bashkir among people he called primitive.[25] Social observers were alarmed by the enfeebled appearance of urban factory workers and their children and the high failure rate in military physical examinations as well as the epidemic of madness. Confronting Morel's hereditary dynamic of ever earlier and more disabling illness, Weinberg explained in 1903 how apparent evidence for it could arise as a statistical artifact. Parents will necessarily have survived their childhood, while children who die young can never manifest diseases of adolescence and adulthood. Also, severe congenital conditions such as idiocy greatly inhibit parenthood. He had believed all along, he said, that Morel's theory was overblown, and now he knew why. "No one can in earnest comprehend facts readily explicable in mathematical terms as a proof of degeneration."[26] German and French understandings of heredity were increasingly distinct. A standard German introductory text on the theory of inheritance in 1913 referred almost exclusively to German and English writings.[27]

The family pedigrees that proliferated from about 1900 in Germany, Scandinavia, Britain, and North America had little in common with Morel's tables of family decay. The German historian Ottokar Lorenz, who did much to launch this form of research, wanted to replace the family tree, or *Stammbaum*, which typically followed a descending line of succession, by the ancestry table, or *Ahnentafel*, which ascended from an individual of interest, the propositus (in German, *Proband*) to the two parents, four grandparents, and so on. Although he favored work on royal and noble families for which there might be centuries of data, his vision extended to family diseases, inspiring a wave of pedigrees of mental heredity.[28] Wilhelm Strohmeyer, a Jena-based specialist in mental abnormalities of children and a vigorous advocate of family research, saw no way that medicine could construct reliable family genealogies even for three generations. Rejecting "mass statistics," he called instead for *Individualstatistik*, records of mental disease compiled by doctors on the families of their patients. In a 1901 lecture, he reiterated Legrand du Saulle's complaint of discrepant measures and appealed for reliable medical data.[29] By 1904 he had encountered Wagner-Jauregg's report on Koller's critique of these percentage measures. With proper pedigree charts based on scrupulous medical classification, he now argued, many apparent cases of polymorphic psychopathology would be shown to exhibit "an exquisitely homogeneous hereditary tendency."[30]

Strohmeyer's confidence presumed that there was something enduring to be inherited, some stability of disease forms. Especially in Germany, such faith reflected a growing confidence in Kraepelin's new diagnostic categories. Grassmann had spoken of reliable classification as a prerequisite for hereditary research, and in 1901 an Alsatian doctor flipped the argument over, invoking patterns of hereditary transmission in support of newly identified disease forms. A medical dissertation at the University of Geneva in 1903 cited all of these authors to challenge "French" degeneration theories, which could not stand up to the empirical finding that dementia praecox and manic-depressive illness were generally stable in transmission.[31]

Sites of Centralized Data

Especially in Germany, the last prewar years brought a frenzy of projects for centralized data collection. Heron had appealed in 1907 for "a General Register of the Insane for preservation in the office of the English Lunacy Commissioners."[32] Sommer in 1910 called for a psychiatric division within the Imperial Health Office. He began: "The International Congress for the Care of the Mentally Ill, which took place in Berlin at the beginning of October 1910, showed with great clarity that psychiatry has become a social science." Clinical psychiatry, itself a branch of social medicine, had turned research on causes of insanity into a social study, and psychical hygiene was expanding into eugenics, the care of the *Anlagen* of a whole people. He proposed to split the division into four sections, including one for statistics and institutions, a second for heredity and psychical hygiene, and a third for methods of research on causes. Writing to support the proposal, Alois Alzheimer, a member of Kraepelin's clinic in Munich, emphasized the need for a central office to direct statistical research on causes. The Imperial Health Office supported these sections, rejecting only a fourth, clinical one. Sommer weakly defended it as necessary to keep the other three in touch with real processes of illness.[33]

Strohmeyer declared in 1913 that knowledge of heredity had advanced sufficiently to enable medical counselors to promise healthy offspring with high probability. He had been won over to psychiatric Mendelism by Davenport's figures for epilepsy and Rosanoff's for neuropathic constitution. He now discerned Mendelian ratios in Dahl's pedigree tables, recently reprinted in the catalogue of the 1911 International Hygiene Exhibition in Dresden from Pearson and Heron's *Treasury of Human Inheritance.* Strohmeyer described Mendel's success in explaining dormancy as one more nail in the coffin of "number-crazed statisticians of heredity."[34] But rules, he added, are worth nothing without empirical numbers. The road to hereditary health would be paved with superabundant data from asylums, church books, and registry offices as well as from private medical practice. He called for personalized, printed cards on every kind of mental illness, cards combining kinship tables and pedigree tables with tables of descendants, to be amassed in central bureaus for family research. Since there was so much data around already, the need was to organize it. "From the cradle to the bier, we have records for every citizen on their birth, their baptism, their vaccination, their school and military duty, their place of residence, their income, their insurance payments for sickness and invalidism, their success in examinations, their prior convictions, their causes of death." Now, medicine could move on to the most vital task of all, the certification of fitness to participate in reproducing the race.[35]

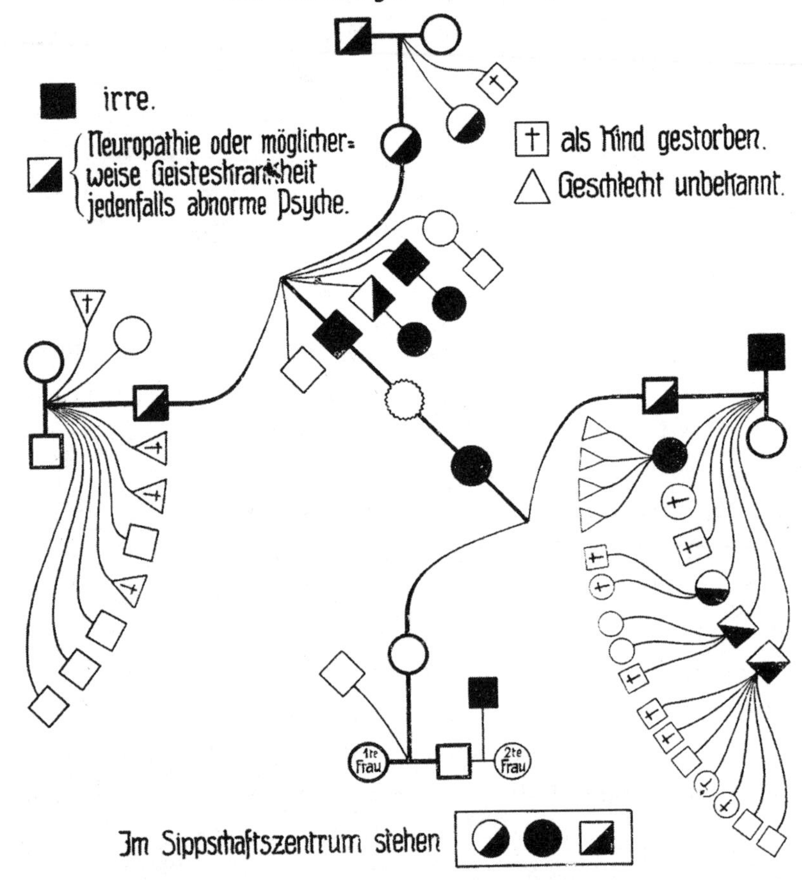

FIGURE 12.6. This 1908 chart of inheritance of mental illness was featured in the 1911 Dresden Hygiene Exhibition and in the catalogue by Max von Gruber and Ernst Rüdin, *Fortpflanzung, Vererbung, Rassenhygiene*, 84. Males are represented by squares in this diagram, females by circles. The fully darkened squares and circles signify two Mendelian factors for madness (*Irre*), while the half-darkened ones signify just one factor. The chart was drawn by Arthur Crzellitzer, a Jewish ophthalmologist, from data in a paper by Strohmeyer. The catalog explains (81–82) that this *Anlage* was not completely recessive, and that a single *Anlage* might appear as a neuropathic abnormality. The image here is from a 1913 reprint in Rüdin's "Einige Wege," 532.

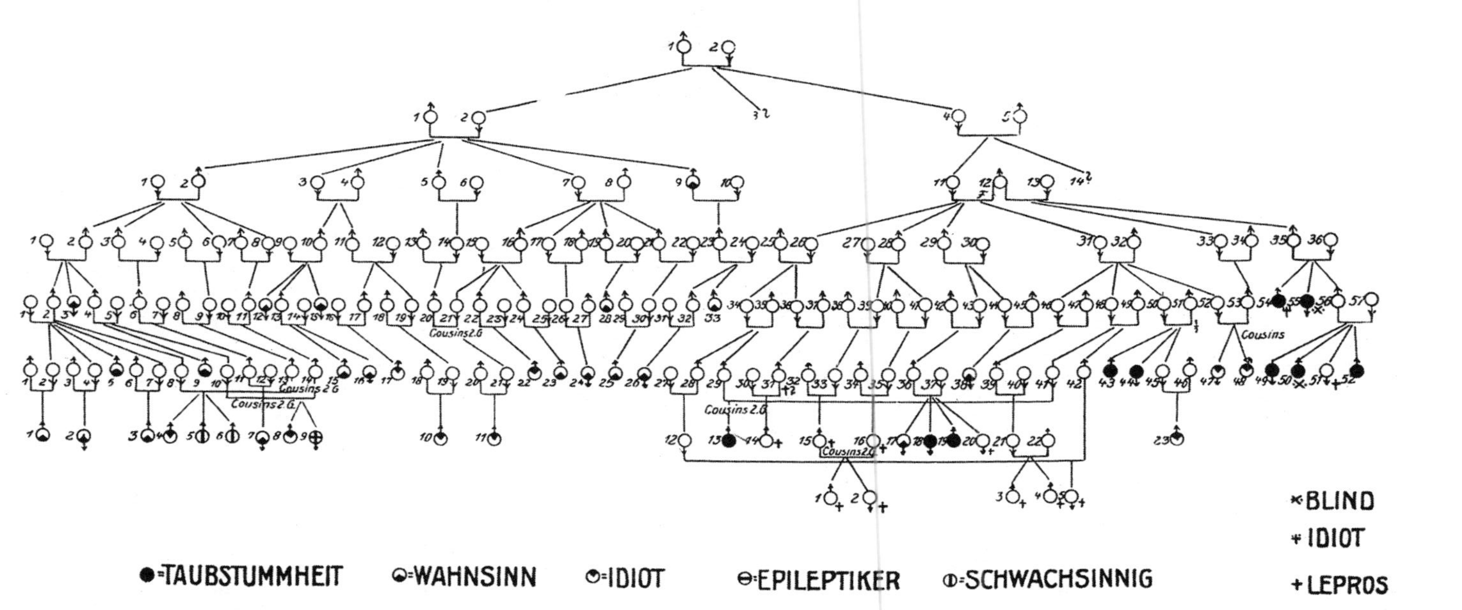

FIGURE 12.7. The 1911 Dresden Exhibition catalog, *Fortpflanzung, Vererbung, Rassenhygiene*, also included this image from Karl Pearson and David Heron's *Treasure of Human Inheritance*, plate 10, figures 59 and 63. They had assembled it from two figures in Ludvig Dahl's *Bidrag* (1859), tables 4 and 5. Dahl, like Pearson and Heron, understood this table as illustrating the diverse defects to be found in a single human kin group. The Dresden catalogue called the figure a table of deaf-mutism among blood relatives, lending implicit support to a Mendelian explanation of the inheritance pattern. This copy is from the larger-format image in Gruber and Rüdin's reprinted exhibition catalogue, *Fortpflanzung, Vererbung, Rassenhygiene*, 2nd, larger-format ed., still 1911, 81.

Another ambitious advocate of Mendelian family research was the Alsatian asylum doctor Ernst Wittermann. He anticipated a Renaissance of hereditary research, and then of humanity itself. While medicine, until recently, had struggled to limit the increase of costly mental defects, it was now becoming possible to direct the forces that reshape hereditary *Anlagen* for the mind. He cautiously endorsed Rosanoff's argument for a complex hierarchy of dominance in the Mendelian transmission of neuropathy. It was necessary now to go beyond mere data accumulation and to explain biologically the conditions leading to the expression of this hereditary factor. Such research should open a whole new perspective on regeneration and racial hygiene, he argued. It required, however, abundant and exact observational materials, data whose analysis would depend on real mathematical expertise.[36]

Was Wittermann a member of this mathematical elect? He proceeded in the usual way of data-intensive medicine, providing case material from 81 kin groups with a total of 2,660 members and giving case descriptions for the 169 of these, whom he diagnosed as mentally ill, including 96 psychopaths and 87 drinkers. The data, he concluded, confirmed Rosanoff's discovery that neuropathy was a Mendelian recessive. Strohmeyer had given reasons why Mendel's rules could not be quantitatively exact in psychiatry, including an absence of fixed traits, the inheritance of predisposition rather than of the disease itself, the effects of germ damage, and the uncertainties arising from the deaths of young persons before they can manifest mental illness. Wittermann countered by invoking those excellent numbers from American research and Hermann Lundborg's massively detailed study of a small district in Sweden.

All these physicians sought justification from correct (Mendelian) ratios, and so did Wittermann. He identified parents as heterozygous from disease states of their children and applied Weinberg's correction for family size. He summed up his case with printed tables giving the numbers of offspring with and without a pathological character for three combinations of recessive (R) and dominant (D) traits in the parents: RR × DD, RD × RD, and RR × RD. When the results were satisfactory, he calculated a percentage. He was

$$\frac{\Sigma x_1}{\Sigma p} = \frac{70}{186} = \frac{24}{100}\,!\,!$$

$$\frac{\Sigma x_2}{\Sigma p} = \frac{89}{286} = \frac{31}{100}\,!\,!$$

$$\frac{\Sigma x_3}{\Sigma p} = \frac{101}{286}$$

$$\frac{\Sigma x_1(x_1 - 1)}{\Sigma x_1(p - 1)} = \frac{24}{293}$$

$$\frac{\Sigma x_2(x_2 - 1)}{\Sigma x_2(p - 1)} = \frac{106}{402} = \frac{26}{100}\,!\,!$$

$$\frac{\Sigma x_3(x_3 - 1)}{\Sigma x_3(p - 1)} = \frac{160}{460}\,.$$

Und aus Tabelle VII:

$$\frac{\Sigma x_1}{p} = \frac{11}{38} = \frac{29}{100}$$

FIGURE 12.8. Ernst Wittermann's data for the proportions of diseased among the offspring of (presumed) heterozygotes. He inserted an enthusiastic "!!" whenever the numbers appeared to manifest a basic Mendelian ratio. There is a misprint in a denominator in the first line, a 186 that should be 286. From "Psychiatrische Familienforschungen," 262.

delighted to get proportions of 31% and 26% when each parent had just one *Anlage* for dementia praecox. He indicated his satisfaction with a double exclamation point and did not withhold it from a differently defined figure at the top of the table that need not have manifested the golden 25% but did. He reached the conclusion of Davenport's dreams, "that the *Anlage* for dementia praecox is a recessive trait in the sense of Mendel's rules."[37]

Weinberg and the Data of Württemberg

In contrast to psychiatrists who identified genetics with family research, Weinberg insisted that Mendelian genetics required the subordination of pedigrees to statistics. He showed how data are biased by a focus on families displaying multiple defects. Lorenz's reliance on tables of ancestors made things still worse. We all descend from parents and grandparents, but those who have no children can never be ancestors and are thus made invisible by tables of this kind. Weinberg, a canny statistician, was the outstanding medical Mendelian of his day.[38] The child of a Jewish father and a Lutheran mother, he was born in 1862 on Christmas Day and baptized into his mother's church. He married a Lutheran wife, who bore five Christian children. By the time he died in November 1937, ancestry was all that counted in Germany, though a biographical essay on which his son consulted suggests that he faced only a little overt antisemitism. Rüdin, who had depended on Weinberg's statistical expertise, marked his passing with one admiring sentence. A laudatory obituary in the *AZP* by Hans Luxenburger, a pioneer of empirical prognosis, concluded: "Psychiatric hereditary research, especially, is indebted to him for the best tools of method that it possesses." But Luxenburger was uncomfortable with Weinberg, describing him as isolated, resentful, and intolerant of statistical shortcomings in others. Franz Kallmann, who escaped to America, remarked that Weinberg ran the risk "of seeming to physicians an excellent mathematician who handled mysterious figures as an artist, and of appearing to mathematicians as a physician who devoted his private life to the study of mathematical problems." He was also an outsider in hereditary research, earning his living from public health work rather than in an elite psychiatric clinic or institute.[39]

Yet Weinberg was a central figure across a range of fields, including medical statistics, psychiatric genetics, and anthropology. His achievements had to be forgotten from about 1936 to 1945, and in the aftermath it was unclear how he should be remembered. A moral disgrace has justly descended on German eugenic and genetic projects in which he participated for a time. He was a founding member of the German Society for Racial Hygiene, and he served for many years as head of its Stuttgart branch. However, he declined to endorse sterilization as effective, appropriate, or necessary to control the propagation of mental illness. As a not-quite-innocent bystander, he plays only a supporting role in histories of Nazi eugenics, though few have wanted to treat him as merely an

FIGURE 12.9. Portrait of Wilhelm Weinberg (1862–1937). Weinberg mainly earned his living as a doctor for the poor while compiling abundant medical and vital statistics for the town of Stuttgart. By 1908, he had acquired impressive skills in medical and hereditary statistics, enabling him to advise Rüdin and others on the methods of Mendelian experimentation on humans. Photograph reproduced from C. Stern, "Wilhelm Weinberg," *Genetics* 47 (1962), 1–5, with permission of the Genetics Society of America.

accomplished scientist.[40] He acquired the fame of eponymy for a fundamental but mathematically trivial result (G. H. Hardy spoke of "a little mathematics of the multiplication-table type") that was long attributed to Hardy alone. Due to the efforts of geneticist Curt Stern, who left Germany and achieved success in the United States, it became the Hardy-Weinberg equilibrium. Weinberg credited it to himself and Pearson.[41]

Data work seems to have provided the basis for Weinberg's quantitative skills. His 1886 dissertation used numbers to reason from cases in a typically medical way. After a few years of travel, he began compiling medical statistics for the city of Stuttgart, where he also delivered babies for over forty years. From 1892 to 1907, he edited and prepared tables of births, deaths, and marriages for the annual Stuttgart medical-statistical report. He was listed there as a physician to the poor. By 1896 he was working with records of the health of physicians, statistics of military recruitment, and insurance numbers, using Poisson's formula to estimate errors. He recognized by then that, due to self-selection, mortality and morbidity statistics of different occupations do not reliably indicate their unhealthfulness.[42]

As a keen connoisseur of data, Weinberg put great emphasis on the quality of medical statistics, not least of mental illness, in the kingdom of Württemberg. The archives there hold bundles of reports for a census of the insane in December 1832, organized by its medical collegium. Wilhelm Köstlin wrote up the results in 1840 as a medical dissertation, which drew on Quetelet's *Sur l'homme* and provided comparisons with numbers from Scotland, New York, France, Westphalia, and Norway. Köstlin inferred causes from tables, using such variables as state, religion, landform, and urban or rural but omitting personal factors such as heredity.[43] In 1852, the Württemberg Statistical-Topographical Bureau helped organize doctors to count the mentally ill and to assess their conditions of care. The results appeared under the authorship of P. Sick. A neat pile of booklets sent in from every district in 1864 includes tabular data, but no names, for every mentally ill individual in the kingdom.[44] About 1893, just as Weinberg assumed responsibility for health statistics in Stuttgart, the medical college of Württemberg commenced a series of annual volumes on the mad, feebleminded, and epileptic for the entire kingdom largely conforming to the Prussian categories.[45]

The most noteworthy census of the Württemberg insane, however, was carried out in 1878 under the direction of Julius Ludwig August Koch, director of the asylum in Zwiefalten. A reviewer in the *AZP* called it "extraordinarily diligent and careful," though too much obsessed with heredity. That obsession was a conscious choice, justified, Koch held, by the scientific importance of heredity and a need to remedy the neglect of his predecessors.[46] Diligent he certainly was. His book began with a historical register of every tally of the insane in German lands going back to 1804, then, less exhaustively, in the rest of the world, and finally in his native Württemberg. In describing his methods, he emphasized all the precautions taken to insure completeness. He enlisted Protestant, Catholic, and Jewish religious leaders to survey the mentally ill in every town and district. The result was 7,953 mentally ill, about equally divided between idiots and mad, in a population of 1,881,505. This gave Württemberg a ratio of 1:236.58, second only to the Swiss canton of Bern. Koch doubted that his kingdom was peculiarly vulnerable to insanity, explaining it instead as the result of his own meticulous planning. He tried to correlate numbers by district with terrestrial and atmospheric factors then warned against spurious trends. It is said that the mad have a longer average life than the healthy, he remarked, but their mean age of death is raised artificially because madness becomes visible only in adulthood.[47]

Koch laid out his hereditary data in a massive table with rows labeled (for example) "epileptic without mental weakness or mental illness," "mentally ill suicide in last 24 months," and "ambiguous mental condition," and columns for different categories of relatives. He declined to draw conclusions based on illness of distant relatives but narrowed his focus to parents and grandparents plus uncles and aunts. Whenever possible he used written asylum records as a check on oral testimonies. His clerical enumerators tended to downplay heredity, he thought, so as not to show their parishes in a bad light and to avoid any hint of materialism.[48] Koch also recognized the fallacies that can arise from the dormancy of the hereditary *Anlage*, which is more likely to be expressed in a large family. For this reason, France, with its low birth rate, may actually have a higher frequency of such factors than other nations despite its favorable statistics.[49] Weinberg is often credited with the recognition of "ascertainment bias" in genetics, for example, the greater visibility of recessive genes in large families. We see that something very similar was apparent to Koch on the basis of asylum data more than two decades before Mendelism.

Weinberg's close attention to local statistics was balanced by his study of Jung, Hagen, Tigges, and Koller. Like Hagen, he was responsible for reviews of English-language publications on statistics and heredity during his most creative period. He looked to modern demographic and statistical conceptions as a basis for genetic analysis in medicine, since real experimentation was not possible. No one before him had been so alert to the fallacies that may arise from thoughtless calculation, especially when there was no proper basis for comparison. When Paul Mayet, in an influential lecture on family marriage, invoked statistics to argue that the *Anlage* for mental illness must sometimes be beneficial, Weinberg countered that this could be an effect of class, since cousin marriages are more common among the prosperous, who also have more complete knowledge of family *Anlagen*. From life insurance work he built up expertise on age effects. He stressed that any conclusions about a condition arising in adulthood, be it marriage, occupational choice, or mental illness, depends on proper age standardization.[50]

Weinberg's focus on heredity did not commence with asylum studies, but with other public-health work. His obituaries all mention the 3,500 births he attended, including 120 or more twin births, for which he kept meticulous records. He calculated the percentage of monozygotic or identical twins from the excess of same-sex twins over boy-girl pairs. He also inferred that mothers who bore nonidentical twins were about three times more likely in subsequent births to have twins again. Having determined that women of German origin have more twins than Latin ones, he inferred that twin birth cannot be a sign of degeneration. However, the very existence of racial difference implied that this ratio is inherited. It was his first topic of Mendelian investigation, providing the stimulus for his algebra of genetic equilibrium that brought him posthumous renown. In context, the purpose of this formula was to show that his data on the twinning tendency were in better accord with expectation for a recessive trait than for a neutral or dominant one.[51]

Weinberg's other early focus of hereditary study was tuberculosis, on which he wrote a report in 1906 for the Stuttgart Medical Association. It was a natural role for this long-term custodian of medical statistics, and he had already touched on the topic in 1903. Although the disease was infectious, there might still be a hereditary predisposition. Ever mindful of the need for a comparison population, he counted up additional cases of tuberculosis in families that already had had one. This study resembled the twin investigation, but since an increase might be due to family environment, he spoke simply of "burden" (*Belastung*) without the adjective "hereditary." Although tuberculosis was a paramount concern of life insurance institutions, he relied mainly on a vast file of data cards from thirty years of Stuttgart death certificates. When he compared the effect of a tuberculous parent on the child to that of a stepfather or stepmother, the difference was negligible. The effects of heredity, he concluded, must be minimal.[52]

Weinberg vaunted his city and kingdom as treasure troves of data for inquiries like this. He worked tirelessly to expand and improve its data reserves, and not only for medical records. In a paper for the statistical journal of Württemberg, he praised the family registers of Württemberg, Baden, and nearby Swiss cantons. Beyond their genealogical uses, these registers enabled researchers to extract from a single source information on family origins, ancestors, marriages, divorces, remarriages, marriage partners, residences, and births, including subsequent events in the lives of children. In 1913, he applied the Stuttgart registers to a task endorsed by the German Society of Racial Hygiene to assess family causes of tuberculosis, this time by "turning off" the hereditary element and comparing the effect of a tuberculous parent on his or her children with that on the spouse.[53] While many of these inquiries were eugenic, he consistently took as much interest in environmental as in genetic causes. In 1907, he called for an alliance of medicine and social science. He dreamed of a comprehensive database in which, along with medical records by physicians, "anthropological, criminal-statistical, recruitment-statistical, school-statistical, and psychiatric data would be collected with the results of mandatory medical reports in a scientific central office." Zola's *Docteur Pascal* would have been humbled.[54]

Data Divides

His most productive period as geneticist extended from 1908 through the First World War. In 1908, he was authorized to review a study by Pearson on the statistics of pulmonary tuberculosis for the *Archiv für Rassen- und Gesellschafts-Biologie* (*ARGB*). "Racial and social biology" in the title reflects the bold ambitions of this journal, whose concern for racial purity grew out of Swiss and German temperance movements. Founded in 1904, it was the first journal of eugenics in the world, though at first it included much general biology. Weinberg criticized Pearson, not for his strong eugenic commitments as such, but for calculating correlations in a way that obscured the social factor. Pearson, he wrote, had failed to notice a petitio principii: that the family correlation already includes environment, since *minderwertig* (inferior) persons usually live in inferior conditions. Yet he recognized biometry as an English science, and "Pearson, one of our first biometricians," as virtually without peer. The *Archiv* had included in its first issue an approving review of one of his papers on the inheritance of mental and moral characters. A few months after Weinberg's review, it published in German translation Pearson's 1907 Boyle Lecture at Oxford, "The Scope and Importance to the State of the Science of National Eugenics."[55]

Pearson, who blew hot and cold on German science, was not gratified by Weinberg's comments, nor by those of other reviewers in this journal. Weinberg repeatedly complained of the English statistician's seeming failure to correct for data interactions. Pearson had exaggerated the dysgenic role of the mentally ill by neglecting their low fertility. He had found raised levels of mental illness, tuberculosis, and crime in first-born children only because they were older on average and hence more likely to have reached the age at which people commit crimes or go mad. He had mistakenly compared the correlation of tuberculosis between spouses with the correlation of parents to children as if blood relationship were the only variable besides environment that could matter.[56]

Or did he? Pearson refused to concede that he had anything to learn from his critical but somewhat intimidated German follower. Having responded angrily in 1911 to a critique of his (skeptical) views on the ties of alcohol to degeneration by the journal's most prolific reviewer, Rudolf Allers, Pearson wrote again in 1912, now in basic German, to protest Weinberg's misconstrual of his research on the tubercular diathesis. "Why should the theme of German critics be always the same, namely that the English biometrician either neglects what everyone knows or writes tendentiously from beginning to end?" No raw mortality number ever gets into my work, he continued, but all are adjusted to a standard population age distribution. Pearson could never commit such an error, chipped in two recent associates in his lab. Curiously, Weinberg's charge of selection bias against him was just the accusation Pearson had recently leveled against Mott's degenerationist theory of "ante-dating" (see chapter 10).[57] In 1909, Pearson accused Weinberg of a related error: using an inappropriate comparison group. Weinberg immediately acknowledged the problem in a letter, adding that no better comparison group was available and that his paper had explained this. As an admiring student of Pearson's statistics, Weinberg recognized that Pearson was as well qualified as anyone to deal with such complications. Still, he was not satisfied with vague assurances. He felt the sting of Pearson's aloofness even more than that of his criticism. With this one exception, Pearson ignored his papers, and in their printed exchanges he referred to Weinberg as *Referent* rather than by name. In a second letter in 1909, Weinberg asked Pearson to report on his work on heredity of twin births for *Biometrika*. He never got a reply.[58]

Mendelism may have had something to do with their bad relations. Weinberg described Mendel's findings in 1908 as enticing in their simplicity but difficult to reconcile with medical experience of inheritance. In contrast to the results of plant breeding, he wrote, Bateson had found only a few discrete dominant or recessive traits in humans, and these were rare. There could be no doubt, Weinberg explained, that the principles of heredity in plants and animals are the same for humans, but experimental investigation of the latter was impossible. Since so few human traits seemed to be transmitted intact from parents to offspring, he had no alternative but to proceed with statistics. His interest in Mendel after 1907 did not require him to reject his earlier methods, and he sometimes treated the *Anlage* as equivalent to a Mendelian unit. He hoped to trace the transmission of *Anlagen* across generations, on the supposition that population variability may reduce to a mixing of "pure lines." (Within a pure line, as the Danish breeder and geneticist Wilhelm Johannsen explained, there is no heritable variation.)[59] In 1908, Weinberg criticized Pearson for rejecting so hastily the possibility of Mendelian inheritance of the tubercular *Anlage* (Pearson called it a diathesis). The 3:1 genetic ratio might simply be disguised by other causes, Weinberg supposed. Neither Pearson nor his critics was yet accustomed to discussing Mendelian ratios in this complex way. In 1912, Weinberg called attention to the deformities pictured in Rischbieth's section of the *Treasury of Human Inheritance* as possibly a dihybrid (two-factor) recessive.[60]

Weinberg's reports for the *ARGB* on the Mendelian discoveries of Davenport, Rosanoff, and Goddard were hesitant but hopeful. The Munich psychiatrists to whom he would soon lend his statistical expertise were more severe, accusing the Americans of insufficient attention to medical proprieties. Rüdin, though an instant convert to the ERO's Mendelian ratios, complained of insufficient diagnostic skill. His own genealogical researches, he told Davenport in a letter, always began with a patient in his own clinic whom he could properly examine. He said of Goddard's first paper on inheritance of feeblemindedness: "It is an extraordinary shame that American works like the one before me, undertaken with so much diligence and such ample philanthropic means, are so often marred by defects on the most important scientific points." What possible use are pedigree tables incorporating so many meaningless categories, for hysterical, St. Vitus Dance, and so on, but no explanation of how these labels are applied? He wanted to understand the basis for diagnosis. He charged the eugenic fieldworkers with contributing little more than colorful descriptions. Only doctors possessed the competence to identify disease entities that might be inherited. He judged David Heron's pedigrees at the Galton Eugenics Laboratory less severely as flawed yet still valuable as evidence of a general hereditary burden.[61]

Rüdin's associate Allers, like Jelliffe, found more promise for psychiatry in English biometry than in American Mendelism. How could neuropathic make-up be a Mendelian recessive when it failed so badly as a medical or biological category? "The American hereditary researchers can hardly be spared the charge of prematurely applying concepts to practice, as if they must correspond directly to biological facts," he complained.[62] Weinberg, who reviewed materials from both sides of Davenport's dispute with Pearson and Heron, supported the search for Mendelian factors but condemned the ERO's handling of data. Rosanoff and Orr's 3:1 ratio looked promising at first, he explained, but it could not possibly be meaningful. The researchers never even mentioned age, though the numbers must evolve as patients grow older. The Americans wavered on whether neuropathy was homozygous or heterozygous and failed to justify the classification of epileptics as feebleminded. The processing of numbers was so slipshod that when Allers recalculated from their pedigree charts, he found 81% where they had 100% and 15.9% instead of 50%.[63]

Weinberg was scathing on American techniques of observation, especially the use of female assistants (*weibliche Hilfskräfte*). These fieldworkers, he complained, seemed to conspire with their subjects to generate noteworthy results that would satisfy the researchers. The arrival of David Heron's damning critique, he continued, reinforced doubts about the data that he had already expressed in the first part of his review. He now endorsed Heron's point that the fieldworkers were given improper instructions, being told, for example, not even to record ancestors who failed to display dominant traits, but to keep searching for recessive ones in collateral relatives if they could not be detected in the parents. Weinberg called it a betrayal of objectivity.[64] He also criticized Davenport's study of racial mixing in Jamaica. Here he chose to lay responsibility directly on the fieldworkers, even indulging in some misogyny. "These labors with *fieldworkers* generally remind us of the types of ants that are fed by others and must be hungry or die when they obtain no food or solely the wrong kind. And how easily these ladies come to see their meal as more important than the ideas of their leaders. At a minimum, the *fieldworkers* must understand that they will be constantly watched." Yet he somehow accepted their evidence of the inherited discontinuity of skin color. "In this way the stronghold of *blending inheritance* has finally fallen."[65]

Weinberg seemed often to care less than Davenport about finding those elemental Mendelian ratios. "The mere fixing of numerical proportions cannot be conclusive," he once declared. Ratios were interesting, but the true basis of Mendelism was the law of segregation, the division of offspring into discrete forms. He concluded his assessment of the American work on psychiatric heredity with a comment on Wittermann's paper, just then being corrected according to Weinberg's methodological suggestions. It seemed to be converging on a correct Mendelian ratio for the general psychopathic *Anlage*. That sounds like Rosanoff's much-ridiculed "neuropathic make-up," so glaringly ill-suited to Kraepelinian psychiatry. Weinberg might insist on the recognition of complex inheritance, but he, too, clearly longed for some clear-cut Mendel numbers. We can detect this also in his reaction to a 1915 Swiss dissertation on hereditary similarity in psychiatry, whose author, William Boven, referenced him on the pointlessness of chasing after Mendelian proportions. Weinberg, as reviewer, noted with satisfaction Boven's application of his research methods but denied that such ratios could never succeed in psychiatry.

Why were they so hard to find? He was as disappointed as Rüdin when the massive study he had helped to design gave figures for inheritance of dementia praecox at least five times too small. Perhaps it involved two recessive genes, or even two recessives and a dominant.[66]

Managing Heredity

Although Mendelian genetics was well established in the laboratory by 1914, its application to vital issues of human heredity now appeared problematical to leading German researchers. Inheritance of mental illness and feeblemindedness, so central to the eugenics movement, could not be reduced to Mendelian explanation. Weinberg and Rüdin had already accepted large-scale studies as the only way forward. As Rüdin wrote in 1916, "Every family tree is only a particular instance in the dice game of heredity and proves, by itself, nothing." Indeed, the most diseased family trees, because they were unrepresentative, led necessarily to false conclusions.[67] The medical ideal of accumulating "observations" as a basis for generalization was now unacceptable to elite hereditary researchers. As Allers remarked in his review of a book-length paper by Phillip Jolly, a psychiatric assistant in Halle, a hundred deeply researched families would not be sufficient, even if his method of purposive selection were acceptable. It was not.[68]

Rüdin designed his large-scale test of the Mendelian theory of dementia praecox according to Weinberg's precepts. He focused, as Davenport often did, on siblings of affected persons whose parents were both normal. On the hypothesis that dementia praecox was a Mendelian recessive, each parent had to have one *Anlage* for the disease in order to give birth to an ill child, but none could have two, or they would themselves be affected. Excluding the diseased propositus to avoid selection bias, he hoped for numbers approaching 25% and 50%. Instead, they came in as 4.48% and 6.12%, or 4.12% and 10.3%. Although Rüdin and Weinberg had supposed from the beginning that other factors were involved, these were devastating results.

Not long after completion of the study, a few days before Christmas 1914, Weinberg began his own investigation of the "family burden of mental illness," using data to be supplied by Württemberg mental hospitals and supported by a grant from the Württemberg Medical Council. For the kingdom to launch this vast project at such a time demonstrates a remarkable devotion to hereditary statistics. The government, in fact, was not thinking of better knowledge someday in the future, but of more efficient health policies now. Upon its completion in 1919, Weinberg would claim that his findings were exemplary for all of Germany and could justify depriving criminals and the mentally ill of their voting rights. In March 1915, when the survey was set in motion, he was preparing for a rapid (scientific) advance at the conclusion of a short war. "While it cannot be expected that the work will progress briskly during the war, it is still possible to make ready the plan of organization and to begin making records" so that the case histories might begin as soon as peace returned. Would it be possible for Württemberg asylums to send their recent *Zählkarte* to Stuttgart for extracts, along with a list so he could avoid duplicate records?[69]

The institutions responded quickly but disappointingly. The hospital in St. Vincent had never introduced these *Zählkarte*, and it lacked the personnel to do the work requested. It did have extracts, which were kept by a former physician, on the hereditary burden of the dementia praecox group. At Weissenau there were records but no personnel to spare. Zwiefalten had full records only since 1914 and worried about confidentiality if the cards were entrusted to patients for copying, but it was willing to send what it had to Stuttgart. And so on. After receiving these poorly standardized documents, Weinberg began ruminating on the advantages of a single location in Stuttgart to store them. Officialdom evidently favored this centralizing plan. In March 1916, a Dr. Camerer commented that the records were needed for practical management of the mentally ill and not merely for research purposes. Camerer was thinking of the newly approved *Zählkarte* that Weinberg had drawn up, whose section on heredity has a familiar look.

Erblichkeit							
	Geisteskrank	Nervenkrank	Trunksucht	Selbstmord	Abnorm. Charakter	Verbrechen	Unbekannt
Vater							
Mutter							
Geschwister							
Halbgeschwister							
Verwandte des Vaters							
„ der Mutter							

FIGURE 12.10. Wilhelm Weinberg drew up this simplified census card for his hereditary survey of asylum patients in Württemberg during World War I, after Ernst Rüdin's tallies of inheritance of schizophrenia yielded results that diverged unacceptably from Mendelian expectations. From Staatsarchiv Ludwigsburg, E 163 Bü 105, Bl. 92, reproduced with permission.

In mid-1916, this mishmash of prior asylum records was still circulating. By then, shortages were growing more severe, and it was hard to find workers. A letter from Weissenau in November proposed that Dr. Weinberg could come there himself if he wanted to consult patient records. In March 1917, the asylum reported having found three women to do the work for 2.50 marks per day and requested funds to pay them. The Winnenthal institution reported in July 1917 that it could not keep women to copy records, since they did not know in the morning if they would have enough bread, potatoes, and milk for the day. And yet the records Weinberg accumulated appeared valuable enough for Rüdin to insist on copies for use in Munich.[70]

By mid-1919, according to Weinberg's own account, he had completed a large work on the foundations of hereditary statistics, which he showed to Rüdin and to the geneticist Max von Grüber. But printing, like bread and potatoes, was in limited supply, and publication of his manuscript was delayed, indefinitely. The next year he published three journal articles of modest length: two on statistics of dementia praecox and one on manic-depressive insanity. They aimed to account for Rüdin's "surprisingly low" figure for inherited dementia praecox through application of Weinberg's sibling method. The papers integrated chromosome work and fruit fly results as well as asylum statistics. They seem to depend on ad hoc explanations. There is, he said, more than one way to match data to theory. While he did not seek justification by Mendel numbers alone, he deployed there a more subtle form of calculation in hope of redeeming somehow these rules of heredity that failed to predict.[71]

His remedies fall into four broad categories. One was to suppose that the genetics involved *Polymerie*, or multiple *Anlagen*. He hoped the chromosome structure might provide a clue. A second, related possibility was to interpret mental illness as sex-linked. This seemed especially promising for manic-depressive insanity, which was diagnosed more often for women than for men, even if their suicide rate was much lower. He also tried out the possibility of assigning women a second, dominant *Anlage*, for this illness.[72] A third idea was to consider the influence of external circumstances, the interactions of heredity and environment. He knew well that some hereditary factors were manifested only in particular environments. Although this relationship could be complex, he formulated it simply as a parameter y, the probability that a genetic tendency would be expressed. The geneticist Fritz Lenz had already tried out such a factor, whose value he estimated on the basis of Rüdin's numbers as 4.5/25. Weinberg, taking issues of ascertainment into account, suggested a lower figure, about 1/7.[73]

Finally, he was drawn to the possibility that the rate of insanity resulted mainly from cousin marriages. If so, the observed rate of mental illness would not require such a high percentage of *Anlagen* in the general population. This also would enhance the effectiveness of eugenic intervention. By 1920, many doctors and statisticians believed that inbreeding had an important role in the perpetuation of feeblemindedness but not of schizophrenia or manic-depressive illness. Weinberg's push for the wartime Württemberg survey was designed primarily to get data on cousin marriages among asylum patients. Such information should enable him to distinguish a recessive *Anlage*, which, if rare, required a raised level of inbreeding to be expressed, from a dominant one, which did not. With Rüdin's data, unfortunately, he couldn't tell. He mentioned a census project at the beginning of his paper on manic-depressive insanity and later commented on the inadequacy of existing data on marriage and disease. "More exact information can only be expected from a complete survey of the mentally ill and their marriages, as is now being produced in Württemberg." He also had hopes of measuring Lenz's parameter y rather than simply calculating a value that made the numbers work. This would require years of continuous collection from every insane asylum in Germany.[74]

While he was optimistic about redeeming Mendelian psychiatry as a basis for prediction, he worried that reliance on multiple *Anlagen* and on environmental interactions implied the blending of some traits rather than clean Mendelian segregation.[75] That was the direction taken, in practice if not in theory, by a group forming around Rüdin in Munich. Their new method relied on surveys and statistics not to identify *Anlagen* but to predict mental defect in offspring purely on the basis of the characteristics of their relatives. Weinberg's Mendelism, admitting so many exceptions to the simple quantitative rules of the classic pea experiments, now appeared unworkable for practical eugenic purposes.

CHAPTER 13

Psychiatric Geneticists Create Colossal Databases, Some with Horrifying Purposes, 1920–1939

"He was a very great scientist, but he made a terrible mistake. Not the way people said, not for any vulgar sordid reasons. To him it must have seemed perfectly logical. He was always as cool and . . . how do you say it? Objectionable?"

"Objective?" Ruth suggested.

"Objective, yes. Finding out the secrets of life was his obsession."

—Wallace Stegner, *The Spectator Bird* (1976)

"*Reading* it?" he said. "I never read the text, I don't care what they think, I only look at the tables."

—Lionel Penrose, as recalled by E. B. Robson (1998)

By now it is clear to scholars that German science survived and even flourished under the Nazis. While the mathematical level of British statistics in the 1930s, including quantitative genetics, was unmatched, German writings on inheritance patterns of mental illness and related conditions were admired internationally right through the Second World War and beyond. Such questions did not yield to Mendelian mathematics. Empirical hereditary prognosis arose from biometric work in Pearson's lab and from Weinberg's collaboration with Rüdin's groups in Munich and Basel. The tradition of hereditary investigation based on institutional data and population surveys persisted almost to the end of the 1930s as a cosmopolitan scientific endeavor. With the outbreak of war, Anglo-German scientific cooperation became impossible, but shared research persisted on both sides of the military divide. Science could not stand above politics, but neither did it collapse into politics. Meanwhile, the scale of research on what were widely supposed to be inherited mental traits expanded hugely. As relations of the science of human heredity to social and medical intervention grew increasingly consequential, boundaries between pure and applied knowledge became more and more difficult to sustain. Yet research on institutionalized populations, now in prisons as well as special schools and mental hospitals, remained as central as ever to genetic study of humans.

Contested Expertise and the Great German Experiment

Karl Pearson's pronouncement at his retirement dinner on 23 April 1934 has become notorious: "In Germany a vast experiment is at hand, and some of you may live to see the results. If it fails it will not be for want of enthusiasm, but rather because the Germans are only just starting the study of mathematical statistics in the modern sense." It is still often misunderstood to imply the continuity of his eugenic science with Nazi crimes.[1] He was referring specifically to the Nazi sterilization law. In contrast to various brilliant contemporaries, from George Bernard Shaw to R. A. Fisher, he refused to the end of his life to endorse forced eugenic sterilization, insisting that too little was known of its potential effects.

Just a few months earlier, Pearson had received a paper in German called "Mathematical Reflections on Racial Hygiene, especially Sterilization," for the *Annals of Eugenics*. The Leipzig-based author, Ewald Bodewig, held a PhD in philosophy for a thesis on the place of mathematics in Saint Thomas Aquinas. Bodewig calculated the population effects of eugenic sterilization on the assumption that the traits it targeted were caused by a single recessive gene (*Erbanlage*). His mathematical biology, if inelegant, was clear enough. He concluded that since the presumed factors must be rare and remain invisible unless an individual has two of them, such restrictions on breeding would accomplish little, perhaps not even enough to counteract the negative eugenic impact of war. Pearson published the paper with a footnote explaining that it held no novelty for English or American readers and that it incorporated false assumptions such as random mating. In the next issue, still dated 1933, Pearson inserted a "Memorandum on Dr. Bodewig's Paper" apologizing for printing such flawed and unoriginal work. His reason for doing so was less scientific than political. The new German sterilization law had come into effect on 1 January 1934, and Bodewig would not be allowed to publish it at home. Although knowledgeable English readers no longer imagined that insanity and mental deficiency could be inherited in a Mendelian way, the paper might have value as an answer to German supporters of this law.[2]

Upon Pearson's retirement, the editorship of his journal passed to Fisher, his bitter antagonist, who graciously offered Bodewig a page to respond. The assumption of causation by a single recessive factor, Bodewig explained, was the best-case scenario for backers of the new law. Multiple recessive factors would slow still further this weeding out. He could correct the calculation, he continued, only if he knew how many factors were involved. Fisher also offered a few pages to a scientific backer of the sterilization law, Siegfried Koller, to reply to Bodewig's "attack." He countered that Bodewig must be ignorant of the scientific literature, since in Germany, at least, calculations like his were well known. Certainly he did not accept the inferiority of German statistics to that of the English. Koller acknowledged that the new law was no panacea but merely the beginning of the great work of healing the nation. He included his own calculation for the elimination of a single-gene recessive mental defect. With nonselective mating, the presence of a recessive trait in 1% of the population implies that 10% of the genes are recessive. In that case, blocking reproduction by defective individuals would lower the rate of feeblemindedness by about 17% in the first generation. Like Fisher, who had published more or less the same calculation in 1924, and like Pearson, Koller believed that nonselective mating would allow the weeding out to proceed somewhat more rapidly.[3]

Koller added, however, that German geneticists were not taken in by naïve one-factor explanations and that empirical hereditary prognosis, as worked out by Ernst Rüdin and his school, was preferable for practical uses, since it gets around all theoretical unclarity regarding hereditary transmission and inbreeding. By the end of his brief rejoinder, he seemed to forget the modesty of his initial claims, promising within a century to reduce by one half the kind of mental weakness the law was designed to combat.[4] Koller, who had studied in Göttingen with Felix Bernstein, the pioneering expert on genetics of blood types, clearly did understand the algebra. The next year he calculated the decrease in successive generations of genetic defects on the basis of varying assumptions as to dominance relations, the number of factors involved, and the thoroughness of selection. In his 1949 textbook of human genetics, Curt Stern, by then a biology professor at the University of California in Berkeley, reproduced in graphical form Koller's results for selection in successive generations against a "rare genotype." Koller was to have a lead role in implementing Nazi racial laws. Bodewig, who was probably a Catholic dissident on sterilization, decamped to the Netherlands.[5]

By 1934, Fisher was more circumspect about Mendelian assumptions. In a letter to Weinberg, he dismissed the exchange as an unfortunate legacy from his editorial predecessor, adding that Bodewig had deserved a chance to respond to the points on which he felt ill-used by Pearson, but that he would publish nothing further. "So far as I can judge, Dr. Bodewig's paper will be taken no more seriously in this country than in Germany." Certainly it was old news for Fisher, but he had another reason to declare the discussion closed. Weinberg, as we will see, had just weighed in with a letter in English for publication, raising demographic and political issues that Fisher could not have considered appropriate for a scientific journal.[6]

Smooth Distributions

By 1910, T. H. Morgan had begun to accumulate genetic variants of the fruit fly *Drosophila* in his Columbia University laboratory. Soon, geneticists there would show how correlated inheritance could be used to tell which genes were on the same chromosome and to map their locations. Agricultural researchers learned to isolate the effects of individual genetic factors in the United States, where the federal government funded a research station in every state. Davenport's ambitious data initiative with his eugenics fieldworkers looked for a time like the most important Mendelian research program of all. It does not appear that Pearson regarded the breeding of *Drosophila* or hybrid corn as pertinent to the problems of human science that mattered most to him. By 1920, however, he had moved well beyond his agnostic view of Mendelism as a promising theory, spoiled by the flawed statistical practices of its advocates. In 1925, in the editorial introduction to *Annals of Eugenics*, he and Ethel Elderton laid out a vision of their journal as vital for the "study of man." They described eugenics as "the supreme form of knowledge," comparable to theology in the Middle Ages (an admiring comment, coming from Pearson). Mendelian eugenics would do best to pursue Galton's vision, a "highly developed and *applied* Anthropology" built on a foundation of probability theory. "The whole development of Mendelism in recent years has been in the direction of a multiplicity of factors even for apparently simple characters." They concluded that discrete hereditary units were of modest interest in themselves, becoming fruitful mainly when summed using continuous integrals, as with molecules in mechanics.[7]

The single-factor mental defects advertised by Davenport, Rosanoff, and Goddard, though subjected to occasional harsh criticism, persisted for about two decades then began to falter in the face of mounting attacks. In 1930, at a moment of intense discussions between German and British eugenicists, Cora Hodson of the (English) Eugenics Society sent Rüdin a pamphlet containing Fisher's calculation of the potential for genetic improvement by sterilizing the feebleminded. She explained in a letter that they were revising the figures on account of the "prejudice" of one or two English members against "the American work," which, she added in parentheses, was "largely discounted in this country." While her wording is peculiar, a reassessment of feeblemindedness as a one-gene trait was clearly underway. Pauline Mazumdar surmises correctly that the refutation of Davenport's program was based on arguments published fifteen years earlier by Heron and Pearson.[8]

Carthage, however, would not be razed in a day. Again and again, geneticists acknowledged the inadequacy of single-gene explanations in one breath and then proceeded in the next as if heredity could mean nothing else. The issue came to a head about 1930. Fisher cosigned a letter to the editor of the *Lancet* that year reaffirming his results on the effectiveness of winnowing the gene for feeblemindedness. Yet in his *Genetical Theory of Natural Selection*, also published in 1930, he reflected: "The fashion of speaking of a given factor, or gene substitution, as causing a given somatic change, which was prevalent among the earlier geneticists, has largely given way to a realization that the change, although genetically determined, may be influenced or governed either by the environment in which the substitution is examined, or by other elements of the genetic composition."[9] The gist of this line was quoted in the Brock Committee's *Report on the Departmental Committee on Sterilisation*, presented to Parliament in December 1933. Fisher, a prominent expert on this committee, must have welcomed the reported work of Torsten Sjögren, a medical superintendent in Sweden, who had uncovered a rare type of amaurotic idiocy (involving loss of sight) that behaved as a Mendelian recessive. Otherwise, apart from Huntington's chorea, the committee found no evidence of "the transmission of mental disorders in Mendelian ratios" and no sharp line separating mental defect from ordinary dullness.[10]

For reasons like these, Goddard's Mendelian findings, which had made such a sensation two decades earlier, now appeared scandalous. The committee report condemned the "dismal chronicles of the Kallikaks, the Jukes, and the Nams" as unworthy of serious attention. It cited as the source for this criticism a 1925 book by the American psychiatrist Abraham Myerson, who had cast a critical eye over a large field of work on inherited insanity, reserving his sharpest barbs for the Eugenics Record Office. Mostly he reiterated points made by Heron and Pearson. If my critique seems exaggerated, he remarked, just compare it with the severity of their reports, which had shown that "Davenport and his followers" were "dogmatic offenders against logic and science." They drew their

Mendelian conclusions in advance then arranged the data to verify them by sending out field workers with inadequate medical training to construct ancestries. "To a medical man the sang-froid with which the social worker makes a diagnosis on people she has never seen or else met in a casual way, is nothing short of appalling." The Brock Committee concluded that Myerson's criticism had never been answered. "Judged by modern standards the technique employed was unscientific and the instructions to the field workers so tendentious that it is not surprising that they succeeded in finding what they were told to seek."[11]

Davenport's program to explain every human trait with a genetic factor was now an embarrassment, not least to the directors of the Carnegie Institution to whom he reported. Their appeal for a purer, less politicized science reflected a shift in the politics of science. This was most forcefully articulated in England, where geneticists like J.B.S. Haldane, Lancelot Hogben, and Lionel Penrose now attacked the old eugenics as a biological expression of class bias. The Carnegie Institution appointed a committee in 1935 to evaluate the ERO's work, led by geneticist L. C. Dunn, who, having recently visited Nazi Germany, was well attuned to the dangers of racialized genetic politics. All those files on ordinary behaviors and personality traits, gathered up in family pedigrees to demonstrate genetic causation, now seemed ridiculous. Davenport and his assistant Harry Laughlin were forced into retirement, and in 1940 the ERO was permanently closed.[12]

Pearson's eugenics projects were not altogether free of such problems. Goddard's pedigrees of feeblemindedness were even included in the first issues of the *Treasury of Human Inheritance*. Yet the *Treasury* as a whole appeared far more scrupulous and less tendentious, a neutral work of reference rather than a proof or refutation of a genetic theory. Julia Bell, who participated from the beginning, gained a reputation for exacting research, in stark opposition to the ERO fieldworkers.[13] A comparative history of human genetics by Daniel Kevles contrasts the eugenic science of Galton and Pearson, flawed but brilliant, to Davenport's shoddy data and naive moralism. He implies a still sharper contrast with a British group that took form about 1930 including Fisher, whose eugenic views remained conservative, as well as Hogben, Haldane, and Penrose, on the political and scientific left. All combined a full commitment to Mendelism with superb statistical skills.[14]

In a way, such work vindicated Pearson's insistent claim that Mendelism depended on the tools of biometry, that insanity and feeblemindedness "are very far from being simple unit characters" and that a continuous distribution was easily generated as the sum of effects of multiple genes. By March 1930, when he reiterated these views in a lecture to the Board of Control (set up to implement the Mental Deficiency Act), many geneticists agreed.[15] Yet his prophecy that the future of human genetics would emphasize multigene traits and continuous distributions was vindicated only in part. Although Penrose, who focused his work on mental deficiency and took his data from an institution for the feebleminded, aligned himself with Pearson's insistence on a "continuous gradation of intelligence," much of the new genetics was built up around population models of genetic units. That research grew out of work on blood groups, which Felix Bernstein had explained in 1924 in terms of three alleles at one locus. The force of his conclusions rested on a match of calculation with frequency distributions.[16] Here at last was a common and significant Mendelian trait in humans, a scientific exemplar for a new human genetics.

The genetics of blood groups, though useful for identifying paternity in courts and reconstructing human migration patterns, provided no clue on the most urgent genetic question of the day, the inheritance of mental defect. Its centrality to a new human genetics meant pushing aside, for the moment, eugenic ambitions and anxieties. Even deeply committed eugenicists like Fisher recognized the advantages of a more detached genetic science. Scientists and even historians have not always resisted the temptation to associate the evils of eugenics with reliance on humdrum, practical tools and to depict empirical approaches, especially in Germany, careening toward moral calamity just as British human genetics set off on a new path of accomplishment. That argument is hard to sustain. Genetics and the statistics of mental and bodily states were not readily separable. Both had a role in race hygiene and eugenic sterilization.

Databases of Hereditary Defect

Mendelian or not, human heredity was a data science. The perfect system of classification and filing was Davenport's dream. He and Laughlin envisioned decimalized eugenic codes, with "General Traits" under 0, 1 for the integumentary system (skin and hair), 2 for the skeletal system, 3 for "Nervous system, Criminality [!]," and so on, with additional digits for more specific characters.[17] Proper filing became all the more crucial as the scale of data ambitions expanded.

The new century brought bold efforts to connect diagnoses of mental illness across the generations. Between the pages of the Worcester case book for women admitted from 1871 to 1874, we find a slip sent from the Boston State Hospital in 1911 requesting information about their patient's aunt, who had been admitted almost three decades earlier. A similar inquiry in 1936 concerned a young man seeking admission to a Benedictine mission, which wanted to know if his grandfather's case from the 1890s had been hereditary. (He had been admitted for general paralysis and his case was not.) The book for men admitted from 1879 to 1881 holds a letter of inquiry from a curator at the Warren Anatomical Museum of the Harvard Medical School about a patient admitted for the second time in 1878. "Could you let us know something of his family history, education, occupation and habits, and a brief account of his two hospital residences? I would be very grateful for this information, as his case is to be included in an atlas of brains of criminals." And there we find it, the brain of a man accused of attempting to rape a little girl in 1878, serial no. 43 in a book of photographs prepared in the 1910s by E. E. Southard and published in 1942. The man was acquitted with a diagnosis of dementia praecox and sent to the Bridgewater State Hospital, where he died about 1918.[18]

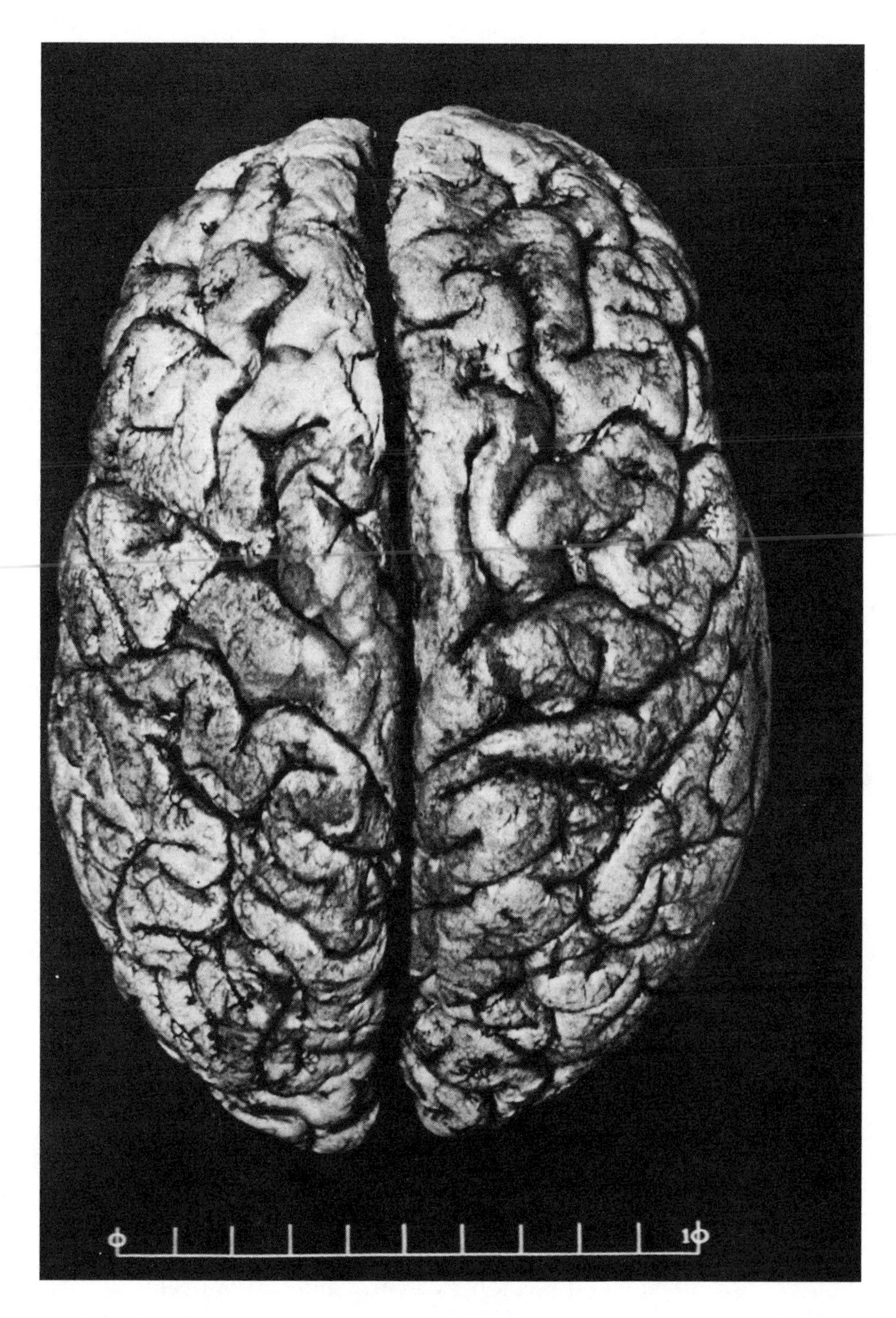

FIGURE 13.1. This photograph of the brain of a one-time Worcester patient, originally taken in the 1910s by the Massachusetts state pathologist Elmer Southard, was reproduced for a retrospective study. The psychiatrists hoped to correlate characteristics of the preserved brain with the actions and behaviors of insane criminals. Reproduced from Canavan and Eisenhardt, *Brains of Fifty Insane Criminals: Shapes and Patterns*, 172–173.

In Germany, the level of hereditary detail on admission forms was again ramped up after the First World War, supporting new efforts to incorporate ancestral data into files of current patients.[19] In archives of the Pforzheim Institution in Baden, we discover a still-empty file for a patient admitted in 1897 and in its place a note indicating that the case history had been removed for use in schizophrenia studies. The files from the Provincial Hospital of Zwiefalten in Württemberg hold many such requests for information, and by 1941, printed forms declaring: "For hereditary-biological comprehension of [name] we require at no cost a complete extract from the family register, including still births, early deaths, about [named relative]. We also request communication of what is known about presence of mental illness, feeblemindedness, alcohol dependence, suicide, criminality, or other abnormal personalities among the kin." If asylum officials lacked this information, they were supposed to inquire immediately to the police chief and to health and parish offices. The abundant mobilization of records of this kind enabled Baden and Württemberg to outpace all other German states in sterilizations up to 1939.[20]

Another such effort is documented in institutional records for the Rhine Province. Otto Löwenstein, a professor at the University of Bonn and chief neuropsychiatrist at the State Hospital for Nervous and Mental Diseases, began assembling an archive of hereditary biology in 1923. He sent a detailed questionnaire to local institutions, and he took advantage of the long history of the Siegburg asylum to assemble a full century of patient records, including 8,000 files from 1825 to 1878, to be organized and made accessible in a cabinet (*Kartothek*). Heredity, he argued, is involved not only in "endogenous" cases (generated from within), but also "exogenous" ones (those with identifiable external causes). He dreamed of interlinking the family trees of asylum patients so that each newly admitted child could be checked straightaway for mentally ill relatives. He hoped that someday his records might match the superb files of the psychiatric institute in Munich. In March 1933, when he fled to Switzerland, his office and data were passed on to Kurt Pohlisch, a psychiatrist untainted by Jewish ancestry, notorious now for giving medical lectures in his SS uniform. Pohlisch carried on the project with redoubled vigor, introducing a new card to facilitate searches for relatives and a central office with records for every mentally ill person over the last century. By May 1938, his cabinets bulged with 750,000 cards from prisons, asylums, and special schools, and in the next few years they doubled again. The Rhine Province was thus ideally positioned to apply the German sterilization law—and worse. Pohlisch issued hundreds of expert hereditary assessments of mental patients, children as well as adults, often authorizing euthanasia.[21]

German investigations of criminal heredity illustrate how the advancing ideal of integrated data centers was linked to a more general and somewhat indiscriminate sense of genetic defect. While the idea of heredity as a factor in crime was nothing new, prisons did not share the medical-administrative culture of mental hospitals, which routinely recorded data on causes. Serious data gathering on criminal heredity, in fact, was initiated in special schools and asylums. Julius Koch, the census taker of Stuttgart, had introduced a concept of *Minderwertigkeit* (inferiority), encompassing a range of moral and intellectual defects less severe than idiocy and madness. His idea was widely discussed at international congresses and applied explicitly to criminals. An assistant to Bleuler at Burghölzli presented data in 1904 showing a hereditary burden among criminals that was almost as high as for mental patients.[22]

Carl Rath, a prison pastor in Siegburg, carried out the first detailed investigation of criminality as a Mendelian recessive in his 1914 dissertation at Bonn. He drew up a massive questionnaire to record family and criminal histories of prisoners and then, on the basis of records of prior sentences, diagnosed in some a criminal disposition. Such a disposition, he supposed, was likely to be associated with "specific *Anlagen*." Although he acknowledged that criminal behavior could not be independent of environment, he was delighted whenever he found a numerical approximation to a Mendelian ratio, in one case 53:52 (close enough to 1:1). Yet it was difficult to match behaviors and genes. Women, he found, often fail to manifest their genetic constitution, so he coded the mothers of habitual criminals as heterozygous. Fathers of criminals were assumed to be homozygous.[23]

The Rhine area soon gave way to Munich as the prime focus of research on criminal heredity. Kraepelin had long nurtured an interest in criminology and, as Richard Wetzell shows, had a role in much early research on criminal biology. During the 1920s, Rüdin positioned himself at the center of data work on criminality, which he, too, regarded as allied to mental illness. He soon joined forces with the physician Theodor Viernstein, who had been authorized to conduct anthropological studies in a Bavarian prison, Straubing. Viernsten established there a Bavarian Criminal-Biological Service, which soon advanced a step beyond the familiar card cabinet for a single institution to become a *Sammelstelle*, a storehouse of data cards from multiple prisons. About 1930 he moved his files to the psychiatric research institute in Munich and began conducting detailed interviews each week with two newly admitted prisoners, a method he preferred to superficial records on every new inmate. His questionnaire took in criminal records, alcohol use, details of the convict's life, and the mental health of relatives. He also performed a full physical examination, assessed intelligence and temperament with a psychological test, and sent out forms to be completed by school, church, and police authorities. Viernstein wanted to go beyond individual diagnosis and to identify the racial stocks from which criminals were recruited. He lent the database to his Munich colleague Johannes Lange for twin research, on the model of psychiatric twin studies developed in Munich and Basel by Hans Luxenburger.[24]

Medicine and Death

German eugenics was unparalleled in its scale and unique in moving beyond incarceration and sterilization to mass murder. It resembled that of other countries, however, in being shaped by notions of public health, social welfare, and budgetary need. The familiar weighing of costs of treating the mad against savings for those cured turned in darker times to arguments that there were better uses for state funds than to house and feed unproductive mental patients. By 1914, the German state was perhaps uniquely unhopeful about the prospects that asylum care could check the increase of insanity. Patient numbers in unified Germany had increased alarmingly, from 47,000 in 1880 to almost 240,000 in 1913.[25] Rüdin spoke of an "avalanche" of medical expenditures and of the duty of psychiatry to contain it "to preserve the state." Asylum patients were last in line for food during World War I, so that 140,000 of them died, many from hunger and disease. By 1929, their number had bounced back to surpass the prewar peak. The cost of maintaining so many ill persons in the straitened circumstances of the early 1930s seemed unacceptable. German racial hygiene was backed up by commanding medical expertise, allied to systems of institutions for those who broke laws or could not care for themselves. With encouragement from these doctors, the Weimar parliament was deliberating on a sterilization law even before the Nazis seized power. Many socialists and communists defended sterilization as a way to alleviate poverty by reducing the social burden of mental illness.[26]

The medical profession had a notoriously high proportion of party members, and National Socialism was merciless in dealing with all those it deemed unworthy. Although it spent generously to preserve and restore the health of "valuable" Germans, it had no regard for those with inherited weaknesses or whose incapacity owed to moral failings such as alcoholic excess or venereal disease. On some matters the party was inflexible, particularly on the evils of racial mixing. Eugen Fischer, head of the Kaiser Wilhelm Institute for Anthropology at Dahlem in Berlin, nearly destroyed his career by failing to toe the line on interracial marriage. Geneticists learned to keep quiet or to endorse this dogma, though many thought it scientifically doubtful. Nazi positions on eugenic weakness were less dogmatic. This did not make them gentler. No eyebrows were raised when a researcher concluded that the wide bounds of (statistical) error made it impossible to demonstrate the psychiatric hereditary burden of prisoners. It hardly mattered, since criminals by then were routinely sterilized anyway. In 1941, Heinrich Wilhelm

Kranz and Siegfried Koller forsook the winding path of genetics to defend the sterilization of criminals simply on the basis of their irredeemable unsociability.[27]

The Nazi takeover enabled genetic researchers to press forward with existing projects, now with more generous funding. Rüdin's work on psychiatric heredity had long been integrated with eugenic policy aims. Already in 1924, as an expert consultant, he lectured on the genetic basis of reproductive decisions. He has been described as forsaking genetic science for simplified and politically expedient research methods. This argument presumes that the Mendelian orientation of British and American geneticists was the true path of science. In fact, even the Anglophones recognized by the early 1930s what Rüdin's colleagues had reluctantly admitted a decade earlier, that Mendelian genetics offered no easy foothold for studies of common mental and psychiatric conditions. The turn to blood groups was a diversion from the pressing issue of inheritance of mental illness, where a need to rely on medical and psychological data in place of Mendel numbers was at last winning acceptance. For Rüdin, who had taken up genetics for the sake of psychiatry, medical and demographic statistics now appeared to be the only way forward.

Genetic research was a leading site of murderous science. Yet science never set the agenda on matters so integral to Nazi dreams as the extirpation of groups they called degenerate. Surpassing by far the standard of forced sterilization set in North America, Switzerland, and Scandinavia, German officials did not insist on a demonstration of hereditary defect. One finds data forms authorizing sterilization on which not a single person is indicated on the space for mentally ill relatives. When war broke out, the Nazi state revved up its engines of mass murder, with mental patients among the first targets. Euthanasia depended on euphemism, yet the killers were sincere about "lives unworthy of living." They targeted people leading disorderly lives, perhaps dependent on alcohol or drugs, who had failed in school and work or could not take care of themselves. Grounded in revulsion as well as a rough budgetary calculation, decisions to sterilize or to kill were eugenic in a loose sense only.

Historians no longer claim that the researchers involved were scientifically incompetent or that Nazi moral failings depended on scientific ones. Scientists like Fischer and his successor, Otmar von Verschuer, were at the top of their fields. They pursued *Phänogenetik*, or developmental genetics, as an alternative to the more reductive and, to Anglophones, more familiar "transmission genetics," which mostly put aside the issue of bodily mechanisms. While the scientific questions they asked have come to be recognized as fundamental, their most interesting experiments are radically unethical when performed on human subjects. National Socialism, which treated millions of persons as unworthy of moral regard, cleared the way for transgressive interventions. These were in part immoral experiments in service of an evil state, in part the exploitation of moral breakdown in pursuit of recognized scientific goals. A rich historical literature has grown up on this terrible topic. The most notorious of the genetic experiments involved twin research by Verschuer and his former PhD student Josef Mengele. It was convenient for research to be able to dissect at will individuals showing unusual traits. Under some other ethics of research, it would not be possible to examine simultaneously and in depth the developmental state of identical or fraternal twins. Mengele, the Angel of Death, picked out newly arrived Sinti prisoners at Auschwitz to be killed immediately so that Karin Magnusson at Dahlem could use their eyes to develop an "iris table" as a racial indicator.[28]

Hereditary Counseling

Family or marriage counseling (*Familienberatung, Eheberatung*) implied no fixed political orientation but was addressed in diverse Weimar publications on psychiatry and heredity. Rainer Fetscher, an outspoken anti-fascist, created a network of centers for hereditary counseling around Dresden in the 1920s. He described the work as a utopian dream made real, the first consciously eugenic program and a welfare service that every state should provide at no cost to its citizens.[29] He assembled a hereditary database of prisoners in order to make Saxony a center of data on criminal heredity and even, or so he claimed, carried out (voluntary) sterilizations three years before the Nazis made them legal and mandatory. He also developed a *Kartei* of the inferior, which Rüdin soon incorporated into his data storehouse in Munich.[30] Rüdin, like Fetscher, insisted that eugenic sterilization should be strictly voluntary until he was given the opportunity to help interpret and implement a law that made it compulsory, the notorious Law for the Prevention of Hereditarily-Diseased Offspring. While his dedication to eugenic policies was sincere and unflagging, he acted opportunistically to expand his research empire. Toward the bitter end, as he found it necessary to seek funds from increasingly dubious sources, he gained access to corpses for research and lost control of the work.[31]

The conditions in Munich for research on inheritance of mental illness had been steadily improving in the twentieth century. Kraepelin's nerve clinic in Munich, where Rüdin made much of his early career, not only survived the terrible demands of the First World War but flourished. Thanks to a major donation by the American banker James Loeb, it was reorganized in 1917 as the German Institute for Psychiatry. In 1925, Rüdin added a well-funded chair in Basel to his appointment at the Munich institute. He returned full-time to Munich in 1928 as the institute was upgraded, once again, to become the Kaiser Wilhelm Institute for Psychiatry, with a new building and munificent research support from the Rockefeller Foundation, which ranked him with Penrose as the best in the business.[32]

Rüdin never gave up on Mendelian psychiatry. He and his associates repeatedly scrutinized their data for new categories that might reveal the underlying Mendel numbers. In 1924, he called for new genealogical groupings to refute the charge that schizophrenia lacked genetic unity. If the numbers never worked out, the validity of Mendelism was still assured on the basis of experimental results such as Mendel's own. Meanwhile, empirical correlations based on Jenny Koller's and Otto Diem's methods of comparison appeared to be the best way forward. Some may claim that probability methods reduce the biology of heredity to a throw of the dice,

Rüdin remarked (echoing his own earlier dismissal), but "honest empirical work" is precise enough for reliable prognosis.[33]

Under this banner he assembled a research group of his own students. Eugen Kahn was one of the first, working in Munich until 1930, when the Rockefeller Foundation endowed a chair to bring him to Yale. Rüdin's associates Luxenburger, Carl Brugger, and Bruno Schulz did much of the work in Basel and Munich to develop this empirical program.[34] Medical and behavioral data were at the heart of the project, requiring a room full of skilled clerical workers. In negotiating his return to Munich, Rüdin requested a medical statistician and was granted a car and driver for travel to distant interviews. The funding was sufficient to carry out complete local surveys, and in this way to circumvent problems of inference associated with family pedigree research. During the last years of the Weimar Republic, his team gathered data on hereditary burden in five communities in Allgäu (in the German Alps) based on 5,000 medical examinations, and from a population near Rosenheim in eastern Bavaria. Brugger, meanwhile, examined tax lists, consulted local officials, and recruited doctors, mayors, teachers, preachers, and older inhabitants to provide information on the psychopathologies of 37,651 persons from the cities of Jena and Stadtroda.[35]

The Nazi state pushed forward this pioneering information society. In their classic history of German eugenics, Peter Weingart, Jürgen Kroll, and Kurt Bayertz estimate that by 1942 there were 10 million file cards on the hereditary worth of Germans. These records, organized into families and sorted by form of disease or abnormality, were the not-so-raw data of empirical hereditary prognosis. In the classic way of asylum studies, they relied on tables in place of mathematics, yielding results in a relatively accessible form. Yet statistics provided no easy solution to the problem of winning over political actors with little experience of science. Powerful Nazi geneticists such as Koller and Fritz Lenz criticized the reliance of the Munich group on mere percentages. In popular lectures, Rüdin invoked the place of chance in modern physics and chemistry as a defense against such charges. He also emphasized that empirical correlations were undergirded by simple laws

FIGURE 13.2. Data preparation at the Kaiser Wilhelm Institute for Psychiatry, 1930s. Data work was central to Rüdin's hereditary investigations and required a considerable staff, mostly of women, to coordinate researchers, arrange interviews, keep files, and analyze data. From Max Planck Institut für Psychiatrie, Historisches Archiv, Bildersammlung, reproduced with permission.

of the *Erbanlage*, or gene. For more than a century, that unit has far outpaced every rival in giving the public a false sense of genetic understanding.[36]

Rüdin's correspondence with the Eugenics Society in England sheds light on the interplay of sophistication and simplification. The collapse of Davenport's reputation had left a vacuum, and Rüdin, after so much instruction by Weinberg, understood well enough why pedigree tables should give way to demographic statistics. His streamlined version of empirical prognosis, an introduction to hereditary correlation at a moment of dissatisfaction with the old methods, dazzled the officers of the Eugenics Society in England. It helped to define a new direction for English eugenics in relation to the pressing problems of insanity and feeblemindedness.

What won them over was an offprint of a popular lecture that Rüdin had published in 1929 in *Das kommende Geschlecht* (The

Coming Race), a journal for women. It came to the unsurprising conclusion that eugenic progress required ample breeding by women and men of good quality and sterilization of those with heritable defects of brain and behavior.[37] The Eugenics Society, having received the offprint in 1930, immediately issued a partial translation under the title *Psychiatric Indications for Sterilisation*. The pamphlet emphasized the relevance of probability to every disease, even Huntington's chorea, a Mendelian dominant, since it is a question of chance whether an adult child of a diseased parent carries the *Anlage*, and again for whether it is passed to each successive child. A risk of 25% appeared as reason enough to Rüdin to bar reproduction. All the more common hereditary defects, he explained, have a complex and unknown genetic structure, leaving empirical numbers as the only workable basis for practical eugenics. Although the Munich research program extended to correlations with grandparents and aunts and uncles, Rüdin's statistics addressed the simplest case. A single schizophrenic parent raised the probability of schizophrenia in a child to 9% or 10%, while with both parents affected it increased to 53%. Manic depressive illness had a still higher heritability: 30% to 33% in the case of one affected parent, and twice that for both.[38]

Although these numbers seemed already to make a sufficient case for sterilization, mental illness was now only the tip of the iceberg. Insanity and idiocy, he wrote, are uncommon and so debilitating that those affected can barely survive on their own, much less reproduce their type. Eugenics must shift its attention to the "vast army of psychopaths," all those inefficient persons with a distaste for life or suicidal tendencies, manifesting cruelty, sex perversions, or grave criminality. Taking psychopathy into account, the hereditary burden of a marriage with just one schizophrenic or manic-depressive partner rose to 60%.

This inclusion of general nervous and behavioral defects was a radical departure from the psychiatric ideals of Rüdin's prior career and a betrayal of the founder Kraepelin's insistence on disease specificity. His speech that year to honor Kraepelin seemed rather to bury him. Like Davenport two decades earlier, Rüdin could no longer maintain diagnostic walls around conditions that were interwoven on so many pedigree charts in so many file cabinets. As he explained on the occasion of his own festschrift in 1934, it no longer mattered if the assumption of genetic homogeneity for schizophrenia should prove incorrect. Although it would be nice if the empirical numbers conformed to Mendelian ones, that too was unnecessary. The data practices of human heredity had already taken a different turn. Like so many statisticians, including Galton (with his "residuum"), Pearson, and Fisher, Rüdin now focused attention on a degraded stratum where genetic defects accumulate and whose bad heredity produces a bad environment. Beneath it all, he supposed, were discrete disease entities, but all mixed together. Defective *Erbanlagen* of diverse types combined to make general psychopathy.[39]

Rüdin welcomed the Nazi triumph in 1933. His lectures and popular essays turned obsequious, and he refocused his work on putting the sterilization law into action. He relied on school and medical reports as the basis for expert diagnoses of low social worth, and hearsay regarding a relative as sufficient evidence of hereditary defect.[40] Yet the claim that his whole program was corrupted by ideology and not up to international standards is difficult to defend. To researchers in psychiatric genetics across northern Europe and North America, the Munich institute remained an inspiration. Rockefeller funds continued to flow to it almost up to the outbreak of World War II, enabling psychiatrists from Scandinavia and Britain to visit the Munich institute and to learn its methods. His students helped to reshape psychiatric genetics in other lands. Franz-Josef Kallmann, who learned genetics with Rüdin's group, was welcomed in 1936 as a psychiatric exile to the United States. The experience of National Socialism did not cool his eugenic ardor.[41] In 1947, in an open letter, he defended Rüdin as an outstanding scientist and a "passionate dreamer." Kallmann's scientific legacy was appreciatively recalled in 1972 by a prominent American genetic psychiatrist, who commended Rüdin as the source of his methods and bemoaned the flight from Rüdin's science owing to unfortunate associations with forced sterilization and murder.

In 1981, a researcher at the National Institute of Mental Health exalted Kallmann as a scientific hero and a Jewish model for the redemption of genetic psychiatry from "recrudescent Lysenkoism" on the left. "Perhaps Franz Kallmann's courageous perseverance and dedication to truth will serve to help human behavior genetics survive its next political buffeting."[42]

Sterilization Debates

In 1929, Rüdin had called on advocates of sterilization to be reasonable. In a passage omitted from the Eugenics Society translation, he worried that Laughlin's proposal to sterilize as much as 10% of the American population in a single year, unthinkable in Germany, was provoking needless fear. His tune abruptly changed in 1933, when there arose a real possibility to halt the reproduction of those teeming psychopaths. Here, again, the lesson of empirical prognosis retained more than a whiff of Mendelism. He divulged a suspicion that psychopathy arose from heterozygous defect and incomplete dominance. A volume prepared to explain the sterilization law, appearing under the names of Arthur Gütt, Rüdin, and Falk Ruttke, began with an introduction to Mendelian heredity, illustrated by idealized family pedigrees in which each genetic factor, whether healthy or flawed, was fully known. These gave way to charts of data from hereditary studies involving filled-in and half-filled circles to give the impression that real mental illness involved homozygosity (two defective *Anlagen*), and less severe or ambiguous defects only one. While acknowledging that basic Mendelism could not explain the transmission of these traits, the authors invoked it to justify sterilization of healthy heterozygotes. They also tried out the idea of inherited *Momente*, or impulses, as a quasi-Mendelian basis for complex empirical results.[43]

Kallmann, still more ambitiously, spoke at the International Congress on Populations Questions, held in Berlin in August 1935, of an *Erbanlage* for schizophrenia. The prevalence of this gene in the population was already receding, he calculated, but much too slowly for its timely elimination. Too many families were formed and children born before the afflicted could be interned, and too many genetic factors were transmitted invisibly by heterozygotes. He proposed to expand the field of sterilization to relatives of schizophrenics whose defective gene could be inferred from psychopathic characteristics. In the discussion, the German geneticist and racial theorist Fritz Lenz and the Austrian hygienist Heinrich Reichel objected that heterozygotes were far too numerous for such a policy to be practical. Kallmann made no concessions. The next year, forced to emigrate on account of his Jewish heritage, he exported his science to America.[44]

Meanwhile, he had attracted the attention of another geneticist of Jewish background, who could not be present at this congress. Although Weinberg's eugenics was comparatively restrained, he found these proposals intriguing, as we know from the letter he sent for publication to Fisher in Australia. After complaining of Bodewig's mistakes, his failure to credit Weinberg's own mathematical contributions, and his "perfectly erroneous description of the state of German biology and eugenics," Weinberg seized the opportunity to raise some big questions regarding sterilization of heterozygotes. He did not see how any European state pursuing such policies could avoid the extinction of 50% to 75% of its population. Without international regulation, there would be new European wars. Other continents would most likely maintain their indiscriminate reproductive practices. The project might still be advantageous if the sterilized 75% could be replaced by sound individuals, assuming the frequency of new mutation-defects was sufficiently low. Here, in short, a dangerous new technology seemed to force a shift toward European if not global government. No wonder Fisher cut off the discussion. Weinberg's radical idea disappeared into his files.[45]

While twin studies, freed of moral constraint, flourished as never before under National Socialism, empirical hereditary prognosis was increasingly focused on the implementation and consequences of the sterilization law. Bruno Schulz's introductory text on methods of medical research on heredity, published in 1936, relied on Weinberg's statistics, and even Fisher's and Haldane's, to model its genetic consequences. His introduction and conclusion included a farrago of arguments for abundant sterilization. Empirical prognosis uses purely statistical methods, he explained, making it possible to proceed without theory. We do not even need to distinguish hereditary from environmental causes, or to determine how defects are passed on. A spirit of caution demands copious interventions, just as we would not take a recreational train ride if there were even one chance in ten of going off the rails. Whether or not schizophrenia is a hereditary unit, expected outcomes are always better when both parents are healthy. He did not overlook the easiest argument of all. It is not the character but the *Anlage* that is inherited, and defective *Anlagen* must be combated.[46]

It fell to Siegfried Koller and his mathematical colleague, Harald Geppert, to write the authoritative work on heredity for national socialists. The mathematical competence of the authors is evident, for example, in their treatment of sampling biases. But the authors displayed the ideology and the anxieties of their time. In discussing blood groups, they avoided any mention of Bernstein, Koller's Jewish teacher. They gave detailed attention to the mixing of Jewish and German blood, offering grounds for hope that purity could over time be restored with appropriate marriage laws. Toward the back, on page 170, they stressed the importance of empirical hereditary prognosis for racial hygiene. Up front, on page 5, they explained how gene loci are lined up along the chromosome, "like beads on a necklace."[47]

British and German Empiricism

The Brock Committee Report in 1934 referred favorably to German work on empirical prognosis. Two of the most prominent British researchers on psychiatric heredity, Aubrey Lewis and Eliot Slater, studied in Munich with Rüdin. They carried on using statistics of inheritance of mental illness in the 1940s and 1950s and taught it to their students. In 1996, a paper on Slater for a historical commemoration of psychiatry praised Rüdin for methods and data that have stood the test of time.[48] Rüdin's approach, as we have seen, did not enter Britain as something alien but had been developing there for a century in the work of alienists and statisticians at asylums and schools. Quantitative geneticists of the 1920s and 1930s remained true to the institutional sources of their data. Fisher collaborated with Cyril Burt, psychologist for the London schools and a pioneer of IQ testing, who proved his devotion to hereditary explanation by relying on data from poorly documented or even invented pairs of identical twins separated at birth. Penrose, too, took an active interest in measures of the intelligence of schoolchildren, especially in debates over its inheritance. After 1944, when, against all eugenic expectations, the mean IQ in Scotland showed a marked increase since 1932, he tried out new relations of genetics and reproductive patterns to explain this result. He was keenly interested in evidence for mental traits that depended on a single gene as well as in discrete human differences that eluded the Mendelian grid. "Mongolian idiocy," or Down's syndrome, was well known to show a non-Mendelian inheritance pattern. His *Colchester Survey* confirmed this point and used the evidence of mean values and error estimates to show it was linked to maternal age rather than birth order. A Munich pediatrician, citing Brugger, had made these points convincingly in 1932 with bar graphs. Penrose, addressing the same topic in 1934 in the *Annals of Eugenics*, deployed advanced statistics, the mathematics of covariance.[49]

"Colchester" meant the Royal Eastern Counties Institution for the Mentally Defective, founded in 1859 as the Eastern Counties Asylum for Idiots and Imbeciles. It was one of the oldest such institutions in England. Penrose's statistical study of its patients, recognized as a landmark of human genetics, shows clear continuities with longstanding traditions of asylum statistics. It drew on German as well as English works of empirical prognosis. His understanding of intelligence as a "graded character" ruled out the direct application of Mendelian analysis and encouraged reliance on non-laboratory sources. Pearson was correct, he said, that mental defect does not reduce to low intelligence but involves more general social failings. Penrose, however, was not content with intelligence as an institutional construct, to be defined by the opinions of schoolteachers, but deployed the Stanford-Binet intelligence test in hope of establishing a measure of true mental ability. These test scores provided a ratio of mental to chronological age, which he made the basis for classification: from 0 to 20 percent meant idiot; 20–49 imbecile; 50–69 feebleminded or (he preferred) simpleton; and 70 to 85 dull. His report itself was packed with correlation tables relating the mental ability of children to that of parents, siblings, grandparents, and aunts and uncles. He did not try to reduce the tabular information to a single measure of hereditary transmission. In other tables he showed how the "mental grade" of children was associated with home conditions and with the professional class of the parents.[50]

Among the authors he cited was Brugger, whom he credited for classifying relatives by degree of defect. The next year, he mentioned Luxenburger and Brugger in a short article on "Eugenic Prognosis" as a basis for advising parents of a defective child whether to try again. Their methods, he wrote, "could be adapted for use in this country if the criteria of mental grade were better standardized." Brugger, who remained in Basel after Rüdin's departure, recorded hereditary data on the Munich forms to calculate how much mental weakness in the siblings of his patients could be explained by the mental grade of their parents, a classic problem of empirical hereditary prognosis. These numbers were high, rising to 93.15% when both parents were mentally deficient. Penrose, who always insisted on environmental as well as genetic causation, argued that high percentages did not prove anything. Yet he was working along the same lines. He felt no need to avoid or disguise this affiliation with the genetics of Munich, even though he despised Nazi eugenic politics.[51]

Data Science, Human Genetics, and History

> When I came into human genetics, I had one, I guess, absolute guiding principle: Try to be as rigorous as I would have been had I remained with *Drosophila*.
>
> —James V. Neel

By no means did the discrediting of the hereditary claims of the Eugenics Record Office, or even exposure of the atrocities of Nazi racial hygiene, bring an end to eugenics. Even the name of eugenics was abandoned only gradually, while the effort to select genetic traits of individuals and of populations lives on. Geneticists were inclined to attribute the abuses of eugenics to bad science, which certainly played its part, and to insist on scientific rigor as the antidote, which seems at best optimistic. Blood groups filled the role of fruit flies for the new human genetics. They were not clearly associated with any compelling medical conditions, however, and the hope of finding genetic solutions for mental and psychiatric disabilities did not fade. Human genetics could never be confined to the laboratory, not even research on blood groups. It generated an ever wider field of data through ethnographic surveys, nuclear radiation studies, and recordkeeping in hospitals and prisons.[1] We are encouraged still by genetic entrepreneurs and by a compliant press to understand the history and to visualize the future of genetics in terms of experimental breakthroughs, along the lines of James Watson's novelistic *Double Helix*, functioning as a fable for our time. It oversimplifies to the point of untruth to suppose that genomics or bioinformatics arose as serendipitous discoveries of disinterested science. Our focus on the data of heredity shows the intertwining of genetic science with health, education, race, law, business, finance, and war.

This book, though concluding in the 1930s, aims to provide a well-documented alternative to the partly implicit historical assumptions that inform much modern commentary. I hope at least to have shown that the history of heredity as a data science is a long one and did not begin with genomics, DNA, or Mendelian genetics. In our own time, the study of human heredity continues to depend on mundane, routinized labor in places like asylums, hospitals, prisons, courtrooms, and schools. Our dazzling technologies for decoding and manipulating nucleic acids do not easily translate into practical solutions to medical and behavioral problems. The fruits of the postgenomic alliance of DNA Mystique with Data Deluge take the form of probabilities, that is, measures of risk and uncertainty. Genomics offers, too, the poisoned fruit of alleles with less predictive value than patients and doctors want to believe, less, sometimes, than the phenotypic characters of parents and siblings. Indiscriminate DNA scans and even focused tests turn up an abundance of alleles or variants "of unknown significance," which yet, on account of legal and medical demands on this genomic astrology, cannot easily be put aside.

Although techniques of hereditary prediction grew more powerful after World War II, there was no quantum leap. When James Neel, appointed to a new professorship at the University of Michigan in 1946, was given the resources to make good on his promise to make human genetics as rigorous as *Drosophila* research, the resources he drew on were familiar ones. Like Davenport, Fisher, and Rüdin, he had ideas for detecting heterozygote "carriers" of recessive characters. He proposed to map genetic traits on human chromosomes, drawing on the same statistical tools employed for mapping mutations of fruit flies. Genes on the same chromosome tend to be inherited together, and the nearer their positions to one another, the higher the probability. Measurement of genetic distances would provide a basis for prediction. Huntington's chorea, for example, does not usually become visible until too late to provide guidance for reproductive decisions, but if a nearby allele on a chromosome of the parent with Huntington's were to manifest in the child, this would suggest a high probability that the gene for Huntington's was inherited along with it. His proposed first phase of the work has a familiar ring: to assemble the largest possible database of genetic traits using family pedigrees. He defended pedigree tables, which had been much criticized in the 1920s and 1930s, as indispensable for human genetics.[2]

Other emerging objects of human genetics included hemophilia and a lengthening list of much rarer genetic traits.[3] *Drosophila* genetics was to provide the baseline against which to assess the more complex aspects of human life and culture. Victor McKusick of Johns Hopkins University came to be recognized as the "father" of the allied field of medical genetics, primarily for his compilation of rare genetic traits, *Mendelian Inheritance in Man*. It was dull, sober work, in deliberate contrast to the overreaching of Davenport's Eugenics Record Office. He did not claim any specific genetic basis for comparatively common and pressing conditions, still widely regarded as hereditary, such as mental deficiency and insanity. Their fuzziness as disease entities has made genetic analysis deeply problematical. Yet social medicine could not ignore them, and it didn't. Curt Stern included a discussion of empirical hereditary prognosis in his pioneering 1949 textbook of human genetics and inserted a table giving probabilities of schizophrenia and of related "psychopathic conditions" for children, siblings, and other relations of a diagnosed schizophrenics. His numbers, compiled from synthetic works published in 1937 by Luxenburger and Verschuer, did not indicate such striking inheritance as Rüdin's, but they still showed greatly elevated rates for near relatives of affected individuals. Stern also drew extensively from German twin studies to assess inheritance of mental disorders, criminality, intelligence, and feeblemindedness.[4]

Genetic counseling, which expanded rapidly in the United States after 1945, was mainly focused on conditions that were not straightforwardly genetic. As Soraya de Chadarevian shows, chromosome research still dominated human genetics during the 1950s and 1960s, notwithstanding Watson and Crick's celebrated publication in 1953. Doctors and scientists examined chromosomal karyotypes drawn from asylum and prison populations and carried out anthropological studies comparing (white) European populations with "remote" or "isolated" ones on other continents.[5] The poster child for genetic counseling was Down's syndrome, identified in 1959 as a chromosomal abnormality that had long been recognized as somehow genetic but not inherited. Counselors gave advice based on the age of the mother, a variable whose relevance to the probability of bearing a child with this condition had been worked out in the 1920s and 1930s. An American pioneer of genetic counseling, Sheldon Reed, in collaboration with J. A. Böök from Sweden, introduced the term "empiric risk" in 1950 for prediction on the basis of statistical evidence. Later, the term was extended to inheritance of mental weakness by another Swedish researcher, Hans Olof Åkesson. Åkesson cited Lionel Penrose's research on intelligence test scores of mentally defective patients in relation to their relatives and Carl Brugger's much-referenced 1935 paper on inheritance of mental weakness. As seems clear even from its name, work on empiric risk perpetuated the intellectual program of empirical hereditary prognosis developed by Brugger, Luxenburger, and Rüdin in Munich and Basel. Scandinavian hereditary counseling, which took off in the interwar years, became a more appealing model to invoke than German versions.[6]

The rise of molecular genetics in the 1960s, and the immense hopes and resources poured into it by investors and pharmaceutical companies as well as states, universities, and foundations, reinforced a narrow view of heredity as genetics, a science concerned with genes as fragments of DNA, described as unmoved movers in biological processes of reproduction and development. Its ambitions were universal. Many practical questions that had appeared so urgent to Pearson, Fisher, Davenport, and Rüdin became research topics for "behavior genetics," which drew from psychiatry, neuroscience, and especially psychology. Behavior genetics ascribed a critical role to (undetectable) genes but performed statistical analysis of phenotypic traits and measures, notably IQ scores, whose relation to genetics became immensely contentious. Geneticists in biology departments did not necessarily object to the politics of psychological heredity, but many disdained hereditary research that relied on test scores and behaviors rather than on model organisms in laboratory experiments. As the Human Genome Initiative took form in the 1990s, geneticists and politicians promised to bypass the inconclusive work of correlating traits and to identify the effects of specific genes on mental abilities and disabilities as well as dread diseases such as cancer. Matching James Wilson's claims for the "fire-proof vaults" of the ERO, "making records of the very souls of our people, of the very life essence of our racial blood," James Watson, who directed the Human Genome Initiative from the same laboratories in Cold Spring Harbor, loved to say that "we used to think our fate was in the stars, now we know it to be in our genes." Daniel Koshland, biochemist and editor of *Science*, promised through genomics to uncover causes of mental illness and in this way to relieve a host of social ills.[7]

To the usual motive for exaggerated scientific promises, the effort to keep research money flowing, genomics and recombinant DNA added high-tech business models for raising capital and filing patents. Informed critics articulated compelling reasons why the promise of finding genes to explain complex maladies and behaviors was unlikely to be fulfilled. There were, however, compelling financial incentives to redirect the genomic gaze beyond rare genetic diseases to complex conditions including "cancer, heart disease, hypertension, schizophrenia, depression, migraines," and so on, which affected vast populations and offered a prospect of unlimited markets.[8] In relation to ambitions like these, the return of genomics has so far been as meager as the skeptics imagined.

The recent experience of genomics confirms what we also find in examining a longer sweep of the past, that the category of genetics is often too narrow for historians. This book draws inspiration from a more encompassing articulation of the topic, "cultural history of heredity," which has taken form as a set of projects associated with the Max Planck Institute for History of Science in Berlin.[9] That effort reaches self-consciously beyond gene mechanisms to take in practices and doctrines of biological transmission over several centuries. Heredity as a historical topic embraces breeding practices and explanations in agriculture, medical study of distinctive family traits, embryological observations, human and animal pedigree charts, and statistical methods. Especially for human heredity, genes and DNA were always only part of the story and rarely if ever independent of the other parts. This history reaches beyond biology and medicine to anthropology, demography, and psychology, which have been so obsessed with the hereditary significance of human difference with regard to skull shapes, birth rates, and IQ scores. I make the case in this book for madness and mental deficiency as medical-social concerns with a key role in shaping this science.

History, which addresses the past, is too often disdained as lacking contemporary relevance. We might just as well dismiss the importance of biological evolution. I would rather say that nothing in our world makes sense apart from history. It tells about how we became what we are and how we misunderstand what we are becoming. Rethinking the past has definite consequences for the present. A more encompassing perspective on the historical sources of hereditary knowledge puts contemporary work in a different light. This account extends to heterogeneous forms of knowledge and practice, their successes not sharply distinguished from their failings, to show how insane asylums and special schools became sites as well as objects of hereditary investigation. These institutional configurations have evolved across the centuries, yet the role of medical and social institutions remains, in our day, central to human and medical genetics. In recent times, such connections have proliferated, extending to finance, insurance, pharmaceutical firms, patient groups, information technologies, diversity movements, and on and on.

History of science, when it deals with topics that matter, cannot be content to treat factors like these as extraneous or external to science. They are not merely elements of context but shape the work itself. Neither can history set aside the particular places and conditions of research, the tools and techniques of gathering, processing, and diffusing information, or the reshaping of research by patients and families, prisons and poorhouses, state ministries and regulators. The history of heredity reaches across national and linguistic boundaries and did so throughout the last two centuries, though regional and national differences have persisted.

Too many simplicities are uttered about science, and the work of science itself is too often presented in microscopic units. Science is quite complex enough to justify dense, wide-ranging narratives, narratives that take account of what scientists discuss among themselves but usually omit from publications as well as their interactions with a variety of other actors. Omniscient narration is not available to the historian, but at almost every point it is possible to examine the surroundings more closely or to dig deeper. As in the work of natural science, attention to new factors does not merely add complexity. Often, it is necessary to widen the investigation to have any hope of closing the circle. Science cannot be performed in isolation from the social world. If we are lucky, a wider investigation may reveal unexpected elements of coherence. My goal in this book has been to put together the pieces that can support a new understanding of the historical development of the far-reaching sciences of human heredity.

Science, like every form of knowledge, depends on strategic simplification, but it is entirely mistaken to suppose that science makes the world thin or simple: that is an illusion created mostly by textbooks. Scientists work outward from graspable elements, which may be created experimentally or technologically and which are invaluable for making sense of a complex reality. Yet ostensible complexity is almost never exposed as mere illusion. Genetics has sometimes pretended to escape the tangled web of phenomena by looking behind them, away from societies, organisms, tissues, and cells to invisible, pearl-like genes or to a disembodied space of pure information. In the brave new era of genomic decoding, scientists again began dreaming that complex human abilities, disabilities, vulnerabilities, and behaviors could be reduced to genetic factors. At present, genomic science seems to be cycling back to the recognition of complexity. This view of life makes the world more interesting, since it draws on a wider range of material and intellectual tools. Complexity is not unknowability. Historians, though resolute defenders of complex interpretation, solve many little problems in the course of framing a comprehensible narrative, while science depends more than is usually realized on interpretation and narrative.[10]

Molecular genetics, now heavily invested in start-ups and takeovers, coalesced around a narrow view of what's "modern" in biology. Ironically, while the promise of genomics to check or cure dread disease and to repair inherited disabilities has had limited success, it offers great promise for the extension of historical knowledge into the more distant past, where documents are scarce or nonexistent, and on certain questions that written sources do not address well. Genomic technologies complement and enrich what can be learned from the study of physical and archeological remains and linguistic traces to reconstruct human migration patterns, diets, diseases, and, more generally, evolution. The press likes this kind of history, with its reliance on new technologies and findings that can be simply described as discoveries. The *New York Times* quoted recently this pronouncement by a geneticist: "For decades we have been trying to figure out what happened in the past. . . . And now we have a time machine."[11] In reality, we know incomparably more about times and places that are documented by archives and printed sources. While natural science has the potential to add information of great value, it would indeed be remarkable if genetic sources could provide historical knowledge rich enough to be articulated as an explanatory narrative.

ACKNOWLEDGMENTS

Sometimes I can't believe how long this book took. At other times it seems a wonder that I ever finished it at all. When to my delight I was nominated for a year at the Berlin Institute for Advanced Study, I decided to put forward the project whose outcome you have before you, but with a bad conscience, since I supposed the book would be done by then. Instead, I was barely halfway through the first draft when the fellowship year ended. I rarely wrote a paragraph without being driven back into the sources, and so it remained during the revisions. Patience, not my best attribute, was thrust upon me. Others had still less choice in the matter. Mary Terrall and Soraya de Chadarevian were my best critics, along with two anonymous reviewers and my editor, Brigitta van Rheinberg (who also had her patience taxed). I can never forget how fortunate I am to work at a university that values research alongside teaching and provides time and resources to do both. UCLA also has superb library facilities and librarians, who managed somehow to get every printed document I asked for.

I list below the archives and a few specialist libraries where I consulted manuscripts and rare books. Whether I visited for a few hours or for weeks, I depended on the generosity and expertise of archivists, first for collecting and cataloguing these precious resources, and then for personal guidance in identifying and using them. I benefited particularly from the knowledge and generosity of Russell Johnson and Teresa Johnson in History and Special Collections of the Louise Darling Biomedical Library at UCLA and Jack Eckert at the Countway Medical Library of Harvard University.

I began the project knowing a lot about the social and historical dimensions of data and statistics, but less about the history of heredity, and not much at all concerning asylums and psychiatry. I sent many queries to friends, colleagues, archivists, and librarians, often out of the blue, for help with things I didn't understand or for advice on sources. Almost everyone responded, and some went to considerable trouble to help me. Other friends provided help with language and translation issues, which I especially needed for the material on Norway. If I tried to thank everyone individually, it would be a very long list, and I would no doubt forget some. I make an exception for Diane Campbell, who, decades ago, taught me much of what I know about measures of phenotypic heredity. Otherwise, it may be appropriate simply to recall how much we depend on one another as colleagues, friends, students, and teachers. You might think we would engage in bitter rivalry, and this is not unknown, but it is overwhelmed by habits of generosity.

Once again, Mary Morgan and Charles Baden Fuller generously put me up in their lovely flat when I worked in London archives. I had the good fortune to be invited for several short-term research appointments in the company of interesting colleagues, and sometimes in proximity to unique archives or library collections. Among them were visits to the ESRC Genomics Policy and Research Forum, University of Edinburgh (April 2007), the Wellcome Trust Centre for History of Medicine and the Wellcome Library in London (June 2009), the Ecole des Hautes Etudes en Sciences Sociales in Paris (November 2008), and the Max Planck Institute for Social Anthropology, Halle, Germany (May 2012). A Norwegian Research Council grant for a project in which I participated, "The Cultural Logic of Facts and Figures" (2013–2015), provided valuable research support as well as stimulating intellectual exchange.

My gratitude to the Max Planck Institute for the History of Science, Berlin, extends beyond its support for a two-month summer residence in the summer of 2008. Lorraine Daston's division of the MPI has provided superb intellectual resources for generations of researchers for over two decades. A project on history of heredity in Hans-Jörg Rheinberger's division helped shape the idea for the approach I take here. My career in history of science has often brought me to Germany, and I take this opportunity to recall the important stimulus I received long ago, and still do, from programs and especially friendships formed at the Center for Interdisciplinary Research of the University of Bielefeld.

For research support on this project I acknowledge gratefully two Scholar's Awards from the National Science Foundation: SES 06-22346 (2006–2010) and SES 10-27100 (2010–2013). Those funds supported much of the travel and leave time that made the work possible. In combination with a grant from the University of California Humanities Society of Fellows, these NSF funds enabled me to spend a happy, collegial, and productive year (2010–2011) as a fellow at the Center for Advanced Study in the Behavioral Sciences, Stanford University. Three years later, in 2013–2014, I passed another wonderful year at an institute for advanced study, this time the Wissenschaftskolleg in Berlin, as part of a highly collegial research group with shared interests in numbers, measures, accounts, and rankings. In Berlin and Stanford, I got to trade in my car for a bicycle and to hang out with new people from diverse backgrounds under circumstances that encouraged conversation. Both also offered dedicated, friendly, and resourceful librarians and staff to make the year rewarding. So much to be grateful for!

My son, David Porter, who was in high school when I began this book, is now wrapping up a dissertation in Chinese history. I learn a lot from him these days.

18 September 2017

NOTES

Introduction. Data-Heredity Madness: A Medical-Social Dream

Epigraphs: *Bericht der zur Verwaltung der Irren-Heil-Anstalt ernannten Abgeordneten der Rheinischen Stände an den 5ten Ausschuss des 3ten Provinzial Landtages* (June 1830), ALVR, 1154, p. 3; Wilhelm Schallmayer, "Grundlinien der Vererbungslehre," in Siegfried Placzek, ed., *Künstliche Fehlgeburt und künstliche Unfruchtbarkeit, ihre Indikationen, Technik, und Rechtslage* (Leipzig: Verlag von Georg Thieme, 1918), 37. Throughout, if no translated edition is referenced, the translation is mine.

1. Soraya de Chadarevian and Harmke Kamminga, *Representations of the Double Helix* (Cambridge: Whipple Museum of the History of Science, 2002); see also Evelyn Fox Keller, *The Century of the Gene* (Cambridge, MA: Harvard University Press, 2000).
2. See especially Staffan Müller-Wille and Hans-Jörg Rheinberger, eds., *Heredity Produced: At the Crossroads of Biology, Politics, and Culture, 1500–1870* (Cambridge, MA: MIT Press, 2007); Staffan Müller-Wille and Christina Brandt, eds., *Heredity Explored: Between Public Domain and Experimental Science, 1850–1930* (Cambridge, MA: MIT Press, 2016).
3. Theodore M. Porter, "Asylums of Hereditary Research in the Efficient Modern State," in Müller-Wille and Brandt, eds., *Heredity Explored*, 81–109.
4. I mention only Horace Freeland Judson's *The Eighth Day of Creation: The Makers of the Revolution in Biology* (New York: Simon and Schuster, 1979), a great success with scientists as well as with a highbrow public.

Chapter 1. Bold Claims to Cure a Raving King
Let Loose a Cry for Data, 1789–1816

Epigraph: William Black, *An Arithmetical and Medical Analysis of the Diseases and Mortality of the Human Species*, 2nd ed. (London: C. Dilly, 1789), ii.

1. *Report from the Committee Appointed to Examine the Physicians who have attended His Majesty During His Illness, Touching the present State of His Majesty's Health* (London: 1789), 20, 25.
2. Andrea A. Rusnock, *Vital Accounts: Quantifying Health and Population in Eighteenth-Century England and France* (New York: Cambridge University Press, 2002); James Cassedy, *American Medicine and Statistical Thinking, 1800–1860* (Cambridge, MA: Harvard University Press, 1984); J. Rosser Matthews, *Quantification and the Quest for Medical Certainty* (Princeton, NJ: Princeton University Press, 1995).
3. Andrew Scull, *The Most Solitary of Afflictions: Madness and Society in Britain, 1700–1900* (New Haven, CT: Yale University Press, 1993).
4. John Strype and John Stow, *A Survey of the Cities of London and Westminster, Containing The Original, Antiquity, Increase, Modern Estate and Government of those Cities, Written at first in the Year MDXCVIII by John Stow . . . Corrected, Improved and very much Enlarged . . . by John Strype*, 2 vols. (London: A. Churchill, 1720). The section "Bethlem Hospital, commonly called Bedlam" is on 192–197.
5. Jonathan Andrews, "Bedlam Revisited: A History of Bethlem Hospital, c1634–c1770," PhD diss. (London University, 1991), 489–490; Jonathan Andrews et al., *The History of Bethlem* (London: Routledge, 1997), 338.

6. William Black, *Observations Medical and Political, on the Small-pox, and the Advantages and Disadvantages of General Inoculation, especially in Cities, and on the Mortality of Mankind at every Age in City and Country* . . . (London: J. Johnson, 1781), 229; Black, *A Comparative View of the Mortality of the Human Species at all Ages; and of the Diseases and Casualties by which they are destroyed or annoyed* (London: C. Dilly, 1788), 235. Black's *Comparative View* appeared in German translation in 1789 but seems not to have made much of an impression. Black, *Arithmetical and Medical Analysis*, 129–130.
7. Black, *Arithmetical and Medical Analysis*, dedication, ii–iv, and 130. The history is recounted in Ida Macalpine and Richard Hunter, *George III and the Mad Business* (New York: Pantheon Books, 1969), esp. 297–299, and dramatized in Alan Bennett's play *The Madness of George III* (1991), on which was based the 1994 film *The Madness of King George*." See also Alan R. Rushton, *Genetics and Medicine in Great Britain, 1600–1939* (Victoria, BC: Trafford Publishing, 2009), 48–49.
8. Black, *Arithmetical and Medical Analysis*, dedication and iii.
9. Black, *Observations Medical and Political*, 229–230.
10. Black, *Arithmetical and Medical Analysis*, 140–141.
11. Black, *Arithmetical and Medical Analysis*, dedication and iii; Andrews et al., *History of Bethlem*, 369; John Haslam, *Observations on Insanity: with Practical Remarks on the Disease, and an Account of the Morbid Appearances on Dissection* (London: F. and C. Rivington, 1798), 114–115. Haslam repeated this in his second edition, *Observations on Madness and Melancholy: Including Practical Remarks on those Diseases, together with Cases: and an Account of the Morbid Appearances on Dissection* (London, 1809), 251–253. On Haslam, see Andrew Scull, Charlotte MacKenzie, and Nicholas Hervey, *Masters of Bedlam: The Transformation of the Mad-Doctoring Trade* (Princeton, NJ: Princeton University Press, 1996), chap. 2.
12. On financial calculations see William Deringer, *Calculated Values: Finance, Politics, and the Quantitative Age* (Cambridge, MA: Harvard University Press, 2018); on the ethic of recordkeeping and publicity, see Edward Higgs, *The Information State in England: The Central Collection of Information on Citizens since 1500* (London: Palgrave Macmillan, 2004); Thomas Crook and Glenn O'Hara, eds., *Statistics and the Public Sphere: Numbers and the People in Modern Britain, c. 1800 to 2000* (London: Routledge, 2011). See also the 1737 report of the newly founded County Hospital for the Sick and Lame at Winchester (available at Eighteenth Century Collections Online, Gale document no. CW106936014).
13. Committee on Madhouses, *Report from the Committee on Madhouses in England* (July 1815).
14. Nancy Tomes, *A Generous Confidence: Thomas Story Kirkbride and the Art of Asylum-Keeping, 1840–1883* (Cambridge: Cambridge University Press, 1984), 294.
15. Thomas Percival, *Medical Ethics, or a Code of Institutes and Precepts, adapted to the Professional Conduct of Physicians and Surgeons* (Manchester: J. Johnson, 1803), 15–17.
16. Black, *Comparative View*, numbered list of causes, 250; Black, *Arithmetical and Medical Analysis*, table of causes, 133. On medical tables from the early modern period, see Rusnock, *Vital Accounts*. Black was as skeptical as Haslam about official Bethlem cure rates.
17. Black, *Comparative View*, 249; Black, *Arithmetical and Medical Analysis*, 143.
18. Black, *Comparative View*, 249; Roy Porter, *Mind-Forg'd Manacles: A History of Madness in England from the Restoration to the Regency* (London: Athlone Press, 1987), 33–34.
19. Haslam, *Observations on Insanity*, 98–99.

20. John Johnstone, *Medical Jurisprudence: On Madness* (Birmingham: J. Johnson, 1800), 8, 10. John A. Paris and John S. M. Fonblanque, *Medical Jurisprudence*, vol. 1 (London: W. Phillips, 1823), 325, also refer to the possibility of latency in one generation and reemergence in the next. I owe these references to John Carson.
21. See Alfred S. Taylor, *Medical Jurisprudence*, 2nd ed. (Philadelphia: Lea and Blanchard, 1850),619, 651; on tables and statistics see the 3rd American ed. (1853, from 4th London ed.), 556; Charles E. Rosenberg, *The Trial of the Assassin Guiteau: Psychiatry and the Law in the Gilded Age* (Chicago: University of Chicago Press, 1968).
22. Haslam, *Observations on Madness and Melancholy*, 225–226, 231, 229.
23. J.E.D. Esquirol, "Folie," *Dictionnaire des sciences médicales*, 60 vols. (Paris: C.L.F. Panckoucke, 1812–22), vol. 16 (1816), 188; see Carlos Lopez Beltrán, "In the Cradle of Heredity: French Physicians and L'Hérédité Naturelle in the Early 19th Century," *Journal of the History of Biology* 37 (2004), 39–72; Laure Cartron, "Degeneration and 'Alienism' in Early Nineteenth-Century France," in Staffan Müller-Wille and Hans-Jörg Rheinberger, *Heredity Produced: At the Crossroads of Biology, Politics, and Culture, 1500–1870* (Cambridge, MA: Harvard University Press, 1984), 155–174.
24. Erwin Ackerknecht, "Diathesis: The Word and the Concept in Medical History," *Bulletin of the History of Medicine* 56 (1982), 317–325; Robert C. Olby, "Constitutional and Hereditary Disorders," in W. F. Bynum and Roy Porter, eds., *Companion Encyclopedia of the History of Medicine*, 2 vols. (New York: Routledge, 2001), 1: 414–416.
25. Table reprinted with corrections in J.E.D. Esquirol, *Des maladies mentales considérées sous les rapports médical, hygiénique et médico-légal*, 2 vols. (Paris: J.-B. Baillière, 1838), 1: 64; Georg Schweig, "Auseinandersetzung der statistischen Methoden in besonderem Hinblick auf das medicinische Bedürfniss," *Archiv für physiologische Heilkunde* 13 (1854), 313, a paper discussed in chap. 6 below.
26. J.E.D. Esquirol, "Mémoire historique et statistique sur la Maison Royale de Charenton," *Annales d'hygiène publique et de médecine légale* 13 (1835), 142–143; reprinted in Esquirol, *Maladies mentales*, 2: 683.

Chapter 2. Narratives of Mad Despair Accumulate as Information, 1818–1845

Epigraphs: Black, *Arithmetical and Medical Analysis*, 149; Gerald Grob, *The State and the Mentally Ill: A History of Worcester State Hospital in Massachusetts, 1830–1920* (Chapel Hill: University of North Carolina Press, 1966), 7. Awl was the founding asylum superintendent in Columbus, Ohio; Woodward, in Worcester, Massachusetts.

1. H. Laehr, "Ueber periodische Berichte aus Irrenanstalten," *Allegemeine Zeitschrift für Psychiatrie* (*AZP*) 32 (1875), 80–82, from a speech given on 14 Dec. 1874. On exchanges of reports, see *AZP* 33 (1877), 56–57.
2. Peter Becker and William Clark, eds., *Little Tools of Knowledge: Historical Essays on Academic and Bureaucratic Practices* (Ann Arbor: University of Michigan Press, 2001). On law as a stimulus to recordkeeping on asylum patients in Germany, see Volker Hess, "Die Buchhaltung des Wahnsinns: Archiv und Aktenführung zwischen Justiz und Irrenreform," in Cornelius Borck and Armin Schäfer, eds., *Das psychiatrische Aufschreibesystem* (Paderborn: Fink, 2015), 55–76.
3. Michel Foucault, *Psychiatric Power: Lectures at the Collège de France, 1973–74*, ed. Arnold Davidson, trans. Graham Burchell (London: Palgrave Macmillan, 2008). See also Andrew Scull's critique of Foucault's *Madness and Civilization* in the introduction to his *Social Order/Mental Disorder: Anglo-American Society in Historical Perspective* (Berkeley: University of California Press, 1989).

4. Samuel Hanbury Smith, "Superintendent's Report," *Thirteenth Annual Report of the Directors and Superintendent of the Ohio Lunatic Asylum to the 50th General Assembly of the State of Ohio for the Year 1851* (Columbus, 1852), 70, 65.
5. Erving Goffman, *Asylums: Essays on the Social Situation of Mental Patients and Other Inmates* (Garden City, NY: Anchor Books, 1961).
6. Andrews, *Bedlam Revisited*, 4; William Farr, *On the Statistics of English Lunatic Asylums and the Reform of their Public Management* (London: Sherwood, Gilbert, and Piper, [1840?]), esp. 42–43.
7. UCLA Louise Darling Biomedical Library, Library Special Collections for Medicine and the Sciences, no. 39, Case Records of Patients Admitted to St. Lukes Hospital for Lunatics 1839–1840, record 7422989. His name was given as Will[m] Shakespear in the entry, but as Shakespeare, W. in the index.
8. Medical Center of New York-Presbyterian/Weill Cornell, New York, NY, Archives of Bloomingdale Asylum, Medical Register of the Bloomingdale Asylum, 31 Dec. 1844—August 1866.
9. Samuel Tuke, *Description of the Retreat, an Institution near York, for Insane Persons of the Society of Friends. Containing an Account of its Origin and Progress, the Modes of Treatment, and a Statement of Cases* (York: W. Alexander, 1813), quote on 208. The printed report, which I found in the National Library of Medicine, Bethesda, MD, is *State of an Institution near York called The Retreat for Persons Afflicted with Disorders of the Mind 1802/03* (Whitby, 1803). See also Ann Digby, *Madness, Morality and Medicine. A Study of the York Retreat, 1796–1914* (Cambridge: Cambridge University Press, 1985), 214.
10. Borthwick Institute for Archives (BIA), York University, Archives of Retreat at York, RET 6/1/1/1A and 6/1/2, Admission Papers 1826–1833. Quotation from 6/1/2, form 104.
11. BIA, RET 6/1/4, Admissions Papers 1826–1833, where the new form is introduced. The printed passage emphasizing the value for patient treatment of all this information was replaced about 1846 with language required (according to the document itself) by an act from the parliamentary session of 1845 (8&9 Victoria).
12. Isaac Ray, "Shakespeare's Delineations of Insanity," *American Journal of Insanity* (*AJI*) 3 (1846–47), 289–332, 291.
13. The ascribed moral causes of insanity were similar over much of the asylumized world. Ann Goldberg, *Sex, Religion, and the Making of Modern Madness: The Eberbach Asylum and German Society, 1815–1849* (New York: Oxford University Press, 1999), explains moral causes such as masturbation as repressive moves directed especially at women, peasants, and religious dissidents.
14. Quotations from BIA, RET 6/1/4. John Thurnam printed other declarations of this kind in *Observations and Essays on the Statistics of Insanity and on Establishments for the Insane to which are added the Statistics of the Retreat near York* (London: Simpkin, Marshall; York: John L. Linney, 1845); see foldout chart between pp. 70 and 71, titled "Recommendations for Filling up the Register of Cases, agreed to at the Annual Meeting of the Association of Medical Officers of Hospitals for the Insane, held at the Asylum, Lancaster, June 2nd and 3rd, 1842." An example from an American asylum of the "circular letter," to be filled out by relatives or guardians, with the assistance of a physician, may be found in *Twenty-Second Annual Report of the Officers of the Retreat for the Insane at Hartford* (Hartford, 1846), 49–51. Question 17 asks about "any constitutional or hereditary predisposition."
15. Thurnam, *Observations and Essays*, 5. On gaming of statistics see Wendy Nelson Espeland and Michael Sauder, *Engines of Anxiety: Academic Rankings, Reputation, and Accountability* (New York: Russell Sage Foundation, 2016), 79–87, 145–148, 194–196.

16. Theodric Romeyn Beck, "Statistical Notices of Some of the Lunatic Asylums of the United States," *Transactions of the Albany Institute* 1, part 1 (1830), 80 (paper read 16 and 29 Apr. 1829); *Report of the Medical Visitors of the Connecticut Retreat for the Insane, Presented to the Society May 13 1830* (Hartford, 1830), 4.
17. Pliny Earle, "Researches in reference to the Causes, Duration, Termination, and Moral Treatment of Insanity," *American Journal of the Medical Sciences* 22 (1838), 348, 339; Beck, "Statistical Notices," 80. On these cure rates and the pride they provoked, see Lawrence B. Goodheart, *Mad Yankees: The Hartford Retreat for the Insane and Nineteenth-Century Psychiatry* (Amherst: University of Massachusetts Press, 2003), 50–56, 62–63.
18. John Curwen, *History of the Association of Medical Superintendents of American Institutions for the Insane from 1844 to 1874, Inclusive* (privately printed, 1875); "Report of the Commissioners Appointed to superintend the Erection of a Lunatic Asylum at Worcester," made 4 Jan. 1832, in *Reports and Other Documents Relating to the State Lunatic Hospital at Worcester, Mass.* (Boston, 1837), 1–36, 26.
19. Quote from Smith, "Superintendent's Report," 19; *Population of the United States in 1860, Compiled from the Original Returns of the Eighth Census under the Direction of the Secretary of the Interior by Joseph C. G. Kennedy, Superintendent of the Census* (Washington, DC: Government Printing Office, 1864), preface, c [100].
20. David Gollaher, *Voice for the Mad: The Life of Dorothea Dix* (New York: Free Press, 1995), 192–193; Andrew Scull, *Social Order / Mental Disorder: Anglo-American Psychiatry in Historical Perspective* (Berkeley: University of California Press, 1989), 112.
21. Thurnam's epigraphs from *Observations and Essays*.
22. Thurnam, *Observations and Essays*, vi–vii, x–xi, 5–6, including quotes from Carlyle and Chalmers. In his pamphlet, *Chartism*, chap. 2, Carlyle dismissed statistics as "like cobwebs, like the sieve of Danaides; beautifully reticulated, orderly to look upon, but which will hold no conclusion."
23. John Thurnam, *Statistics of the Retreat: Consisting of A Report and Tables exhibiting the Experience of that Institution for the Insane from its Establishment in 1796 to 1840* (York, 1841), esp. 18–19.
24. Theodore M. Porter, "Genres and Objects of Social Inquiry from the Enlightenment to 1890," in Theodore M. Porter and Dorothy Ross, eds., *Cambridge History of Science*, vol. 7, *Modern Social Sciences* (Cambridge: Cambridge University Press, 2003), 13–39; Porter, "The Social Sciences," in David L. Cahan, ed., *From Natural Philosophy to the Sciences: Writing the History of Nineteenth-Century Science* (Chicago: University of Chicago Press, 2003), 254–290.
25. Thurnam, *Observations and Essays*, iv–v.
26. BIA, RET 6/2/2/1, Administrative Registers of Cases.
27. The pioneering statistical document from Worcester came out in 1837: *Reports and other Documents Relating to the State Lunatic Hospital at Worcester, Mass.* Bemis expressed his doubts in the *Thirty-First Annual Report of the Trustees of the State Lunatic Hospital at Worcester*, October 1863 (Boston, 1864), 48. On Woodward's devotion to statistics, see Grob, *State and Mentally Ill*, 73–74.
28. Worcester Lunatic Hospital, Admission Books, held by Harvard Countway Library, Center for History of Medicine, box 1, first volume, 1833–1861.
29. On case records and case books in US asylums, see Richard Noll, *American Madness: The Rise and Fall of Dementia Praecox* (Cambridge, MA: Harvard University Press, 2011), 30–31.
30. Countway Library of Medicine, Rare Books and Special Collections, Worcester Lunatic Hospital / Worcester State Hospital Records. MC box 3, Patient Case Books, 1833–1837, books 1 and 2.

31. Woodward's interview with Ezra in this paragraph and the three following comes from Countway Library, Worcester State Hospital Records, Case Book 7.
32. See "Superintendent's Report," 64–102, in *Seventh Annual Report of the Trustees of the State Lunatic Hospital at Worcester*, 1839 (Boston, 1840), 73.
33. *Columbian Star and Christian Index*, vol. 1, 26 Sept. 1829, p. 208, under "Promiscuous Particulars."
34. From the third annual report of the asylum, for 1835, included in *Reports and other Documents*, 92; see also 209–210. These cases were reprinted to assist asylum planning in New Jersey: *Report of the Commissioners Appointed by the Governor of New Jersey to Ascertain the Number of Lunatics and Idiots in the State* (Newark, NJ, 1840), 19–21.
35. Entries from patient tables in *Reports and other Documents*, 46, 78, 104, 141; images are from p. 46 and from *Fifth Annual Report of the Trustees of the State Lunatic Hospital at Worcester, December 1837* (Boston, 1838), 22. Ezra is patient 45. Grob, *State and Mentally Ill*, 84, notes the 1832 Massachusetts law authorizing the removal of dangerous lunatics from jails to mental hospitals.
36. *Fifth Annual Report*, 59–65, summarizes Ezra's story from the case book. Woodward did not mention in print Ezra's prohibition of writing during the interview. In 1844, Issac Ray persuaded the Massachusetts chief justice, Lemuel Shaw, to spell out the doctrine of "irresistible impulse," which the jury then accepted as grounds for acquittal of a prison inmate for the murder of a warden. See Alan Rogers, *Murder and the Death Penalty in Massachusetts* (Amherst: University of Massachusetts Press, 2008), 209–210.
37. Bethlem Hospital Archives, Case Books, CB 12 (1822–25) and various subsequent case books to 1894. The archives contain minute books of admission hearings from 1709 to the twentieth century. For example: "Bethlehem Sub-committee Book Minute Book 1854–1856," HCM 36, box 06/2.
38. London Metropolitan Archives (LMA), H11/HLL/B20/1, Hanwell Lunatic Asylum, Case Book: Males, 1845–1850, all from 1846.
39. LMA, H11/HLL-B20/1, Case Book: Males, 1845–1850, 80–81.
40. LMA, H11/HLL/B19/1–3, Hanwell Lunatic Asylum, Case Book for Woman Patients, 1831–1850, 304.
41. LMA, H11/HLL/B19/1–3, Case Book for Woman Patients, 1831–1850, 13, 319.
42. See ALVR, 1157, "Bericht der standischen Untersuchungs-Commission über die Irren-Heil-Anstalt zu Siegburg," dated 14 June 1843, 11; also Marga Maria Burkhardt, "Krank im Kopf: Patienten-Geschichten der Heil- und Pflegeanstalt Illenau, 1842–1889," Ph.D. diss., Fakultät der Albert-Ludwigs-Universität Freiburg i. Br., 2003, 39.
43. Staatsarchiv Nürnberg, Tit: V No. 2046 4078f Akten der Koeniglich Bayer. Regierung von Mittelfranken, Kammer des Irren. Die Kreisirrenanstalt zu Erlangen, 1851–1897.
44. [Ernst Albert von] Zeller, "Bericht über die Heilanstalt Winnenthal von ihrer Eröffnung den 1. März 1834 bis zum 28. Februar 1837, *Medizinischen Correspondenzblatt des Württembergischen ärztlichen Vereins* 7, no. 30 (5 Aug. 1837), Beilage, 321–335.
45. William Hutcheson, "Physician's First Annual Report," in *Twenty-Eighth Annual Report of the Directors of the Glasgow Royal Asylum for Lunatics* (1841; Glasgow, 1842), 23.

Chapter 3. New Tools of Tabulation Point to Heredity as the Real Cause, 1840–1855

Epigraph: *Report of the medical Superintendent of the Provincial Lunatic Asylum, Toronto, for the Year 1860*, 10.

1. BIA, RET 6/5/1/1A Case Book, patients 1–3.

2. BIA, RET 6/5/1/1A Case Book, patients 9, 18, and 19.
3. BIA, RET 6/4/1/1A Case Book, patients 28 and 83.
4. BIA, RET 6/4/1/1A Case Book, patient 236; BIA RET 6/5/1/2 Case Book 1828–1837, 188.
5. Thurnam, *Statistics of the Retreat*, 18–19.
6. George Chandler, "Superintendent's Report," *Seventeenth Annual Report of the Trustees of the State Lunatic Hospital at Worcester* (Boston, 1850), 47.
7. James Cowles Prichard, *A Treatise on Insanity and Other Disorders Affecting the Mind* (London: Sherwood, Gilbert, and Piper, 1835), 336.
8. Isaac Ray, "Observations on the Principal Hospitals for the Insane," *AJI* 2 (April 1846), 289.
9. Pliny Earle, *A Visit to Thirteen Asylums for the Insane in Europe; to which are Added a Brief Notice of Similar Institutions in Transatlantic Countries and in the United States* (Philadelphia: J. Dobson, 1841), 3, 131; Pliny Earle, *Institutions for the Insane in Prussia, Austria and Germany* (New York: Samuel S. and William Wood, 1854).
10. Heinrich Laehr, "Die Insel San Servolo bei Venedig," *AZP* 27 (1871), 189–199, 189. See also archives at San Servolo, Venice, Italy, holding records of the Manicomio Centrale for San Servolo (male patients) and the Frenicomio Centrale at San Clemente (female patients). Along with patient records there are many statistical documents: "Tavole Statistiche Degle Alienati che ebbero cura nel Morocomio Centrale di San Servolo," with tables for 1850–60; also *Tavolo Statistische Triennali 1865–1866–1867 del Moricomio Centrale Maschile in San Servolo di Venezia*; and *Relazione statistica del Manicomio Centrale Maschile in S. Servolo di Venezie pel Quinquennio 1884–1888*.
11. E. T. Wilkins, *Report of E. T. Wilkins, M. D., Commissioner in Lunacy for the State of California, Made to His Excellency H. H. Haight, Governor, December 2d, 1871* (Sacramento, 1872); G. A. Tucker, *Lunacy in Many Lands* (Sydney, 1887).
12. Lead editor Jules Baillarger's "Introduction," *Annales médico-psychologiques* (*AMP*) 1 (1843), i–xxvii. On census issues, see Ramon de la Sagra, "Statistique des aliénés et des sourds-muets dans les États-Unis de l'Amérique du Nord," *AMP* 1 (1843), 281–288; discussion of the number of insane in France, led by A. Moreau de Jonnès, *AMP* 2 (1843), 300–303, 467; no author given, "De la prédominance des causes morales dans la génération de la folie," *AMP* 2 (1843), 358–371; see also chap. 5 below.
13. H. Aubanel and M. Thore, "Recherches statistiques sur l'aliénation mentale, faites à l'Hospice de Bicêtre," *AMP* 3 (1844), 141–147; "Statistique de l'Asile des Aliénés de Maréville," *AMP* 3 (1844), 434–443; A.J.F. Brierre de Boismont, Review of *Statistique de Bethlem*, by John Webster, *AMP* 3 (1844), 443–448; Le Payen, "Notice historique et statistique de l'Hospice d'Aliéns d'Orléans," *AMP* 4 (1844), 278–286.
14. Notice and review (by A.J.F. Brierre de Boismont) of the new AZP in *AMP* 4 (1844), 156 and *AMP* 5 (1845), 291–297; M. Falret, "Visite á l'Établissement d'Illenau," *AMP* 5 (1845), 419–444 and 6 (1845), 69–106.
15. Notice of Baillarger, "Heredity of Insanity," *AJI* 1 (1844–45), 92–93; "Miscellaneous," *AJI* 1 (1844–45), 286, 285.
16. M. Parchappe, *Recherches statistiques sur les causes de l'aliénation mentale* (Rouen, 1839), 2, 7–9, 60–61; see also Parchappe, "Asile des aliénés de Rouen," *AMP* 3 (1844), 133–135; L. De Boutteville [Deboutteville] and M. Parchappe, "Notice statistique sur l'Asile des Aliénés de la Seine-Inférieure," *AMP* 7 (1846), 133–146.
17. L. Deboutteville and M. Parchappe, *Notice statistique sur l'Asile des Aliénées de la Seine-Inférieure (Maison de Saint Yon de Rouen), pour la période comprise entre le 11 Juillet 1825 et le 31 Décembre 1843* (Rouen, 1845), i–ii, 18–19. See also Frédéric Carbonel, "L'Asile pour aliénés de Rouen: Un laboratoire de statistiques morales de la Restauration à 1848," *Histoire et mesure* 20 (2005), 97–136.

18. Thurnam, *Observations and Essays*, iv.
19. Thurnam, *Observations and Essays*, iii, vii–ix, 5–6; P. W. Jessen, "Aerztliche Erfahrungen in der Irrenanstalt bei Schleswig," *Zeitschrift für die Beurtheilung und Heilung der krankhaften Seelenzustände* 1 (1838), 585, 590, 599, 601.
20. *First Annual Report of the Managers of the State Lunatic Asylum* (of New York, for 1843) (Albany, 1846), 18–24, esp. 21; *Second Annual Report* (for 1844), 24–25; *Third Annual Report* (for 1845), 53–56.
21. *Ninth Annual Report of the Managers of the State Lunatic Asylum* (of New York, for 1851), 21–22. See also the cumulative table in *Twenty-Second Annual Report of the Officers of the Retreat for the Insane at Hartford* (Hartford, 1846), 21–22; Bénédict Augustin Morel, review of reports of the Worcester Lunatic Hospital and the Connecticut Retreat, in "Rapport sur les asiles des États-Unis (anal)," *AMP* 9 (1847), 302–306 and "Rapport sur les établissements d'aliénés des États-Unis et de l'Angleterre (anal)," 10 (1847), 299–304.
22. *Fourteenth Annual Report of the Board of Trustees for the Benevolent Institutions and of the Officers of the Ohio Lunatic Asylum to the General Assembly of Ohio for the Year 1852* (Columbus, 1853), 10; *Fifteenth Annual Report . . . for 1853*, 5, 19; *Sixteenth Annual Report . . . for 1854*, 15; *Seventeenth Annual Report . . . for 1855*, 41.
23. Woodward's reports described masturbation as an avoidable vice and as the one with the lowest cure rate. He wrote a best-selling tract against it: *Hints for the Young in Relation to the Health of Body and Mind* (Boston: George W. Light, 1838). Samuel G. Howe, *The Causes of Idiocy* (Edinburgh: Maclachlan and Stewart, 1848), 3.
24. Smith, "Superintendent's Report," 37.
25. Smith, "Superintendent's Report," 40, 44.
26. Amariah Brigham, "First Annual Report of the Superintendent," *Annual Report of the Managers of the State Lunatic Asylum, made to the Legislature January 18, 1844* (of New York, for 1843), 19, 21.
27. *Eighth Annual Report of the Managers of the State Lunatic Asylum* (of New York, for 1851), 31–32, authored, it would appear, by Dr. N. D. Benedict or Dr. George Cook.
28. Hutcheson in *28th Glasgow Report 1839*, 23–24; report of physician-superintendent Mackintosh in *41st Glasgow Report* (1854), 32.
29. Robert K. Reid, Resident Physician, *Report of the Board of Trustees Stockton State Hospital, 1853*, Document no. 18 of the California State Assembly, 45.
30. *Report of the Board of Trustees of the Insane Asylum of the State of California. Submitted to the Legislature, January 20, 1854*, 23–25.
31. *Biennial Reports of the Directors and Medical Visitors and Fifteenth Annual Report of the Superintendent of the Insane Asylum of California, 1867*, 27–28. On explanations of the high rate of insane commitments in California, see Richard W. Fox, *So Far Disordered in Mind: Insanity in California, 1870–1930* (Berkeley: University of California Press, 1978), 17–27.
32. State of Washington, *Report of the Board of Trustees of the Western Washington Hospital for the Insane, 1898–1900* (Olympia, 1901). More elaborate causes may be found in Washington State Archives, Olympia, Western State Hospital, box 64, Record of Commitment (1891–1899), cases for 1891.
33. Nicolaus Heinrich Julius, *Beiträge zur britischen Irrenheilkunde aus eignen Anschauungen im Jahre 1841* (Berlin: Theod. Chr. Fr. Enslin, 1844), esp. 213–214; Samuel Tuke, "Introduction" and "Statistics of Insanity," in Maximilian Jacobi, *On the Construction and Management of Hospitals for the Insane: with a Particular Notice of the Institution at Siegburg*, trans. John Kitching (London: John Churchill, 1841), iii–xliv and xliv–lxxii.

34. Heinrich Damerow, "Einleitung," *AZP* 1 (1844), i–xlviii, xiv; Damerow, "Die Zeitschrift. Ein Blick rückwärts und vorwärts," *AZP* 3 (1846), 17–19; Salina Braun, *Heilung mit Defekt: Psychiatrische Praxis an den Anstalten Hofheim und Siegburg, 1820–1878* (Göttingen: Vandenhoeck & Ruprecht, 2009), 74–75; Eric J. Engstrom, *Clinical Psychiatry in Imperial Germany* (Ithaca, NY: Cornell University Press, 2003), 35–44. See also Winfried Berghof, *Heinrich Damerow (1798–1866)—Ein bedeutender Vertreter der deutschen Psychiatrie des 19. Jahrhunderts*, Dissertation zur Erlangung des akademischen Grades Dr. med., Karl-Marx-Universität Leipzig, 1990, 96–108.
35. Damerow, "Einleitung"; C. F. Flemming, "Einladung an die Irrenanstalts-Direktoren zur Benutzung gemeinschaftlicher Schemata zu den tabellarischen Uebersichten," *AZP* 1 (1844), 430–440, quotes at 430–431. Although this paper appeared under Flemming's name, he later explained that he was writing as a reporter for all three editors: "Betreffend die Aufstellung eines Normal-Schemas für irrenstatistische Uebersichten," *AZP* 3 (1846), 665–675, 665, fn. On the archival vision, see Lorraine Daston, ed., *Science in the Archives: Pasts, Presents, Futures* (Chicago: University of Chicago Press, 2017).
36. Thurnam, *Observations and Essays*, introduction, iii–xii; Dr. Bernhardi, "Irrenstatistische Bemerkungen zu dem Vorschläge eines Normalschemas tabellarische Uebersichten," *AZP* 2 (1845), 269–271;
37. Bernhardi, "Irrenstatistische Bemerkungen," 277–278. The editors of the *AZP* revised their own recommendations in 1846 in response to this paper; see "Betreffend die Aufstellung." On causes, see Thomas Schlich, "Die Konstruktion der notwendigen Krankheitsursache: Wie die Medizin Krankheit beherrschen will," in Cornelius Borck, ed., *Anatomien medizinischen Wissens. Medizin Macht Moleküle* (Frankfurt am Main: Fischer Taschenbuch Verlag, 1996), 201–229.
38. Pliny Earle, "On the Causes of Insanity, as exhibited by the Records of the Bloomingdale Asylum from June 16th, 1821, to December 31st, 1844," *AJI* 4 (1847–48), esp. 186–192; Earle, *History, Description and Statistics of the Bloomingdale Asylum for the Insane* (New York, 1848), 76–100; Earle, "Causes of Insanity," esp. 76–82.
39. On his asylum tour, see Gerhart Zeller, ed., *Albert Zellers medizinisches Tagebuch der psychiatrischen Reise durch Deutschland, England, Frankreich und nach Prag von 1832 bis 1833* (Zwiefalten: Verlag Psychiatrie und Geschichte der Münsterklinik, 2007); Thomas Müller, Bobo Rüdenburg, and Martin Rexer, eds., *Wissenstransfer in der Psychiatrie: Albert Zeller und die Psychiatrie Württembergs im 19. Jahrhundert* (Zwiefalten: Verlag Psychiatrie und Geschichte des Zentrums für Psychiatrie, 2009), 27–32.
40. Ernst Albert von Zeller, "Bericht über die Wirksamkeit der Heilanstalt Winnenthal von ihrer Eröffnung den 1. März 1834 bis zum 28. Februar 1837," *Medizinischen Correspondenzblatt des Württembergischen ärztlichen Vereins* 7, no. 30 (5 Aug. 1837), Beilage, 231–335, 334; Zeller, "Zweiter Bericht über die Wirksamkeit der Heilanstalt Winnenthal vom 1. März 1837 bis zum 29. Febr. 1840," *Medizinisches Correspondenz-Blatt des Würtembergischen ärztlichen Vereins* 10 (1840), no. 17–18–19, 129–147, 132, 144.
41. Zeller, "Bericht über die Wirksamkeit der Heilanstalt Winnenthal vom 1. März 1840 bis 28 Febr. 1843," *AZP* 1 (1844), 74–76; reprinted from *Medizinisches Correspondenzblatt des Württembergischen Ärztlichen Vereins* 13, no. 38 (1843), 297–322.
42. Staatsarchiv München, RA 57430 Irrenanstalten Allgemeines, "Satzungen für die oberpfälzisch-regenburgische Kreis-Irren-Anstalt zu Karthaus-Prüll," issued by the Königliche Regierung der Rezatkreises, Kammer des Innern.
43. Archiv des Bezirks Oberbayern (ABO): *Satzungen der oberbayerischen Kreis-Irren-Anstalt zu München* (1859), Beilage, 18–20; *Revidirte Satzungen der Kreis-Irren-Anstalt für Oberbayern in München* (Munich, 1876), Beilage, 23.

44. Ernst Albert von Zeller, "Vorrede," J. Guislain, *Phrenopathien oder neues System der Seelenstörungen gegründet auf praktische und statistische Beobachtungen und Untersuchung der Ursachen . . .*, translated from French by Dr. Wunderlich (Stuttgart: L. F. Rieger, 1838), 28; C. F. Flemming, "Aphorismen zur Prognostik der Geistes-Verwirrung," *Zeitschrift für die Beurtheilung und Heilung der krankhaften Seelenzustände* 1 (1838), 395–407, 396.
45. Zeller, "Bericht 1840 bis 1843," 53.
46. Ernst Albert von Zeller, "Bericht über die Wirksamkeit der Heilanstalt Winnenthal vom 1. März 1846 bis 28 Februar 1854," *Medicinisches Correspondenz-Blatt des Württembergischen Ärztlichen Vereins* 24, no. 38 (October–November 1854), 297–318, 298.
47. L.F.E. (Emile) Renaudin, "De la statistique appliquée à l'étude des maladies mentales . . . , Lettre à M. Baillarger . . . ," *AMP* 7 (1846), 468.
48. Zeller, "Bericht 1846 bis 1854," 302–303. *Seelenorgan* refers to the link between the mind and the external world; see Michael Hagner, *Homo Cerebralis: Der Wandel vom Seelenorgan zum Gehirn* (Berlin: Berlin Verlag, 1997).

Chapter 4. The Census of Insanity Tests Its Status as a Disease of Civilization, 1807–1851

Epigraph: Étienne Esquirol, "Remarques sur la statistique des aliénés, et sur le rapport du nombre des aliénés à la population. Analyse de la statistique des aliénés de la Norwège," *Annales d'hygiène publique et de médecine légale* 4 (1830), 340.

1. From a series of articles in *Literary Panorama* 2 (1807), 1255–1263, including "Lunatics. Thoughts on the State of Criminal and Pauper Lunatics, in England and Wales: Excited by the Measures Pursued in the Late Parliaments," 1255–1256; and "Report from the Select Committee, appointed by the Honourable House of Commons, to enquire into the State of the Criminal and Pauper Lunatics, in England and Wales, and of the Laws relating thereto—July 15, 1807," 1256–1261; quotations from 1255.
2. "Copy of a Letter from Dr. Halliday to Mr. William Wynn," followed by population table of districts of England "for Erection of Lunatic Asylums," in *Literary Panorama* 2 (1807), 1261–1263; Richard Powell, "Observations upon the comparative Prevalence of Insanity, at different periods," *Transactions of the College of Physicians* (1 Mar. 1814), 150–151.
3. Andrew Halliday, *A General View of the Present State of Lunatics and Lunatic Asylums in Great Britain and Ireland and in Some Other Kingdoms* (London: Thomas and George Underwood, 1828). See also Halliday, *A Letter to Lord Robert Seymour with a Report of the Number of Lunatics and Idiots in England and Wales* (London: Thomas and George Underwood, 1829).
4. Dora B. Weiner, *Comprendre et soigner: Philippe Pinel (1745–1826). La médecine de l'esprit* (Paris: Fayard, 1999), 143.
5. J.E.D. Esquirol, *Des passions considérées comme causes, symptômes, et moyens curatifs de l'aliénation mentale*, thèse de médecine de Paris, no. 574, présenté et soutenue à l'École de Médecine de Paris le 7 nivose an 14 (Paris, An XIV [1805]), 14–15, 20.
6. Esquirol, "Folie," 177–180; John Carr, *A Northern Summer, or Travels Round the Baltic* (Hartford: Lincoln and Gleason, 1805), 208–209; French translation as *L'Été du nord* (1808). Carr explained this low level of insanity in terms of climate, not Russian primitivism. For a few more examples, see Andrew Scull, *Madness in Civilization: A Cultural History of Insanity* (Princeton, NJ: Princeton University Press, 2015), esp. 224–229. On Rush, see n. 7 below.
7. Marianne Berg Karlsen, "Den første norske telling av sinnsvake," *Nytt Norsk Tidsskrift* 17, no. 3 (2000), 276–293; Frederik Holst, *Beretning, Betankning og Indstilling*

fra en til at undersøge de Sindsvages Kaar i Norge og gjøre Forslag til deres Forbedring i Aaret 1825 (Christiania, 1828), 72; J.E.D. Esquirol, "Remarques sur la statistique des aliénés, et sur le rapport du nombre des aliénés à la population. Analyse de la statistique des aliénés de la Norwège," *Annales hygiene publique et de médecine légale* 4 (1830), 335–338; quote on 338.

8. Holst, *Beretning*, foldout table K at back of volume.
9. Benjamin Rush, *Medical Inquiries and Observations upon the Diseases of the Mind* (Philadelphia: Kimber & Richardson, 1812), 65–66, 69.
10. *Third Annual Report of the Managers of the State Lunatic Asylum* (of New York, for 1845) (Albany, 1846), 54–56.
11. Untitled note in *AJI* 1 (1844–45), 287–288.
12. "Lunatics," *Literary Panorama* 2 (1807), 1255–1256; Esquirol, "Remarques," 338.
13. Holst, *Beretning*; Esquirol, "Remarques," 346–347, 355; Prichard, *Treatise on Insanity*, 347–350 seems to have been following Esquirol's review when he recited Holst's ratio and others, then endorsed the argument for insanity as a disease of civilization.
14. Pierquin mentioned Roger's *Fables sénégalaises* (Paris: Firmin Didot, 1828) but quoted from a private letter (he says). See Claude-Charles Pierquin, *De l'arithmétique politique de la folie, ou considérations générales sur la folie, envisagée dans ses rapports avec l'ignorance, les crimes et la population des diverses régions du globe*, 2nd ed. (Paris, 1831), 12–15. I find no indication of a first edition.
15. Pierquin, *De l'arithmétique politique*, 8, 23, 37, 49, 58. On French debates regarding education and crime, see Theodore M. Porter, *The Rise of Statistical Thinking, 1820–1900* (Princeton, NJ: Princeton University Press, 1986), 28–29.
16. Adolphe Quetelet, *Sur l'homme et le développement de ses facultés, ou essai de physique sociale*, 2 vols. (Paris: Bachelier, 1835), 2: 120–132.
17. Scull, *Most Solitary of Afflictions*, 157–160; Scull, *Madness in Civilization*, 228–229.
18. Sir James Coxe, "On the Causes of Insanity, and the Means of Checking its Growth; being the Presidential Address," *Journal of Mental Science* (*JMS*) 18 (1872), 316–318, 332–333.
19. On societies and journals, see Scull, *Most Solitary of Afflictions*, 232–235; Engstrom, *Clinical Psychiatry*, 35–44; Jan Goldstein, *Console and Classify: The French Psychiatric Profession in the Nineteenth Century* (Cambridge: Cambridge University Press, 1987), 340–341; Ian Dowbiggin, *Inheriting Madness: Professionalization and Psychiatric Knowledge in Nineteenth-Century France* (Berkeley: University of California Press, 1991), chap. 4.
20. Patricia Cline Cohen, *A Calculating People: The Spread of Numeracy in Early America* (Chicago: University of Chicago Press, 1982), chap. 6; Gerald Grob, *Edward Jarvis and the Medical World of Nineteenth-Century America* (Knoxville: University of Tennessee Press, 1978), 69–75. See also Paul Schor, *Compter et classer: Histoire des recensements américains* (Paris: Éditions de l'École des Hautes Études en Sciences Sociales, 2009). The census publication is *Compendium of the Enumeration of the Inhabitants and Statistics of the United States . . . from the Returns of the Sixth Census . . .* (Washington, 1841), p. 7 for the insane in Maine.
21. Ramon de la Sagra, "Statistique des aliénés et des Sourds-Muets dans les États-Unis de l'Amérique," *AMP* 1 (1843), 281–288.
22. John Butler, physician and superintendent, in *Twenty-Second Annual Report of the Retreat for the Insane at Hartford* (Hartford, 1846), 24.
23. Ministre de l'Agriculture et du Commerce, *Statistique de la France. Administration Publique* (Paris, 1843), 305–370, "Aliénés"; Alexandre Moreau de Jonnès, "Statistique," *Comptes-rendus hebdomadaires des séances de l'Académie des sciences* (1843), 65–67.

24. Debates on Moreau de Jonnès, "Statistique," *Comptes rendus hebdomadaires des séances de l'Académie des sciences* (1843), 134–136 and 231–235, sessions for 17 July and 7 Aug. 1843. This debate was excerpted in *AMP* 2 (1843), 300–303. Also Maximien Parchappe, *Recherches statistiques sur les causes de l'aliénation mentale* (Rouen, 1839), 60. The alienist François Leuret criticized the census numbers, and especially the neglect of heredity as cause of insanity, in "Quelques observations sur la statistique des aliénés en France," *Annales d'hygiène publique et de médecine légale* 31 (April 1844), 444–449.
25. [Amariah Brigham], "Statistics of Insanity," *AJI* 6 (1849–50), 141–145.
26. Heinrich Damerow, "Statistique de la France. Section III. Aliénés," *AZP* 2 (1845), 723.
27. Scull, *Most Solitary of Afflictions*, 334.
28. Smith, "Superintendent's Report," 18–22.
29. Quotes from *Twentieth Annual Report of the Trustees of the State Lunatic Asylum at Worcester, Dec. 1852* (Boston, 1853), 40, and *Twenty-Second Annual Report of the Officers of the Retreat for the Insane at Hartford*, 7–8, 12. See also Gerald N. Grob, "Introduction" to Edward Jarvis, *Insanity and Idiocy in Massachusetts: Report of the Commission on Lunacy, 1855* (1855; repr., Cambridge, MA: Harvard University Press, 1971), 1–71; David J. Rosner, *The Discovery of the Asylum: Social Order and Disorder in the New Republic* rev. ed. (New York: Alding de Gruuyter, 1990), 272–276, shows how the changing character of institutionalized populations in America caused respectable families to avoid public asylums.
30. Wilhelm Griesinger, *Die Pathologie und Therapie der psychischen Krankheiten, für Ärzte und Studirende* (Stuttgart: Adolph Krabbe, 1845), 386; repeated in Dr. Kelp, "Irrenstatistik des Herzogthums Oldenburg," *AZP* 4 (1847), 585–633, 617.
31. "Report of the Managers of the Temporary Lunatic Asylum at Quebec, January 1849," in *Reports of the Proprietors and Managers*, 5–30, 10.
32. "The Willard Asylum and Provision for the Insane," *AJI* 22 (1865–66), 192–212, 208; "Annual Report of Medical Superintendent for 1866," in *Annual Reports of the Provincial Lunatic Asylum, Toronto, for the Years 1866 & 1867* (Toronto, 1868), 5–38, 19.
33. ALVR, 1154, Acta 1830, *Bericht der zur Verwaltung der Irren-Heil-Anstalt zu Siegburg während der Jahre 1827, 1828, und 1829* (Koblenz, 1830), 28; ALVR, 1156, Acta 1837, *Bericht über die Verwaltung der Irren-Heil-Anstalt zu Siegburg während der Jahre 1833, 1834, 1835 und 1836* (Koblenz, April 1837), 13; ALVR, 1157, *Bericht über die Verwaltung der Irren-Heil-Anstalt zu Siegburg während der Jahre 1837, 1838, 1839, und 1840* (Koblenz, March 1841). The report for 1830–32 in ALVR, 1155 mentioned heredity in the same way as its successor.
34. [C.F.W. Roller], review of *Rechtfertigung der Erfahrungs-Heillehere der alten scheidekünstigen Geheimärzte*, by Johann Gottfried Rademacher, *AZP* 4 (1847), 169.
35. *Twenty-Second Annual Report of the Directors of James Murray's Royal Asylum for Lunatics* (Perth, 1849), 8.
36. B. A. Morel, "Rapport médical sur l'Asile de Maréville (Meurthe)," *AMP*, 2nd ser., 2 (1850), 353. See also his reviews of American asylum reports in *AMP* 8 (1846), 299–304; 9 (1847), 302–306; 10 (1847), 299–304.
37. Edward Jarvis, "On the Supposed Increase of Insanity," *AJI* 8 (1851–52), quotes on 333–334. He read this paper to a meeting of American asylum directors on 21 May 1851.
38. Jarvis, *Insanity and Idiocy*, 11.
39. Jarvis, "Supposed Increase of Insanity," 347–349, 355–362.
40. *Fifteenth Report of the Commissioners in Lunacy to the Lord Chancellor* (1861), 75. Here I follow Scull, *Most Solitary of Afflictions*, 338–344. See also D. J. Mellett, "Bureaucracy and Mental Illness: The Commissioners in Lunacy 1845–1890," *Medical History* 23 (1981).

41. This count was extended to Canada in 1861; Kathrin Levitan, *A Cultural History of the British Census: Envisioning the Multitude in the Nineteenth Century* (New York: Palgrave Macmillan, 2011), 34, 51, 155.
42. *Fifteenth Report*, 78–79.
43. J. Mortimer Granville, *The Care and Cure of the Insane: Being the Reports of The Lancet Commission on Lunatic Asylums, 1875–76–77*, 2 vols. (London: Hardwicke and Bogue, 1877), 1: 331.
44. Rosner, *Discovery of the Asylum*, 112–122.
45. Richard Dunglison, "Statistics of Insanity in the United States," *North American Medico-Chirurgical Review* 4 (1860), 656–692, 677.

Part II. Tabular Reason

Epigraph: Isaac Ray, "The Statistics of Insane Hospitals," *AJI* 6 (1849), 25–26.

1. *Twenty-Fifth Annual Report of the Trustees of the State Lunatic Hospital at Worcester* (Boston, 1857); John Gray, review of Worcester Hospital report for 1860, *AJI* 16 (1859-60), 463–465 and 17 (1860-61), 450–452. See also Grob, *State and Mentally Ill*, 201.
2. Luther V. Bell, "Thirty-Second Annual Report of the Physician and Superintendent of the McLean Asylum for the Insane," in *Report of the Trustees of the Massachusetts General Hospital, presented to the Corporation at their Annual Meeting, January 23, 1850* (Boston: Eastburn's Press, 1850), 14–20, 15. Bell had been arguing this way in his annual reports since 1840.
3. Ray, "Statistics of Insane Hospitals," 25–26.

Chapter 5. French Alienists Call Heredity Too Deep for Statistics While German Ones Build a Database, 1844–1866

Epigraphs: Ulysse Trélat, "Des causes de la folie," *AMP*, 2d ser., 6 (1856), 189; Wilhelm Jung, "Noch einige Untersuchungen über die Erblichkeit des Seelenstörungen," *AZP* 23 (1866), 243 (emphasis in second epigraph added).

1. On printed forms and information, see JoAnne Yates, *Control through Communication: The Rise of System in American Management* (Baltimore: Johns Hopkins University Press, 1989).
2. [Jules Baillarger], "Introduction," *AMP* 1 (1843), i–xxvii; Baillarger, "Recherches statistiques sur l'hérédité de la folie," *AMP* 3 (1844), 328–339.
3. On cases and observation see Lorraine Daston and Elizabeth Lunbeck, eds., *Histories of Scientific Observation* (Chicago: University of Chicago Press, 2011), especially Gianna Pomata's article there, "Observation Rising: Birth of an Epistemic Genre," 45–80.
4. Baillarger, "Recherches statistiques."
5. His eight-page dossier, "Liste et analyse succincte des travaux anatomiques, physiologiques et pathologiques," prepared for his candidacy, is available at the Bibliothèque numérique, gallica.bnf.fr.
6. "Les chiffres pour quelques hommes, même très éclairés, sont une expression irréfragable de la vérité. Mais . . . les chiffres, aussi bien que les mots, n'ont qu'une valeur représentative; la statistique, qui les rassemble et les fait mouvoir, n'est, par elle-même, qu'une aveugle qui ne raisonne pas: aussi, pour arriver à cette vérité qu'elle nous promet, faut-il, de tout nécessité, percer toute cette draperie, et aller droit aux choses représentées, à travers les signes qui les représentent." Hippolyte-Louis Royer-Collard, "Rapport: *Recherches statistiques sur l'hérédité de la folie* par M. Bail-

larger . . . ," *Bulletin de l'Académie Royale de Médecine* 12 (1846–47), 765. Note that the *elle* who promises truth can be either the blind woman or statistics. The other committee members were Jean-Pierre Falret and François Mélier.

7. Royer-Collard, "Rapport," 766–767.
8. Royer-Collard, "Rapport," 769–770, 774, 776.
9. Trélat, "Des causes," 10–14.
10. Trélat, "Des causes," 14–15.
11. Trélat, "Des causes," 7.
12. Trélat, "Des causes," 188–189.
13. Carl Hohnbaum, "Ueber Erblichkeit der Geisteskrankheiten" (translation of and commentary on Trélat, "Des causes"), *AZP* 5 (1848), 540–568, Hohnbaum's commentary, 558–568.
14. Jules Baillarger, "De la statistique appliquée à l'étude des maladies mentales," *AMP* 7 (1846), 163–168.
15. L.F.E. Renaudin, *Notice statistique sur les aliénés du Département du Bas-Rhin, d'après les observations recueillies à l'Hospice de Stéphansfeld, pendant les années 1836, 1837, 1838, 1839* (Paris: J.-B. Baillière, 1840).
16. L.F.E. Renaudin, "Administration des asiles d'aliénés," in *AMP* 5–6 (1845), especially chap. 10, "Compte Administratif," *AMP* 6 (1845), 401–408.
17. Baillarger, "De la statistique," 163; unnamed translator, "On Statistics Applied to the Study of Mental Disease," *AJI* 5 (1848–49), 322–327, 322.
18. [Amariah Brigham], "Statistics of Insanity," *AJI* 6 (1849–50), 141–145; Elllen Dwyer, *Homes for the Mad: Life inside Two Nineteenth-Century Asylums* (New Brunswick, NJ: Rutgers University Press, 1987), 61–62, discusses Brigham's statistics of causes of insanity.
19. Renaudin, "De la statistique appliquée." Honoré Aubanel of the asylum at Marseille added his voice to Renaudin's in support of public statistics in a continuation on 469–472. Jan Goldstein discusses the circumstances of these letters in *Console and Classify*, 340–341.
20. Renaudin's critique appeared under "ER" in a review of three issues of *AZP* from 1846 to 1847, "Journaux Allemands," *AMP* 12 (1848), 133–138, 136.
21. Damerow, "Die Zeitschrift," 17.
22. This report, printed with no named author, was presented as the shared view of the editors, Damerow, Flemming, and Roller. It consisted of a single page of commentary and a set of sixteen specimen tables: "Betreffend die Aufstellung eines Normal-Schemas für irrenstatistische Uebersichten," *AZP* 3 (1846), 665–675, 671.
23. Rudolf Leubuscher, "Bemerkungen über die Erblichkeit des Wahnsinns," *Archiv für Anatomie und Physiologie und für klinische Medicin* 1 (1847), quote on 73; translated into English as "Remarks on the Hereditary Transmission of Insanity," *Journal of Psychological Medicine and Mental Pathology* 1 (1848), 264–277.
24. Wilhelm Griesinger, *Die Pathologie und Therapie der psychischen Krankheiten für Aerzte und Studirende* (Stuttgart: Adolph Krabbe, 1845), 101, 112–115. Still, he ascribed special significance to heredity, particularly in his third edition (Braunschweig: Friedrich Wreden, 1871), 155.
25. C. F. Flemming, "Aerztlicher Bericht über die Heilanstalt Sachsenberg aus dem 10jährigen Zeiträume von 1840–1849," *AZP* 9 (1852), 377–378.
26. Georg Schweig, "Auseinandersetzung der statistischen Methoden in besonderem Hinblick auf das medicinische Bedürfniss," *Archiv für physiologische Heilkunde* 13 (1854), esp. 313–314. He cited Esquirol's table in German translation from Carl Christian Schmid's Leipzig-based digest of medical writings, the *Jahrbücher der in- und ausländische gesammte Medicin*, first supplement vol. (1836), 458–463, 461.

27. Moritz Martini, "Uebersichtliche Darstellung der Resultate, welche die Zählung der i. J. 1830 in der Provinz Schlesien vorhandenen Gemütskranken gewährt hat," *Streit's Schlesische Provinzial-Blätter* 95 (January–June 1832), 401–408 and 96 (July–Dec. 1832), 115–130 and 210–215. His calculation is in the final section.
28. [Heinrich] L[aehr], "Uebersicht der Resultate der ärztlichen Wirksamkeit der Provinzial-Irren-Heilanstalt zu Leubus in den Etaatsjahren 1856, 1857, 1858, 1859, und kurzer Rückblick auf dreissig Jahre Bestehens," *AZP* 19 (1862), 327–329; L[aehr],"Bericht über die Resultate der ärztl. Verwaltung der Provincial-Irren-Heilanstalt zu Leubus in den Jahren 1861–62 und über die Ergebnisse der Irrenzählung am Schlusse d. J. 1862, von Dr. Martini," *AZP* 22 (1865), 311–312. The neighboring kingdom of Saxony attached a triennial census to its asylum report: see, for example, *Erster Jahresbericht des Landes-Medicinal-Collegiums über das Medicinalwesen im Königreich Sachsen auf das Jahr 1867* (Dresden: In Commission bei C. Heinrich, 1869).
29. "Es soll diese Arbeit eine rein statistische sein, sie wird deshalb sich fern halten von Raisonnements, nur die nackten Thatsachen sprechen lassen, aus diesen die nackten Schlüsse ziehen." Wilhelm Jung, "Untersuchungen über die Erblichkeit der Seelenstörungen," *AZP* 21 (1864), 535–536.
30. Jung, "Untersuchungen," 576, 569; Moritz Martini, "Berichte über die Resultate der ärztlichen Verwaltung der Provinzial-Irren-Zählung in den Jahren 1863, 1864, 1865, und über die Irren-Zählung am Schlusse des Jahres 1865–1866," in *XX. Provinzial-Landtag* [Silesia], 1868, 20.
31. Jung, "Untersuchungen," e.g., 581–585.
32. Jung, "Untersuchungen," 579. He refers to Martini's results, which he combined with his own, on 626–627. The reports may be found in Geheimes Staatsarchiv Preussischer Kulturbesitz. Bestand: I HA Rep. 76 Kultusministerium / Ministerium der Geistlichen- u. Medicinal-Angelegenheiten. Abtheilung für die Medicinal-Angelegenheiten, vol. 20. For Martini's numbers see Laehr, "Uebersicht," 328.
33. Jung, "Untersuchungen," 586–587; following table on 622. See also Katharina Banzhaf, "Vorläufer der psychiatrischen Genetik: die psychiatrischen Erblichkeitsforschung in der deutschsprachigen Psychiatrie im Spiegel der Allgemeinen Zeitschrift für Psychiatrie 1844 bis 1911," Inauguraldissertation zur Erlangung des Grades eines Doctors des Medizin des Fachbereichs Medizin der Justus-Liebig Universität Gießen, 2014, 43–56.
34. Jung, "Untersuchungen," 630–631, 638–639, 641, 651–652.
35. Hugh Grainger Stewart, "On Hereditary Insanity," *Journal of Mental Science* 10 (1864), 51–52, 54–55. The French translation by E. Dumesnil appeared with the author's name misspelled as Hugh Grainger-Steward, "De la folie héréditaire," *AMP*, 4th ser., 4 (1864), 356–378.
36. Stewart, "On Hereditary Insanity," 59; Jung, "Noch einige Untersuchungen," 220. He gave the result in his 1864 study, "Untersuchungen," on 624.
37. Gerhard Lötsch, *Von der Menschenwürde zum Lebensunwert: Die Geschichte der Illenau von 1842–1940* (Kappelrodeck: Achertäler Verlag, 2000), 50.
38. *Illenauer Wochenblatt* 1, no. 10 (7 Sept. 1867), 37–38, and 2, no. 52 (26 Dec. 1868), 209–210, 215–216, 221–222, preserved at the Heimatmuseum Achern. The journal was edited from 1867 to 1896 by the Lutheran preacher, Theodor Achtnich.
39. *Beiträge zur Statistik der inneren Verwaltung des Großherzogthums Baden, Zweiundzwanzigstes Heft. Die Heil- und Pflegeanstalt Illenau* (Carlsruhe: Chr. Fr. Müller'sche Hofbuchhandlung, 1866), esp. ix, xxxiii–xxxiv, 42–53.
40. Heinrich Damerow, "Zur Statistik der Provinzial-Irren-Heil- und Pflege-Anstalt bei Halle vom 1 November 1844 bis Ende Dezember 1863 nebst besonderen Mitthei-

lungen und Ansichten über Selbsttödtungen . . . ," *AZP* 22 (1865), 233–235. Some statistical records from the Halle asylum are preserved in the Landeshauptarchiv Sachsen-Anhalt, Abteilung Magdeburg, C92 No. 1322 and C20 I No. 776. See also Adolph Wagner, *Die Gesetzmässigkeit in den scheinbar willkührlichen menschlichen Handlungen vom Standpunkte der Statistik* (Hamburg: Boyes und Geisler, 1864). On Wagner, see Porter, *Rise of Statistical Thinking*, 168–171.

41. Table in Jung, "Untersuchungen," 630.
42. Jung, "Noch einige Untersuchungen," 235–236, 241–242; Henry Thomas Buckle, *History of Civilization in England* (New York: D. Appleton, 1858), 127n12; Damerow, "Zur Statistik," 236–237; Friedrich Rolle, *Darwin's Lehre von der Entstehung der Arten im Pflanzen- und Thierreich in ihrer Anwendung auf die Schöpfungsgeschichte* (Frankfurt: Joh. Christ. Hermann, 1863), 63.
43. Jung, "Noch einige Untersuchungen," 243–244 (emphasis added); Hermann Eberhard Richter, "Zur Darwin'schen Lehre," [Schmidt's] *Jahrbücher der in- und ausländische Medicin* 126 (1865), 249.

Chapter 6. Dahl Surveys Family Madness in Norway, and Darwin Scrutinizes His Own Family through the Lens of Asylum Data, 1859–1875

Epigraphs: [John P. Gray], "Statistics of Insanity," *AJI* 18 (1861–1862), 13; Ludvig Dahl, "Fortsatte Bidrag til Kundskab om de Sindssyge i Norge," *Norsk Magazin for Lægevidenskaben* 16 (1862), 548–549.

1. [John P. Gray], "Statistics of Insanity in Europe," *AJI* 17 (1860–61), 348–349; Bénédict Augustin Morel, *Traité des maladies mentales* (Paris: Librairie Victor Masson, 1860), 79–83; Jean-Christophe Coffin, *La transmission de la folie 1850–1914* (Paris: L'Harmattan, 2003), 7–8.
2. *Report by Her Majesty's Commissioners appointed to inquire into the state of Lunatic Asylums in Scotland* (Edinburgh: HMSO, 1857), 31–35, 38. On Scottish asylum statistics see Jonathan Andrews, *They're in the Trade . . . of Lunacy. They "cannot interfere"—they say: The Scottish Lunacy Commissioners and Lunacy Reform in Nineteenth-Century Scotland*, Wellcome Institute for the History of Medicine, Occasional Publications No. 8 (London: Wellcome Trust, 1998).
3. "Notice of Dr. Dahl's Report Respecting the Insane in Norway," *AJI* 17 (1860–61), 342–344; from *Dublin Quarterly Journal of Medical Science* 30 (1860), 193–195. These censuses are briefly discussed in Einar Lie and Hege Roll-Hansen, *Faktisk Talt: Statistikkens historie i Norge* (Oslo: Universitetsforlaget, 2001), 123–134. Holst also published a summary of one of these censuses in German: "Ueber die Anzahl der Geisteskranken, Blinden und Taubstummen in Norwegen im Jahre 1835," *AZP* 4 (1847), 479–487.
4. [Gray], "Statistics of Insanity," 1.
5. The review of Legoyt's census compilation is in *AJI* 16 (1859–60), 436–461 and *AJI* 17 (1860–61), 421–443, see 444 and 454. The earliest source I can find for these claims, clearly a later publication than what Gray reviewed, is Alfred Legoyt, *La France et l'Etranger: Études de statistique comparée* (Paris: Veuve Berger-Levrault et fils, 1864), 121–122.
6. [Gray], "Statistics of Insanity," 2–3, 13 (see epigraph).
7. Ludvig Dahl, "Beretning om en med Kgl. Stipendium foretagen Reise i Danmark, Holland, Belgien og Storbritannien," *Norsk Magazin for Lægevidenskaben* 11 (1857), 387–411; 12 (1858), 1–19, 81–111.
8. W. Lauder Lindsay, "On Insanity and Lunatic Asylums in Norway," *Journal of Psychological Medicine and Mental Pathology* 11 (1858), 246–247, 251–252; A.J.F. Brierre de

Boismont, "Sur l'aliénation mentale et les asiles d'aliénés en Norwège," *AMP*, 3rd. ser., 5 (1858), 441–442.

9. Brierre de Boismont, "Sur l'aliénation," 442–443; Dahl, "Beretning," 102. Lindsay, "Insanity in Norway," 276, describes this treatment and its successful use in Edinburgh, without any mention of French sources. But see A.J.F. Brierre de Boismont, "De l'emploi des bains prolongés et des irrigations continues dans le traitement des formes aigues de la folie, et en particulier de la manie," *Mémoires de l'Académie de Médecine* 13 (1847), 527–599, where he explains that he had communicated his method orally to other doctors before publishing it in order to confirm its value, for the sake of humanity, even if it undercut his claim to priority. The use of water to warm the body and chill the head was not new.
10. The Danish word *Anlæg* (*Anlegg* in modern Norwegian) derives from the German *Anlage*, according to Hjalmar Falk and Alf Torp, *Etymologisk Ordbog over det Norske og det Danske Sprog*, 2 vols. (Kristiania: Forlagt af H. Aschehoug, 1903), 1: 19. Its meanings with respect to heredity seem to be identical.
11. Ole Sandberg, Direktor, *General beretning fra Gaustad Sinsdssygeasyl for Aaret 1856* (Christiania: Steenske Hogtrykkeri, 1857), and similarly for subsequent reports; Lindsay, "Insanity in Norway." See also David Cornelius Danielssen and Wilhelm Boeck, *Traité de la Spédalskhed ou Éléphantiasis des Grecs*, trans. L. A. Cosson (Paris: J.-B. Baillière, Librairie de l'Académie Royale de Médecine, 1848); Ludvig Dahl, *Bidrag til Kundskab om de Sindssyge i Norge* (Christiania: Steenske Bogtrykkeri, 1859), 76. There was another Norwegian precedent for these statistics of heredity: a government-commissioned report on the Bergen hospital for lepers in relation to others outside of Norway, jointly authored by a physician there and a medical professor from Christiania. It was published in 1848 in Danish and almost simultaneously in a French translation by the Royal Academy of Medicine. Dahl discussed these researches on leprosy in his 1859 book. For an introduction to the topic of leprosy and heredity, I thank Svain Atle Skålevåg, who wrote his 1998 doctoral thesis at the University of Bergen on insane asylums in Norway. See also Lorentz M. Irgens, "The Fight against Leprosy in Norway in the 19th Century," *Michael Quarterly* 7 (2010), 307–320.
12. Peter Jessen, mentioned in chap. 3 above, worked in Schleswig, a Danish territory until 1862. See also J. R. Hübertz's report of the exemplary Danish census, "Statistik der Irrenwesens in Dänemark," *AZP* 1 (1844), 457–479; also Holst, "Ueber die Anzahl."
13. Dahl, *Bidrag*, 76; Gerhard von dem Busch, review of *Bidrag til Kundskab om de Sindssyge i Norge*, by L. Dahl, *AZP* 18 (1861), 474–518, 474–482. Dahl later argued that this was a fundamentally important division and should be registered even by nonmedical census takers: "Om Tilveiebringelse af en fælles Sindssygestatistik for Sverige, Danmark og Norge," *Norsk Magazin for Lægevidenskaben* 17 (1863), 459–460.
14. Dahl, *Bidrag*, 76–77; von dem Busch, review of *Bidrag*, 482–483.
15. A near-exception is Pliny Earle's "genealogical chart" of color blindness in four generations of his own family in "On the Inability to Distinguish Colors," *American Journal of the Medical Sciences* 9 (April 1845), 346–354.
16. Dahl, *Bidrag*, 76–96; von dem Busch review, 483; Ireland, *Idiocy and Imbecility*, reprinted tables at end of volume.
17. Dahl, *Bidrag*, 81; Ireland, *Idiocy and Imbecility*, 15.
18. Dahl, *Bidrag*, 77–78, 89; von dem Busch, review of *Bidrag*, 483–484.
19. Dahl, "Fortsatte," 524–530. Von dem Busch's review is "Fortgesetzter Beitrag zur Kenntniss über die Geisteskranken in Norwegen," *AZP* 21 (1864), 283–306, 285–287.
20. Søren Christian Sommerfelt, *Physisk-oeconomisk Beskrivelse over Saltdalen* (Trondheim: Kgl. Norske Vidsksabers Selskab, 1827), 109. For basic biographical informa-

tion on Sommerfelt see entry in Norsk biografisk leksikon, accessed 25 Oct. 2017, https://nbl.snl.no/.

21. Sommerfelt, *Physisk-oeconomisk*, 100, 103–106.
22. Dahl, "Fortsatte," 542–544; von dem Busch, "Fortgesetzter," 294–295. Von dem Busch read Dahl as attributing the description of the healthy immigrants from Ranen to Sommerfelt, but Dahl speaks merely of information from "the priest," and Sommerfelt's text never mentions Ranen. I do not understand how as many as *et par Hundrede* (a couple hundred) settlers from Ranen can be reconciled with the rest of Dahl's (or Sommerfelt's) population history of Saltdal, or why, if they arrived recently, they are described so vaguely.
23. Dahl, "Fortsatte," 544, 548–549. On causes of insanity among the peoples of Finnmark, see 601–605.
24. Cathy Gere, "Evolutionary Genetics and the Politics of the Human Archive," in Daston, ed., *Science in the Archives*, 211.
25. W. W. Ireland, "Dr. Ludvig Dahl," *JMS* 37 (1891), 334.
26. *Thirty-First Annual Report of the Directors of James Murray's Royal Asylum for Lunatics near Perth* (June 1858), 24–25. See also *Twenty-Seventh Annual Report* (June 1854), 8; *Thirty-Second Annual Report* (June 1859), 50–51; and *Thirty-Third Annual Report* (June 1860), 27–28.
27. The second was that cures depend on early treatment. *First Annual Report of the General Board of Commissioners in Lunacy for Scotland* (Edinburgh: HMSO, 1859), x.
28. [John P. Gray], review of (US edition) *History of Civilization in England*, by Henry Thomas Buckle, *AJI* 15 (1858–59), 237.
29. Alban Stolz, *Ueber die Vererbung sittlicher Anlagen* (Freiburg: Universitäts-Buchdruckerei von Hermann Meinhard Poppen & Sohn, 1859), 10–13, 17, 23.
30. Isaac Ray, *Mental Hygiene* (Boston: Ticknor & Fields, 1863), 19, 22, 24–25; Butler Hospital for the Insane, *Annual Report for 1863* (1864); summarized with quotations in *AJI* 21 (1864–65), 238–242, quote on 242.
31. Ray, *Mental Hygiene*, 18; Ray, "Reports of the Trustees . . . ," in Butler Hospital for the Insane, *Annual Report* for 1853, published January 1854; final quotation from a summary, "Reports of the Trustees and Superintendent of the Butler Hospital for the Insane, presented to the Corporation at their Annual Meeting, Jan. 25, 1854," in *AJI* 11 (1854–55), 181–186, 185–186.
32. "Report to the Commissioners of the Lower Canada Lunatic Asylum at Quebec," January 1858, by James Douglas, Joseph Morrin, and Charles-Jacques Frémont, in *Reports of the Proprietors and Managers of the Lower Canada Lunatic Asylum, to the Commissioners, Quebec* (Quebec, 1858), 49–64, 58–59.
33. William W. Ireland, *On Idiocy and Imbecility* (London: J. & A. Churchill, 1877), 16–17; Scottish Lunacy Commission, *Report by Her Majesty's Commissioners Appointed to Inquire into the State of Lunatic Asylums in Scotland* (Edinburgh: HMSO, 1857), 186–187. Ireland altered the quotation to remove the hypothetical: "all fatuous females should be restricted."
34. Adam Kuper, "Incest, Cousin Marriage, and the Origin of the Human Sciences in Nineteenth-Century England," *Past and Present* 174 (2002), 158–183; Diane B. Paul and Hamish G. Spencer, "Eugenics without Eugenists? Anglo-American Critiques of Cousin Marriage in the Nineteenth and Early Twentieth Centuries," in Müller-Wille and Brandt, *Heredity Explored*, 40–79; Arthur Mitchell, "On the Influence which Consanguinity in the Parentage exercises upon the Offspring," *Edinburgh Medical Journal* 10, no. 2 (1865), 784, 790–791, 903, 1078–1079.
35. Mitchell, "Influence of Consanguinity," 1080–1082. See Farr to Darwin, 21 May 1868, 6 Aug. 1870, and 16 July 1871; Darwin to Lubbock, 17 and 21 July 1870; Lubbock to

Darwin, 23 and 26 July 1870; Letters 6197, 7279, 7281–82, 7286–88, and 7296 in Darwin Correspondence Project, accessed 25 Oct. 2017, www.darwinproject.ac.uk.

36. Kuper, "Incest," 170; Darwin to Shuttleworth, 12 Feb. 1874, letter 9299A in Darwin Correspondence Project, www.darwinproject.ac.uk; George Darwin, "Marriages between First Cousins in England and their Effects," *Journal of the Statistical Society of London* 38 (1875), Galton comment on 184; George Darwin, "On Beneficial Restrictions to Liberty of Marriage," *Contemporary Review* 22 (1873), 412–426.

Chapter 7. A Standardizing Project out of France Yields to German Systems of Census Cards, 1855–1874

Epigraphs: A. Legoyt, ed., *Compte rendu de la deuxième session du Congrès International de Statistique réuni à Paris les 10, 12, 13, 14, et 15 Septembre 1855* (Paris, 1856), 370; Ludwig Wille, "Ueber Einführung einer gleichmässigen Statistik der schweizerischen Irrenanstalten," *Zeitschrift für Schweizerische Statistik* (1872), 249–254, 249.

1. Ludvig Dahl, "Om Tilveibringelse af en fælles Sindssygestatistik for Sverige, Danmark og Norge," *Norsk Magazin for Lægevidenskaben* 17 (1863), 448–461, for the 1863 Skandinavska Naturforskara-Sällskapets.
2. Ludvig Dahl, "Ueber einige Resultate der Zählung der Geisteskranken in Norwegen, den 31. December 1865," *AZP* 25 (1868), quote on 839.
3. Dahl, "Ueber einige Resultate." For statistics of alcohol he cited an early work of the Norwegian statistician Eilert Sundt, *Om Ædrueligheds-Tilstanden i Norge* (Christiania: J. Chr. Adelsted, 1859). Dahl summarized the results of the 1865 census in "De Sindssvage i Norge den 31te December 1865," *Norsk Magazin for Lægevidenskaben* 23 (1869), 705–724.
4. W. Charles Hood, *Statistics of Insanity; Being a Decennial Report of Bethlem Hospital from 1846 to 1855 Inclusive* (London: David Batten, 1856); Hood, *Statistics of Insanity: Embracing a Report of Bethlem Hospital from 1846 to 1860 Inclusive* (London: David Batten, 1862). See also Andrews et al., *History of Bethlem*, 496–498; Akihito Suzuki, "Framing Psychiatric Subjectivity: Doctor, Patient, and Record-Keeping at Bethlem in the Nineteenth Century," in Joseph Melling and Bill Forsythe, eds., *Insanity, Institutions, and Society, 1800–1914: A Social History of Madness in Comparative Perspective* (New York: Routledge, 1999), 115–136.
5. LMA, H11/HLL/A5/1: John Conolly, "[Third] Resident Physician's Report," in *Fifty-Fourth Report of the Visiting Justices of the County Lunatic Asylum at Hanwell* (London, 1840), 5; H11/HLL/A5/2: Conolly, "Fourth Report of the Resident Physician of the County of Middlesex Pauper Lunatic Asylum at Hanwell, October 1st, 1842," 22; and Conolly, "Sixth Report of the Physician," in *Seventy-Second Report of the Visiting Justices of the County Lunatic Asylum at Hanwell*, 37.
6. J.C.B. Bucknill, "The Annual Reports of the County Lunatic Asylums and Hospitals of the Insane in England and Wales, published during the Year 1855," *Asylum Journal of Mental Science* 2 (1855–56), 257–285, 258; Bucknill, "Annual Reports . . . for the Year 1856," *Asylum Journal of Mental Science* 3 (1856–57), 464–506, 479, quoting Dr. D. C. Campbell of the Essex Asylum.
7. *Report of the Metropolitan Commissioners in Lunacy to the Lord Chancellor* (London, 1844), 177–195; C. Lockhart Robinson, "Suggestions towards an Uniform System of Asylum Statistics (With Tabular Forms)," *JMS* 7 (1860–61), 195–211, quote on 197.
8. "Extraordinary Meetings of the Medico-Psychological Society of Paris," *JMS* 13 (1867), 285.
9. AN F20 [Statistique]/282/46.
10. AN F20/282/46–47 and 49.

11. AN F20/282/46 and 48.
12. Dr. Ant. Ritti, "Éloge du Dr. L. Lunier," *AMP*, 8th ser., 20 (1904), 5–47, 27.
13. Ludger Lunier, "Recherches statistiques sur les aliénés du Département des Deux-Sèvres," *Mémoires de la Société de Statistique du Département des Deux-Sèvres* 16 (1853), 27–53; Lunier, *Exposé des titres et travaux scientifiques de Docteur L. Lunier . . .* (Paris, 1869), 4–19; both of these documents available online from Gallica, BNF, accessed 23 Nov. 2017, http://www.sudoc.fr/164317112.
14. Ludger Lunier, review of "Rapport statistique et critique sur l'Asile des Aliénés de la Grave," by Gérard Marchant, *AMP*, 2nd ser., 12 (1848), 147–151, 148; Lunier, *Exposé des titres*, 16–17.
15. Ludger Lunier, "Asile départemental d'aliénés de Blois (Loir et Cher)," *Compte-rendu du Service Médical pour l'année 1863* (Blois, 1864), 6; table of causes, 8–9.
16. Ludger Lunier, *De l'influence des grandes commotions politiques et sociales sur le développement des maladies mentales, Mouvement d'aliénation en France pendant les années 1869 à 1873* (Paris: F. Savy, 1874). On madness and political disorder see Jean-Claude Caron, *Les feux de la discorde: Conflits et incendies dans la France du XIXe siècle* (Paris: Hachette, 2006), chap. 15; Laure Murat, *The Man Who Thought He Was Napoleon: Toward a Political History of Madness*, trans. Deke Dusinberre (Chicago: University of Chicago Press, 2014), esp. 207.
17. Ludger Lunier, "De l'augmentation progressive du chiffre des aliénés et de ses causes," *AMP*, 6th ser., 3 (1870), 20–34; see also same title in *Journal de la Société de statistique de Paris* 15 (1874), 35–40. The official report is Augustin Constans, Ludger Lunier, and Dr. Dumesnil, *Rapport générale à M. le Ministre de l'Intérieur sur le Service des Aliénés en 1874* (Paris, 1878), see esp. 63–69. For a sense of the flow of statistics in the asylum reports, see Archives de la Ville de Paris D2X3, e.g., *Rapport sur le Service des Aliénés du Département de la Seine pour l'année 1871* (Paris, 1872).
18. Alfred Legoyt, ed., *Compte Rendu de la deuxième Session du Congrès Internationale de Statistique, réuni à Paris, Sept. 1855* (Paris, 1856), xxvi, 10, 116–120, 370–375, 383–387. Parchappe's contribution was reprinted as "Rapport sur la statistique de l'aliénation mentale fait au Congrès International de Statistique," *AMP*, 3rd ser., 2 (1856), 1–6.
19. L.F.E. Renaudin, "Observations sur les Recherches Statistiques relatives à l'aliénation mentale," *AMP*, 3rd ser., 2 (1856), 486–504. On Trélat, see chap. 6 above.
20. Heinrich Damerow," Kritisches zur Irrenstatistik aus der Anstalt bei Halle," *AZP* 12 (1855), 440–467, 440–441.
21. "Reviews of American Asylums," a summary of Workman's 1859 report for the Canada West Provincial Lunatic Asylum, *AJI* 17 (1860–61), 309–315, 310. The archives of the Toronto asylum are held in the Archives of Ontario at York University, Toronto, under Queen St. Mental Health Centre. RG10–272, microfilmed as MS 640, reel 14, "Records of Medical Superintendent," contains admission forms for the 1840s and early 1850s. These include a query for the "exciting cause" but nothing on heredity.
22. Theodore M. Porter, "*Irrenärzte aller Länder!* Tabular Unity and the Nineteenth-Century Struggle to Comprehend Insanity," *Soziale System* 18 (2012), 211.
23. Ludger Lunier, "Rapport de M. Lunier . . . ," *AMP*, 4th ser., 9 (1867), 284–286; Achille Foville, "Rapport sur la proposition de M. Lunier relative à une réunion des médecins aliénistes de tous les pays," *AMP*, 4th ser., 9 (1867), 286–294. The content of the invitation is from "Extraordinary Meetings of the Medico-Psychological Society of Paris," *JMS* 13 (1867), 285.
24. Ludger Lunier, report, "Congrès Alieniste International," *AMP*, 4th ser., 10 (1867), 428–430; Ludger Lunier, report, "Statistique des aliénés," *AMP*, 4th ser., 10 (1867), 512–514.

25. Ludger Lunier, "Projet de statistique applicable à l'étude des maladies mentales arrêté par le Congrès Alieniste International de 1867. Rapport et exposé des motifs," *AMP*, 5th ser., 1 (1867), 32–59. The commission of doctors was given as Borrel, John Ch. Buchnill [Bucknill], J. Falret, W. Griesinger, Lombroso, L. Lunier, J. Mundy, Pujadas, Roller, Harrington Tuke, and Motet.
26. Lunier, "Projet," 42; compare table in Lunier, "Asile de Blois," 21.
27. Ludwig Wille, *Referant* on statistics of Swiss asylums, "Vierte Jahresversammlung des Vereins schweizerischer Irrenärzte am 13. und 14. September 1867 in Münsterlingen, Kanton Thurgau," *AZP* 25 (1868), 416–419.
28. See "Bericht über die Sammlung deutscher Irrenärzte zu Eisenach am 12. und 13. September 1860," *AZP* 17 (1860), Anhang; Braun, *Heilung mit Defekt*, 74–76, 95–96; Engstrom, *Clinical Psychiatry*, 42–44.
29. It developed haltingly in the late 1840s and 1850s as a branch of the Schweizerische gemeinnützige Gesellschaft. See J. M. Hungerbühler, *Ueber das öffentliche Irrenwesen in der Schweiz* (St. Gallen and Bern: Von Huber und Komp., 1846); Ludwig Binswanger, "Bericht über das Irrenwesen der Schweiz; der Schweizerischen Naturforscherversammlung zu Glarus erstattet," *Zeitschrift der schweizerische Naturforschenden Gesellschaft bei ihrer Versammlung in Glarus den 4., 5., und 6. August 1851*, 111–117; Hans Jakob Ritter, "Von den Irrenstatistiken zur 'erblichen Belastung' der Bevölkerung: Die Entwicklung der schweizerischen Irrenstatistiken zwischen 1850 und 1914," *Traverse: Zeitschrift für Geschichte* 10 (2003), 59–70; Ritter, *Psychiatrie und Eugenik: Zur Ausprägung eugenischer Denk- und Handlungsmuster in der schweizerischen Psychiatrie, 1850–1950* (Zürich: Chronos Verlag, 2009), 59–62.
30. "Berliner medicinische-psychologische Gesellschaft, Sitzung vom 30. July 1867," *Archiv für Psychiatrie und Nervenkrankheiten* 1 (1867–68), 211–216; Heinz-Peter Schmiedebach, *Psychiatrie und Psychologie im Widerstreit: Die Auseinandersetzungen in der Berliner medicinisch-psychologischen Gesellschaft (1867–1899)* (Husum: Matthiesen, 1986), 75.
31. Wille, "Ueber Einführung," 249–250.
32. "A Project of a System of Statistics Applicable to the Study of Mental Disease, approved by the International Congress of Alienists of 1867," *AJI* 26 (1869), 49–80.
33. Friedrich Wilhelm Hagen, Bayerisches Hauptstaatsarchiv, MInn 61955; Friedrich Wilhelm Hagen, "Ueber Statistik der Irrenanstalten mit besonderer Beziehung auf das im Auftrage des internationalen Congresses vom Jahre 1867 vorgeschlagene Schema," *AZP* 27 (1871), 267–294.
34. "Sitzung vom 21. December 1869," *Archiv für Psychiatrie und Nervenkrankheiten* 2 (1868–69), 517–519, 518.
35. Friedrich Koster and Wilhelm (Guilelmus) Tigges, *Geschichte und Statistik der westfälischen Provinzial—Irrenanstalt Marsberg, mit Rücksicht auf die Statistik anderer Anstalten*, supplemental issue to vol. 24 of *AZP* (Berlin: August Hirschwald, 1867), 117–475; Wilhelm (Guilelmus) Tigges, "Die Lunier'schen Vorschläge für die Statistik der Geisteskrankheiten," *AZP* 26 (1869), 667–668.
36. [Carl] Pelman (secretary), "Psychiatrischer Verein der Rheinprovinz. Sechste ordentliche Sitzung vom 18. Juni, 1870," *AZP* 27 (1871), 595–597; [Heinrich] L[aehr], "Kritik der Zählblättchen der Berliner medicinisch-psychologischen Gesellschaft, betreffend die Geisteskranken der Anstalten," *AZP* 27 (1871), 626–637. Also, "Versammlung der Mitglieder des Vereins der Irrenärzte Niedersachsens und Westphalens am 2. Mai 1870," *AZP* 27 (1871), 709–711.
37. Werner Nasse, *Aerztlicher Bericht über die Wirksamkeit der Irren-Heil-Anstalt zu Siegburg während der Jahre 1867, 1868, und 1869* (Cologne: Druck Franz Greven, 1871), 6–7.

38. "Bericht über die Sitzung des Vereins des deutschen Irrenärzte zu Leipzig am 13. August 1872," *AZP* 29 (1873), 458–460. The members were Hagen, Nasse, Roller, Wilhelm Sander, and Tigges. Hagen, from Erlangen in Bavaria and Roller from Illenau in Baden, were seen as representing the south German provinces of the new Reich.
39. Wille, "Ueber Einführung," 249, excerpted in "Verhandlungen psychiatrischen Vereine. VIII. Versammlung der schweizerischen Irrenärzte am 25. und 26. September d. J. in der Irrenanstalt Burghölzli bei Zürich," *AZP* 29 (1873), 579–587; Ritter, *Psychiatrie und Eugenik*, 65–66.
40. Anton Bumm, "Nekrolog. Friedrich Wilhelm Hagen," *AZP* 45 (1889), 298–306, 300.
41. Friedrich Wilhelm Hagen, review of *Observations and Essays on the Statistics of Insanity*, by John Thurnam, *AZP* 3 (1846), 677–718; Ritter, "Von den Irrenanstalten," 62.
42. Quote from Friedrich Wilhelm Hagen review of *Statistical Appendix to Report of the Metropolitan Commissioners in Lunacy, 1844*, *AZP* 2 (1845), 523–539, 538; see also his review of the 1844 report itself on 87–141.
43. *Further Report of the Commissioners in Lunacy to the Lord Chancellor. Presented to Both Houses of Parliament by Command of Her Majesty* (London, 1847), 177–222, quotes on 189, 186–187.
44. See Hagen's review in *AZP* 6 (1849), 315–333, 325–326.
45. Friedrich Wilhelm Hagen, *Der goldene Schnitt in seiner Anwendung auf Kopf- und Gehirnbau, Psychologie, und Pathologie* (Leipzig: Wilhelm Engelmann, 1857); G. Specht, "Friedrich Wilhelm Hagen (1814–1888)," in Theodor Kirchhoff, ed., *Deutsche Irrenärzte: Einzelbilder ihres Lebens und Wirkens*, 2 vols. (Berlin: Julius Springer, 1921–1924), 1: 253–260.
46. I saw the drafts of these reports at the Klinikum am Europakanal, Erlangen, thanks to Dr. Phil. Hans Siemens. The report for 1852–53 is Inventar-Nr. 2240 and for 1856–57 is Inventar-Nr. 2270. His final reports from Erlangen to the Bavarian state are in Bayerisches Hauptstaatsarchiv MInn 62092 and 61941. For development of the basic tables to their definitive form, see MInn 61936.
47. Friedrich Wilhelm Hagen, "Aerztlicher Bericht aus der Kreis-Irrenanstalt Irsee," *AZP* 10 (1853), 1–72, esp. 4, 11–12. The statute, dated 30 Jan. 1850, is from Staatsarchiv München RA 56641.
48. Hagen, "Aerztlicher Bericht Irsee," 15–16; Staatsarchiv Nürnberg, Außenstelle Lichtenau, Patientenakten, Irrenanstalt Erlangen, Acta der Kgl. Verwaltung der Kreisirrenanstalt Erlangen, Regina B, admitted 8 Apr. 1878; see also Joseph H., admitted 25 July 1863.
49. Hagen, "Aerztlicher Bericht Irsee," 22, 38–39, 41. In 1858 he wrote a longer-term report, directed mainly to the public: *Bericht über Bestand und Wirken des Kreis-Irrenanstalt Irsee vom 1. September 1849 bis 30 September 1858* (printed report; no publication information), which I found in the Bayerisches Hauptstaatsarchiv MInn 62102. It was published in a different form in *Der Irrenfreund* 1 (August 1859), 96–101, 110–112, 118–120.
50. Hagen, "Ueber Statistik der Irrenanstalten," esp. 269, 272–273, 278–279, 293. See also Porter, "*Irrenärzte aller Länder*," 213–215.
51. Friedrich Wilhelm Hagen, *Statistische Untersuchungen über Geisteskrankheit, nach den Ergebnissen der Ersten Fünfundzwanzig Jahre der Kreis-Irrenanstalt zu Erlangen* (Erlangen: Eduard Besold, 1876), 10, 14, 20–23; Lunier, *De l'influence*.
52. Nasse, *Aerztlicher Bericht Siegburg 1871*; Nasse, "Vorlage für eine deutsche Irren-Anstalt-Statistik," *AZP* 30 (1874), 240–248; Braun, *Heilung mit Defekt*, 98, 143.
53. *Zählkarten und Tabellen für die Statistik der Irrenanstalten aufgestellt von dem Verein der deutschen Irrenärzte*, dated 1874. This was bound as a supplement to the sixth issue of *AZP* 30 (1874). Volker Roelcke, "Unterwegs zur Psychiatrie als Wissenschaft:

Das Projekt einer 'Irrenstatistik' und Emil Kraepelin's Neuformulierung des psychiatrischen Klassifikation," in Eric J. Engstrom and Volker Roelcke, eds., *Psychiatrie im 19. Jahrhundert. Forschungen zur Geschichte von psychiatrischen Institutionen, Debatten und Praktiken im deutschen Sprachraum* (Mainz: Akademie der Wissenschaften und der Literatur, 2003), 169–188, discusses the relation of the international project and Sander's Prussian project. In "Sitzung vom 21 November 1871," *Archiv für Psychiatrie und Nervenkrankheiten* 3 (1870–71), 503, Sander complained of personal attacks by Tigges in regard to the *Zählblättchen*. Albert Guttstadt, "Die Geisteskranken in den Irrenanstalten während der Zeit von 1852 bis 1872 und ihre Zählung im ganzen Stadt am 1. December 1871 nebst Vorschlägen zur Gewinnung einer deutschen Irrenstatistik," *Zeitschrift des Königlichen Preussischen Statistischen Bureaus* (*ZKPSB*) (1874), 248a, also refers to the severity of this dispute.

54. Tigges, "Lunier'schen Vorschläge," 677–681; Braun, *Heilung mit Defekt*, 301, notes the increasing focus on heredity in revisions of the statistics in the 1870s.
55. *Zählkarten und Tabellen*, table 5a.
56. *Zählkarten und Tabellen*, 11.
57. "Statistische Nachrichten über die im Preussischen Staate bestehenden öffentlichen und Privat-Irren-Heilanstalten für das Jahr 1850" and "Ueber die Irrenheilanstalten und die Anzahl der Irren im Preussischen Staate," *Mittheilungen des Statistischen Bureaus in Berlin* 5 (1852), 94–131 and 328–331. Damerow held missing institutions responsible for only a fraction of the undercount: *AZP* 9 (1852), 330–344. On Engel's early ambitions for Prussian statistics see Ernst Engel, "Die Methoden der Volkszählung, mit besonderer Berücksichtigung der im preussischen Staate angewandten," *ZKPSB*, no. 7 (1861), 157, 194; Michael C. Schneider, *Wissensproduktion im Staat: Das königlich preußische statistische Bureau 1860–1914* (Frankfurt: Campus Verlag, 2013), 70–73.
58. "Die Gutachten der königlichen Regierungen über die Ausführung der Volkszählung am 1. December 1871," *ZKPSB* (1874), 153–196, 161–162; "Die Verhandlungen der Vorstände deutscher statistischer Centralstellen bezüglich der Volkszählungen vom 1. December 1875 im Deutsche Reiche," *ZKPSB* (1874), 197–200h, 200e. See also Schneider, *Wissensproduktion* and Schmiedebach, *Psychiatrie*, 76.
59. This account is based on Christine von Oertzen, "Machineries of Data Power: Manual versus Mechanical Census Compilation in Nineteenth-Century Europe," *Osiris* 32 (2017), 129–150. The sales notice is titled "Zählkarten" in *AZP* 30 (1874), 716.
60. Guttstadt, "Geisteskranken," esp. 248a–b. He again asserted the crucial importance of a *Centralstelle* in the presentation "Statistik der Irrenanstalten in Preussen," given at a meeting of the Berlin Psychiatric Association on 15 June 1874: *AZP* 31 (1875), 609. See also Michael C. Schneider, "Medizinalstatistik im Spannungsfeld divergierender Interessen: Kooperationsformen zwischen statistischen Ämtern und dem Kaiserlichen Gesundheistamt/Reichsgesundheitsamt," in Axel C. Hüntelmann, Johannes Vossen, und Herwig Czech, eds., *Gesundheit und Staat: Studien zur Geschichte der Gesundheitsämter in Deutschland, 1870–1930* (Husum: Matthiesen Verlag, 2006), 49–62.
61. [Fischel], review of *Statistische Untersuchungen*, by Friedrich Hagen, in *AZP* 34 (1878), 112–116.

Chapter 8. German Doctors Organize Data to Turn the Tables on Degeneration, 1857–1879

Epigraph: "Le plus souvent, la maladie que se transmet se transforme." Henri Legrand du Saulle, *La folie héréditaire: Leçons professés à l'École Pratique* (Paris: Adrien Delahaye, 1873), 9.

1. Koster and Tigges, *Geschichte und Statistik*, 264, and graph, 265.
2. Note the enthusiastic reception of Daniel Pick's book *Faces of Degeneration: A European Disorder c. 1848–c. 1918* (Cambridge: Cambridge University Press, 1989).
3. B. A. Morel, *Traité des dégénérescences physiques, intellectuelles et morales de l'espèce humaine* (Paris: J.-B. Baillière, 1857), 323–324, 343–346, 565 and plates 1–5, 7–11; Richard von Krafft-Ebing "Zur Prognose der Geistesstörungen," *Irrenfreund* 13, no. 3 (1871), 33–43, brought out some dimensions of this contrast.
4. See epigraph by Legrand du Saulle; Morel, *Traité des maladies mentales*, 503–573; Pick, *Faces of Degeneration*, 47–52; Koster and Tigges, *Geschichte und Statistik*, 213.
5. Morel, *Traité des maladies mentales*, 634–646; L. F. Calmeil, *Traité des maladies inflammatoires du cerveau*, 2 vols. (Paris: J.-B. Baillière, 1859), 2: 650. Baillarger remarked in 1867 on the importance of hereditary prognosis for families. Robert Nye, *Masculinity and Male Codes of Honor in Modern France* (New York: Oxford University Press, 1993), 76, speaks of this language of marital hygiene as "reproductive eugenics *avant la lettre*."
6. Report on the meeting of the Société Médico-Psychologique for 30 Dec. 1867 in *AMP*, 4th ser., 11 (1868), 272–274.
7. Oscar Schmidt, *Descendenzlehre und Darwinismus* (Leipzig: F. A. Brockhaus, 1873), esp. 164; Tigges, "Bericht über die Irren-Heilanstalt Sachsenberg vom Jahre 1871–1875 mit vergleichender Statistik," *Beiträge zur Statistik Mecklenburgs vom Grossherzoglichen statistischen Bureau zu Schwerin* 8 (1876), 80–81.
8. See the census of mental illness taken in 1865 by Tigges's predecessor, Dr. Löwenhardt, *Die Zählung der Geisteskranken im Grossherzogthum Mecklenburg-Schwerin im Jahre 1865* (Berlin: August Hirschwald, 1866), special supplement to *AZP* 23.
9. Koster and Tigges, *Geschichte und Statistik*, 210–214 and table 6, 162–165.
10. Koster and Tigges, *Geschichte und Statistik*, 214 and table 10, 215, 5.9%. Because almost all of the patient records held by this asylum were for mental illness, the very low percentages in the top two rows (labeled Geisteskrankheit, or mental illness) correspond to absolute numbers larger than the sum of the bottom four rows. Poisson's formula was known to doctors mainly from Jules Gavarret, *Principes généraux de statistique medicale* (Paris: Bechet Jeune et Labé, 1840), available also in an 1844 German translation.
11. Tigges, "Lunier'schen Vorschläge," 681.
12. Wilhelm Griesinger, *Die Pathologie und Therapie der psychischen Krankheiten für Aerzte und Studirende* (Stuttgart: Adolph Krabbe, 1845), 112–115, cited some highly discrepant figures but didn't challenge the validity of the measure.
13. Legrand du Saulle, *Folie héréditaire*, 4–6.
14. Legrand du Saulle, *Folie héréditaire*, 3, 9, 11–12.
15. Jacques-Joseph Moreau de Tours, *La psychologie morbide dans ses rapports avec la philosophie de l'histoire, ou de l'influence des névropathies sur le dynamisme intellectuel* (Paris: Librairie Victor Masson, 1859), 116–117. On the uses of medical heredity in French psychology see Martin Staum, *Nature and Nurture in French Social Sciences, 1859–1914, and Beyond* (Montreal: McGill-Queens University Press, 2011), chap. 4.
16. Gabriel Doutrebente, "Etude généalogique sur les aliénés héréditaires," *AMP*, 5th ser., 2 (1869), 203, 205, 208, 227; table on 213; Legrand du Saulle modified these tables and printed two of them in *Folie héréditaire*, 41, 42. The enthusiastic English review is "French Psychological Literature," *JMS* 16 (1871), 612–620.
17. Pelman review of *La folie héréditaire*, *AZP* 30 (1874), 697–700; Legrand du Saulle, *Die erbliche Geistesstörung*, trans. Dr. Stark (Stuttgart: H. Lindemann, 1874).
18. Carl Pelman, "Protokoll der Sitzung des Vereins der deutschen Irrenärzte am 27. Mai 1874 zu Eisenach," *AZP* 32 (1875), 195.

19. Richard von Krafft-Ebing, "Ueber die prognostische Bedeutung der erblichen Anlage im Irresein," *AZP* 26 (1869), 438–456. His reasoning was statistical, based on dividing 292 patients into three categories: hereditary latency, active psychical illness, or anomalies.
20. Hagen, *Statistische Untersuchungen,* Einleitung, 1–77, 10, 38–42. See also discussion of Hagen in chap. 7 above.
21. Hagen, *Statistische Untersuchungen,* 76, 10–11; Heinrich Ullrich, "Ueber Erblichkeitsverhältnisse," in Hagen, *Statistische Untersuchungen,* 175–244, 179. On issues of homogeneity in the history of statistics see Stephen Stigler, *The History of Statistics: The Measurement of Uncertainty before 1900* (Cambridge, MA: Harvard University Press, 1986).
22. Hagen, *Statistische Untersuchungen,* 48–51, 58. On performances in the asylum, see Friedrich Wilhelm. Hagen, *Bericht über die Kreisirrenanstalt Erlangen in den Jahren 1877–1883* (Erlangen, n.d.), in Bayerisches Hauptstaatsarchiv MInn 61941.
23. Hagen, *Statistische Untersuchungen,* 59–60.
24. Ullrich, "Ueber Erblichkeitsverhältnisse," 176–179, 189–191, 208–210; Ullrich, "Bericht über die psychiatrische Literatur," "13. Statistik," *AZP* 38 (1882), 552–555.
25. Ullrich, "Ueber Erblichkeitsverhältnisse," 175–177, 181.
26. Ullrich, "Ueber Erblichkeitsverhältnisse," 178–185, 243.
27. Hagen, *Statistische Untersuchungen,* 64–75.
28. B. A. Morel, *Rapport médical sur l'Asile des Aliénés de Saint-Yon* (Rouen, 1870), 8–10, 16–18.
29. Tigges, "Bericht über Sachsenberg," 80. Reviews by Dr. Kelp in *AZP* 34 (1878), 109–112; Max Huppert in *Schmidt's Jahrbücher der in- und ausländische gesammten Medicin* 173 (1877), 181–184.
30. Tigges, "Statistik der Erblichkeit, betreff. die Kinder und die Geschwister der in die Anstalt Aufgenommenen," *AZP* 35 (1879), 487.
31. Tigges, "Statistik der Erblichkeit," 486–487, 501–502. On Weinberg, see chap. 12 below.
32. Tigges, "Statistik der Erblichkeit," 487, 501–502, 507, summarized in Dr. Kelp, "Bericht über die Irrenheilanstalt Sachsenberg vom Jahre 1871–75 mit vergleichender Statistik von Med.-Rath Dr. Tigges," *AZP* 34 (1878), 109–112, 111.

Chapter 9. Alienists Work to Systematize Haphazard Causal Data, 1854–1907

Epigraphs: Paul Samt, *Die naturwissenschaftliche Methode in der Psychiatrie: Vortrag gehalten in der Berliner medicinisch-psychologischen Gesellschaft* (Berlin: Verlag von August Hirschwald, 1874), 41; Thomas Mann, *Bekentnisse des Hochstaplers Felix Krull* (1954), book 2, chapter 5, an episode set in the mid-1890s. On these grounds, a duped military doctor exempts Krull from conscription.

1. Scull, *Madness in Civilization,* 232; *AZP* 7 (1850), 534.
2. James H. Mills, *Madness, Cannabis, and Colonialism: The "Native-Only" Lunatic Asylums of British India, 1857–1900* (New York: St. Martin's Press, 2000), 50, 58–60; *Report on the Lunatic Asylums in the Central Provinces* for years 1874–1879 and 1895–1899; *Triennial Report of the Lunatic Asylums under the Government of Bombay for the Years 1918–1920,* both at National Library of Medicine.
3. On heredity in California, see chap. 4 above. On Georgia, see especially *Biennial Report of the Trustees of the Georgia Lunatic Asylum for the Fiscal Years from October 1, 1890, to October 1, 1891 and from October 1, 1891, to October 1, 1892* (Augusta, GA, 1892) and *Annual Report of the Trustees of the Georgia Lunatic Asylum for the Fiscal*

Year from September 1, 1896, to September 1, 1897 (Augusta, GA, 1897). Even the asylum report was segregated, and all dissections involved black patients.

4. See reports of the physician at Hanwell for 1853 and 1854, LMA, H11/HLL/A5/4. In 1853, 12 new male patients and only 1 female showed heredity as cause. In 1854, a new physician in the female department recorded causes for only 18 of 82 new patients, compared to 73 of 87 admitted males. For 21 of the males, two or more causes were recorded.
5. Dr. Kreuser, *Die Heil- & Pflegeanstalt Winnenthal. Fünfzigjähriger Anstaltsbericht* (Tübingen: Franz Fues, 1885), 44, 69. He explained the divergence of his numbers from Hagen's at Erlangen and Julius Koch's at Zwiefalten in terms of patient composition.
6. Bethlem Hospital Archives, Case Book: Females, 1854, 29, 153; Bethlem Subcommittee Book, 1854–1856, 9 June 1854.
7. *General Report of the Royal Hospitals of Bridewell and Bethlem and of the House of Occupations for the Year Ended 31st December 1852* (London: David Batten, 1853), 43, held in Bethlem archives.
8. LMA, H11/HLL/B19/24, Female Case Book, 1873–74 and H11/HAA/B20/13, Male Case Books, 1873–74.
9. LMA, H11/HLL/B19/24, Female Case Book 1873–74, cases 5313, 5392, 5405; H11/HAA/B20/13, Male Case Book 1873–74, cases 4665, 4693, 4758, 4796.
10. LMA, H11/HLL/B19/39 Female Case Book 1893–94, patients 9281, 9296l; J. Peeke Richards, "Report of the Medical Superintendent of the Female Department" for 1876 MLA H11/HLL/AF/9, 24–25.
11. Bethlem Hospital Archives, Case Book Males 1894. On relatives in asylums see LMA, H11/HLL/B20/20 for Hanwell, 1889–1890, e.g., cases 7555, 7568, 7668, and 7738.
12. LMA, H64/B6/2 St. Luke's Hospital Case Book 1875–76, 365.
13. "Report of the Committee on the Statistical Tables of the Medico-Psychological Association," *JMS* 28 (1882), 463–464; LMA, H12/CH/B2/2, Register of Admissions for Colney Hatch, 1888–1907.
14. These remarks are based on Metropolitan Asylums Board (London) (hereafter, MAB), *Report of the Statistical Committee for the Year 1897* (London: McQuordale, 1898), 122–123; also MAB, *Annual Report for the Year 1901*, vol. 2, 174–175; *Annual Report for the Year 1904*, 223, 270a; MAB, *Annual Report for the Year 1913*, foldout table on 230c. The statistical committee of the MAB is briefly described in Gwendoline Ayers, *England's First State Hospital and the Metropolitan Asylums Board, 1867–1930* (London: Wellcome Institute of the History of Medicine, 1971), 135–136.
15. State of New York, State Commission on Lunacy, *Second Annual Report* (for 1890; Albany, 1891), 221.
16. See essays in M. Norton Wise, ed., *The Values of Precision* (Princeton, NJ: Princeton University Press, 1995); Silvana Patriarca, *Numbers and Nationhood: Writing Statistics in Nineteenth-Century Italy* (Cambridge: Cambridge University Press, 1996); Schneider, *Preußische Statistische Bureau.*
17. The lecture was promptly published in the leading German economic journal: Ludwig Wille, "Die Aufgaben und Leistungen der Statistik der Geisteskranken," *Jahrbücher für Nationalökonomie und Statistik* 35 (1879), 307–331, see 331. Heinrich Ullrich excerpted from it in his bibliographic report on asylum statistics in "Bericht über die psychiatrische Literatur im 2. Halbjahre 1880," *AZP* 37 (1881), 133–135.
18. For asylum administration and statistics in Belgium, see *Rapport de la Commission Supérieure d'Inspection des Établissements d'Aliénés, instituée par arrêté royal du 18 November 1851* (dated 30 Jan. 1852), *Bulletin de la Société de Gand* 20 (1853), 40–75; then, e.g., *Deuxième rapport de la Commission Permanent d'Inspection des Établisse-*

ments d'Aliénés, instituée par arrêté royal du 17 mars 1853 (Bruxelles, 1854) and so on in mostly annual volumes to the eighth for 1862, and thereafter less frequently.

19. Ferdinand Lefebvre, "Des Bases d'une bonne statistique international des maladies mentales: Rapport," *Bulletin de la Société de Médecine Mentale de Belgique* 37 (1885), 55–60; "Congress of Psychiatry and Neuro-Pathology at Antwerp," *JMS* 31 (1886), 613–626; J. Christian, "Chronique. Le Congrès de phréniatrie et de psychopathologie, tenu à Anvers du 7 au 9 septembre, 1885," *AMP*, 7th ser., 2 (1885), 371–385; Albert Guttstadt, "Zur internationalen Irrenstatistik," *AZP* 43 (1887), 139–140. Ullrich summarized these reports in "Bericht über die psychiatrische Literature im 2. Halbjahre 1885," *AZP* 43 (1887), 115–117.
20. See Karl Westphal, "Vorschläge zur Abänderung der amtlichen Zählkarten für die Irrenanstalten," *AZP* 38 (1882), 717–719 and 39 (1883), 612–616. For paradoxes of standardization, see Andrew Lakoff, *Pharmaceutical Reason: Knowledge and Value in Global Psychiatry* (Cambridge: Cambridge University Press, 2005).
21. Prof. Dr. Hasimi Sakaki, "Erläuterungen zu den statistischen Tabellen aus der Städtischen Irrenheilanstalt zu Tokio, Japan," *AZP* 48 (1892), 109–133.
22. Carl Pelman, "Ueber Irre und Irrenwesen," *Centralblatt für allgemeine Gesundheitspflege* 1 (1882), 16–24, 45–68. For debates on the data cards and on disease classification at meetings of the *Verein der Deutschen Irrenärzte* see *AZP* 35 (1879), 529–534; *AZP* 38 (1882), 717–728.
23. Engstrom, *Clinical Psychiatry*, 141. Hagen, also a Bavarian, kept a file of *verzettelte Krankengeschichten*, case histories on individual sheets, which allowed them to be rearranged at will. See Universität Erlangen Handschriftabteilung, Nachlaß F. W. Hagen, A15–16. He also kept records on the insane of the region who were not in asylums. Stadtarchiv Erlangen 156/24, Die Benutzung der Irrenanstalt und die Behandlung der Irren, 1852–1894.
24. Emil Kraepelin, *Psychiatrie. Ein Lehrbuch für Studirende und Aerzte*, 5th ed. (Leipzig: Johann Ambrosius Barth, 1896), 318; he spoke here of *Geistesstörungen des Rückbildungsalters*, mental illnesses of age reversion. See also Matthias M. Weber and Eric J. Engstrom, "Kraepelin's 'Diagnostic Cards': The Confluence of Clinical Research and Preconceived Categories," *History of Psychiatry* 8 (1997), 375–385; Engstrom, "Die Ökonomie klinischer Inskription. Zu diagnostischen und nosologischen Schreibpraktiken in der Psychiatrie," in Cornelius Borck and Armin Schäfer, eds., *Psychographien* (Zürich/Berlin: Diaphanes, 2005), 219–240; Volker Roelcke, "Quantifizierung, Klassifikation, Epidemiologie: Normierungsversuch des Psychischen bei Emil Kraepelin," in Werner Sohn and Herbert Mehrtens, eds., *Normalität und Abweichung: Studien zur Theorie und Geschichte der Normalisierungsgesellschaft* (Opladen: Westdeutscher Verlag, 1999), 183–200.
25. Emil Kraepelin, *Die psychiatrischen Aufgaben des Staates* (Jena: Fischer, 1900), trans. by Stewart Paton as *The Duty of the State in the Care of the Insane* in *AJI* 57 (1900–1901), 235–280, 249; Eric J. Engstrom, "Organizing Psychiatric Research in Munich (1903–1925): A Psychiatric Zoon Politicon between State Bureaucracy and American Philanthropy," in Volker Roelcke, Paul Weindling, and Louise Westwood, eds., *International Relations in Psychiatry: Britain, Germany, and the United States to World War II* (Rochester, NY: University of Rochester Press, 2010), 48–66.
26. Samt, *Naturwissenschaftliche Methode*, 59.
27. "Reviews and Notices of Books," *Lancet* 1 (28 Apr. 1866), 459–460; Granville, *Care and Cure*, 2: 212, 206. Granville also rejected the distinction between exciting and predisposing causes.
28. *Supplement to the Thirty-Sixth Annual Report of the General Board of Commissioners in Lunacy for Scotland* (Edinburgh: HMSO, 1895), letter of transmittal by Sir George Otto Trevelyan, vii.

29. ALVR, 1157, *Bericht der ständische Commission*, 14 June 1843, 29–31; *Bemerkungen des Ober-Medicinal-Raths Jacobi zu dem Berichte der ständischen Untersuchungs-Commission über die Irren-Heilanstalt zu Siegburg* (undated), 16 pp., 2–3, 5. On these issues see Braun, *Heilung mit Defekt*, 78.
30. Pliny Earle, *The Curability of Insanity. A Series of Studies* (Philadelphia: J. B. Lippincott Company, 1887), "Study First (written in 1876)," 7–63, 19–29, 10. See also Gerald Grob, *Mental Illness and American Society* (Princeton, NJ: Princeton University Press, 1983), 39.
31. Medical Center of New York-Presbyterian / Weill Cornell, New York, NY, Medical Register of the Bloomingdale Asylum, 31 Dec. 1844—Aug. 1866.
32. Earle, *Curability*, 38–39, 60–61; D. Hack Tuke, "Presidential Address, delivered at the Annual Meeting of the Medico-Psychological Association . . . August 2nd, 1881," *JMS* 27 (1881), 336; see also T. C. Shaw, "The New Statistical Tables," *JMS* 29 (1883), 324–325.
33. Franklin Sanborn, *Memoirs of Pliny Earle with Extracts from his Diary and Letters (1830–1892) and Selections from his Professional Writings (1839–1891)* (Boston: Damrell and Upham, 1898), 158; John G. Park, "Superintendent's Report," *Forty-Seventh Annual Report of the Trustees of the State Lunatic Hospital of Worcester*, for year ending October 1879 (Boston: Rand Avery, 1880), 14.
34. Pliny Earle, "Curability of Insanity," *AJI* 34 (1877), 101–102 and comments on Vermont asylum, 536–530; [John Gray] review of Earle's 1877 report, *AJI* 35 (1877), 542–544.
35. Weber and Engstrom, "Kraepelin's 'Diagnostic Cards.'"
36. See Kraepelin's letter to the minister of justice, religion, and instruction, 12 Nov. 1902, in Wolfgang Burgmair, Eric J. Engstrom, and Matthias M. Weber, eds., *Emil Kraepelin in Heidelberg, 1891–1903*. Vol. 5 of *Edition Emil Kraepelin* (Munich: Belleville Verlag, 2005), 150–152.
37. I thank Nikolaus Braun of the Archiv des Bezirks Oberbayern (ABO) in Munich for explaining this procedure and for helping me with the files there.
38. ABO, Akten der Heil- und Pflege-Anstalt Eglfing No. 5916, patient 2518, Zugang Buch (Frauen) No. 212. For relations to the census in Bavaria, see Georg Mayr, comp., *Die Verbreitung der Blindheit, der Taubstummheit, des Blödsinns und der Irrsinns in Bayern, nebst einer allgemeinen internationalen Statistik dieser vier Gebrechen* (Munich: Commissionsverlag von Adolf Ackermann, 1877), esp. 1–5.
39. ABO, Akten Patient 2413 and corresponding *Zugang* volumes. The *Aufnahmszählkarte* used here is for Johann A. and is dated 14 Aug. 1891. On the keeping of patient files see Sophie Ledebur, "Schreiben und Beschreiben: Zur epistemischen Funktion von psychiatrischen Krankenakten, ihrer Archivierung und deren Übersetzungen in Fallgeschichten," *Berichte zur Wissenschaftsgeschichte* 34 (2011), 102–124.
40. *Annual Report of the Trustees of the Taunton State Hospital*, 1893 (Boston: Rand, Abery), 5.
41. Adolf Meyer, "A Review of the Signs of Degeneration and of Methods of Registration," *AJI* 52 (1895–96), 344–363; Meyer, "The Treatment of the Insane" (from a report to the governor of Illinois, 2 Aug. 1894), reprinted in Eunice E. Winters, ed., *The Collected Papers of Adolf Meyer*, 4 vols. (Baltimore: Johns Hopkins University Press, 1951), 2: 37–57, 45.
42. *Sixty-Third Annual Report of the Trustees of the Worcester Lunatic Hospital*, October 1895 (Boston, 1896), 15; *Sixty-fourth Annual Report . . . 1896*, 16–18; S. D. Lamb, *Pathologist of the Mind: Adolf Meyer and the Origins of American Psychiatry* (Baltimore: Johns Hopkins University Press, 2014), 41–58.

43. State of New York, State Commission in Lunacy, *Thirteenth Annual Report*, 1900–1, 20. Noll, *American Madness*, 159, notes that Meyer focused on patient data.
44. A. H. Newth, M. D., "Systematic Case-taking," *JMS* 46 (1900), 255–256.
45. Newth, "Case-taking," 257–259. See also the Annual Reports of James Murray's Royal Asylum from 1886 to 1909.

Part III. A Data Science of Human Heredity

Epigraph: Francis Galton, *Natural Inheritance* (London: Macmillan, 1889), 66.

1. Two textbook histories are Robert Olby, *Origins of Mendelism*, 2nd. ed. (Chicago: University of Chicago Press, 1985) and Peter Bowler, *The Mendelian Revolution* (Baltimore: Johns Hopkins University Press, 1989). Will Provine, *The Origins of Theoretical Population Genetics* (Chicago: University of Chicago Press, 1971) judges Pearson's biometric school favorably, as does Daniel J. Kevles, *In the Name of Eugenics: Genetics and the Uses of Human Heredity* (New York: Knopf, 1985), while (for example) Ernst Mayr, *The Growth of Biological Thought: Diversity, Evolution, and Inheritance* (Cambridge, MA: Harvard University Press, 1982), treats it as misguidedly anti-Mendelian.
2. Bert Theunissen, "Breeding without Mendelism: Theory and Practice of Dairy Cattle Breeding in the Netherlands, 1900–1950," *Journal of the History of Biology* 41 (2008), 637–676.

Chapter 10. The Human Science of Heredity Takes On a British Crisis of Feeblemindedness, 1884–1910

Epigraph: David Heron, *A First Study of the Statistics of Insanity and the Inheritance of the Insane Diathesis*. Eugenics Laboratory Memoirs 2 (London: Dulau, 1907), 3.

1. University College, London (UCL), Francis Galton Papers, box 37/122/1M "Insanity in Twins." This text is accurately translated, perhaps by Galton himself, from Moreau de Tours, *La psychologie morbide*, 172–173.
2. UCL, Galton Papers 37/122/1M, Clouston to Galton, July–August 1875.
3. UCL, Galton Papers 37/122/1M, Clouston to Galton, July–August 1875; 37/122/1J, London Orphan Asylum; Francis Galton, "The History of Twins, as a Criterion of the Relative Powers of Nature and Nurture," *Fraser's Magazine*, n.s. 12 (November 1875), 566–576, 571. Mathew Thomson, "Disability, Psychiatry, and Eugenics," in Alison Bashford and Philippa Levine, eds., *The Oxford Handbook of the History of Eugenics*, 116–133 (New York: Oxford University Press, 2010), 117, notes the emphasis on heredity in asylum records going back to the early in the nineteenth century.
4. Henry Maudsley, "On the Causes of Insanity," *JMS* 12 (1867), 488–502. See also Maudsley, "Insanity and its Treatment," *JMS* 17 (October 1871), 311–334.
5. Henry Maudsley, "Considerations with Regard to Hereditary Influence," *JMS* 8 (January 1863), 482–512 and 9 (January 1864), 506–530, 525, 492; 488; Francis Galton, "Hereditary Talent and Character," *Macmillan's Magazine* 12 (1865), 57–66 and 318–327; Galton, *Hereditary Genius* (London: Macmillan, 1869).
6. Buckle, *History of Civilization*, 127n12; [Gray], review of Buckle, 236–237; Francis Galton, "Three Generations of Lunatic Cats," *Spectator*, 11 Apr. 1896, 514–515. Charles Rosenberg, "The Bitter Fruit: Heredity, Disease, and Social Thought," in Rosenberg, *No Other Gods: On Science and American Social Thought* (Baltimore: Johns Hopkins University Press, 1976), 25–53, discusses medical origins of hereditarian attitudes in America, and in fn. 27 extends the argument to Galton.
7. Galton, *Hereditary Genius*, prefatory chapter, 2nd ed. (London: Macmillan, 1892).

8. Richard A. Soloway, *Demography and Degeneration: Eugenics and the Declining Birthrate in Twentieth-Century Britain* (Chapel Hill: University of North Carolina Press, 1990); Heinrich Hartmann, *Der Volkskörper bei der Musterung. Militärstatistik und Demographie in Europa vor dem Ersten Weltkrieg* (Göttingen: Wallstein Verlag, 2011); Theodore M. Porter, "Statistics and the Career of Public Reason: Engagement and Detachment in a Quantified World," in Tom Crook and Glen O'Hara, eds., *Statistics and the Public Sphere: Numbers and the People in Modern Britain, c. 1800–2000* (New York: Routledge, 2011), 32–47.
9. Galton, *Hereditary Genius* (1869), vi. On the normal curve, see Porter, *Rise of Statistical Thinking*, part 2. John Waller challenges the familiar story of eugenics as Galton's brainchild in "Ideas of Heredity, Reproduction, and Eugenics in Britain, 1800–1875," *Studies in History and Philosophy of Biological and Biomedical Sciences* 32, 2001, 457–489.
10. Mathew Thomson, *The Problem of Mental Deficiency: Eugenics, Democracy, and Social Policy in Britain c. 1870–1959* (Oxford: Clarendon Press, 1998). The French had a large role in initiating this discussion: see Jacqueline Gateaux-Mennecier, *Bourneville et l'enfance aliénée* (Paris: Centurion, 1989); Gateaux-Mennecier, *La debilité légère: Une construction idéologique* (Paris: Éditions du CNRS, 1990); Monique Vial, *Les enfants anormaux à lécole: Aux origines de l'éducation spécialisée, 1882–1909* (Paris: Armand Colin, 1990).
11. Noel A. Humphreys, "Statistics of Insanity in England, with Special Reference to its Alleged Increasing Prevalence," *Journal of the Royal Statistical Society* 53 (June 1890), 201–252; Humphreys, "The Alleged Increase of Insanity," *Journal of the Royal Statistical Society* 70 (June 1907), 203–241. See Simon Szreter, *Fertility, Class, and Gender in Britain, 1860–1940* (Cambridge: Cambridge University Press, 1996), 79–85.
12. *Report on the Scientific Study of the Mental and Physical Conditions of Childhood . . .* (London: The Committee, Parkes Museum, 1895).
13. R. H. Vetch, "Galton, Sir Douglas Strutt (1822–1899)," rev. David F. Channell, *Oxford Dictionary of National Biography* (Oxford University Press, 2004); online ed., October 2009, accessed 10 Dec. 2014, http://www.oxforddnb.com/view/article/10317.
14. Porter, *Rise of Statistical Thinking*, 136–137. Nicholas Wright Gillham, *A Life of Sir Francis Galton: From African Exploration to the Birth of Eugenics* (Oxford: Oxford University Press, 2001), 61–62, 267, mentions two occasions when the cousins interacted. In one of these, Douglas Galton provided Florence Nightingale with an introduction to Francis Galton.
15. About 1862, the Institute of Actuaries used a card-based data system for a mortality survey involving multiple life insurance offices. Nine years later, the Prudential became the first company in the world to replace bound ledgers with a card filing system. See Timothy Alborn, *Regulated Lives: Life Insurance and British Society, 1800–1914* (Toronto: University of Toronto Press, 2009), 113–115; Alborn, "Quill-Driving: British Life-Insurance Clerks and Occupational Mobility, 1800–1914," *Business History Review* 82 (2008), 40.
16. *Report on the Scientific Study*, 16.
17. *Report on the Scientific Study*, 12, 45, 51–53. On intelligence and inequality, see John Carson, *The Measure of Merit: Talents, Intelligence, and Inequality in the French and American Republics, 1750–1940* (Princeton, NJ: Princeton University Press, 2007).
18. Douglas Galton was a member of the executive committee. See *International Health Exhibition 1884. Official Catalogue*, 2nd ed. (London: William Clowes and Sons, 1884); Francis Galton, "On the Anthropometric Laboratory at the late International Health Exhibition," *Journal of the Anthropological Institute of Great Britain and Ireland* 14 (1884–1885), 205–219, read at the meeting of 11 Nov. 1884.

19. Francis Galton, *Record of Family Faculties. Consisting of Tabular Forms and Directions for Entering Data, with an Explanatory Preface* (London: Macmillan, 1884).
20. Karl Pearson, "Mathematical Contributions to the Theory of Evolution.—III. Regression, Heredity, and Panmixia," *Philosophical Transactions of the Royal Society of London A* 187 (1896), 269.
21. Francis Galton, *English Men of Science: Their Nature and Nurture* (London: Macmillan, 1874); Theodore M. Porter, *Karl Pearson: The Scientific Life in a Statistical Age* (Princeton, NJ: Princeton University Press, 2004), chap. 9, and David Alan Grier, *When Computers Were Women* (Princeton, NJ: Princeton University Press, 2005), chap. 7.
22. Edgar Schuster and Ethel M. Elderton, *The Inheritance of Ability: Being a Statistical Study of the Oxford Class Lists and of the School Lists of Harrow and Charterhouse* (London: Dulau, 1907); Edgar Schuster, *The Promise of Youth and the Performance of Manhood: Being A Statistical Inquiry into the Question Whether Success in the Examination for the B. A. Degree at Oxford is Followed by Success in Professional Life.* Eugenic Laboratory Memoirs 3 (London: Dulau, 1907).
23. John L. Myres, "A Bureau of Biometry," *Oxford and Cambridge Review* 1 (1907), 131–144; and follow-up by E. Lyttleton, "More about Biometry," *Oxford and Cambridge Review* 2 (1907), 36–40, pointing out that such a program had already been set up at Eton.
24. Edward Allen Fay, *Marriages of the Deaf in America* (Washington, DC: Gibson Bros., 1898); Edgar Schuster, "Hereditary Deafness. A Discussion of the Data Collected by Dr. E. A. Fay in America," *Biometrika* 4 (1905–6), 465–482. David Buxton of the Liverpool School for the Deaf and Dumb challenged the importance of heredity in "On the Marriage and Intermarriage of the Deaf and Dumb," *Liverpool Medico-Chirurgical Journal* 1 (1857), 172.
25. Karl Pearson, "On the Correlation of Intellectual Ability with the Size and Shape of the Head," *Proceedings of the Royal Society of London* 69 (1901), 333–342.
26. W. R. Macdonell, "On Criminal Anthropometry and the Identification of Criminals," *Biometrika* 1 (1901–2), 187, 190; J. G. Garson, "The Metric System of Identification of Criminals, as Used in Great Britain and Ireland," *Journal of the Anthropological Institute of Great Britain and Ireland* 30 (1900), 161. See also Simon A. Cole, *Suspect Identities: A History of Fingerprinting and Criminal Identification* (Cambridge, MA: Harvard University Press, 2001), 92–93, and the obituary notice for J. G. Garson in *Nature* 129 (25 June 1932), 931.
27. Karl Pearson, "On the Correlation of Intellectual Ability with the Size and Shape of the Head," *Proceedings of the Royal Society of London* 69 (1901), 333–342; Pearson, "On the Relationship of Intelligence to Size and Shape of the Head, and to Other Physical and Mental Characters," *Biometrika* 5 (1906–7), 105–146. He used inter-observer correlations to defend his reliance on the judgment of teachers in Pearson, "On the Value of Teachers' Opinion of the General Intelligence of School Children," *Biometrika* 7 (1910–11), 542–548.
28. Karl Pearson, "On the Laws of Inheritance in Man: II. On the Inheritance of the Mental and Moral Characters of Man, and its Comparison with the Inheritance of the Physical Characters," *Biometrika* 3 (1904–5), esp. 133–135.
29. Scull, *Most Solitary of Afflictions*, 328.
30. Surrey History Centre, Royal Earlswood Asylum, 392/11/1/1 Registry of Admissions; 392/11/4/1 General Register: Female Case Book 1859–1868; 392/11/4/12 Male Case Book admissions 1882–1886; 392/11/4/4 Female Case Book admissions 1888–1899. See also David Wright, *Mental Disability in Victorian England: The Earlswood Asylum, 1849-1901* (Oxford: Clarendon Press: 2001). On the category of

feeblemindedness, see James W. Trent, Jr., *Inventing the Feeble Mind: A History of Mental Retardation in the United States* (Berkeley: University of California Press, 1994).

31. George Edward Shuttleworth, *Mentally-Deficient Children: Their Treatment and Training* (London: H. K. Lewis, 1895), 35; George E. Shuttleworth and Fletcher Beach, "Idiocy and Imbecility, Aetiology of," in Daniel Hack Tuke, ed., *Dictionary of Psychological Medicine*, vol. 1, 659–665 (London: J. & A. Churchill, 1892).
32. David Heron, *A First Study of the Statistics of Insanity and the Inheritance of the Insane Diathesis*, Eugenics Laboratory Memoirs 2 (London: Dulau, 1907), 7, 12 and 14–15; Pearson also emphasized the concept of diathesis.
33. Heron, *First Study*, 23, 32; A. R. Urquhart, "The Morison Lectures—On Insanity, with Special Reference to Heredity and Prognosis," *JMS* 53 (April 1907), 252–253, 276.
34. John MacPherson to Karl Pearson, 3 Jan. 1904, Karl Pearson Papers 11/1/13/21 and 13 Feb. 1907, Karl Pearson Papers 7/33, UCL.
35. J. F. Tocher to Karl Pearson, 15 June 1904, Karl Pearson Papers 11/1/19/43, UCL; Tocher, "Pigmentation Survey of School Children in Scotland," *Biometrika* 6 (1908–9), 130–235; Tocher, "The Necessity for a National Eugenic Survey," *Eugenics Review* 2 (July 1910), 124–141.
36. David Heron, *The Influence of Defective Physique and Unfavourable Home Environment on the Intelligence of School Children, Being a Statistical Examination of the London County Council Pioneer School Survey*, Eugenics Laboratory Memoirs 8 (London: Dulau, 1910), quotes on 1–3.
37. Karl Pearson and Ethel M. Elderton, *A Second Study of the Influence of Parental Alcoholism on the Physique of the Offspring. Being a Reply to Certain Medical Critics of the First Memoir and an Examination of the Rebutting Evidence Cited by Them*, Eugenics Laboratory Memoirs 13 (London: Dulau, 1910), 30.
38. *Treasury of Human Inheritance*, part 4, Eugenics Laboratory Memoirs 15 (London: Cambridge University Press, 1910), section 12a, by Harold Rischbieth, "Hare-Lip and Cleft Palate," 79–123, 88. This memoir soon became part 4 of Karl Pearson, ed., *Treasury of Human Inheritance*, vol. 1 (London: Cambridge University Press, 1912). Rischbieth (1876–1943), as we learn, from the online Cambridge alumni directory and the Wikipedia article on his father, Charles Rischbieth, was an Australian son of a German immigrant father. Pearson's ally and colleague W.F.R. Weldon framed a theory of heredity in terms of subchromosomal determinants; see Pearson, "On a Mathematical Theory of Determinantal Inheritance, from suggestions and Notes of the late W.F.R. Weldon," *Biometrika* 6 (1908), 80–93.
39. *Treasury of Human Inheritance*, parts 1 and 2 (London: Cambridge University Press, 1909), section 6a by Jobson Horne. "Dahl's cases" are figures 59 and 63 on plate 10 and explanations of figures on 29. Bound into Pearson, *Treasury* (1912). Rischbieth included on pp. 93–107 the "Table of Associated Deformities," with six columns for abnormalities of the vertebral column, the senses, the trunk and viscera, etc. For basic information on Horne (1865–1953), see the online Cambridge University alumni directory. On the German appropriation of Dahl, see chap. 12.
40. On Pearson's conflicts see especially Donald MacKenzie, *Statistics in Britain, 1865–1930: The Social Construction of Scientific Knowledge* (Edinburgh: Edinburgh University Press, 1981); Matthews, *Quantification*.
41. For example, A. F. Tredgold, *Mental Deficiency (Amentia)* (London: Baillière, Tindall and Cox, 1908), 16, on germ plasm; John MacPherson (as his name was written here), *Mental Affections: An Introduction to the Study of Insanity* (London: Macmillan, 1899), 9–10, on alcoholism. Macpherson interpreted the trajectory leading to extinction of these degenerate lines as consistent with "triumphal progress" after all.

42. Ethel M. Elderton, with the assistance of Karl Pearson, *A First Study of the Influence of Parental Alcoholism on the Physique and Ability of Offspring*, Eugenics Laboratory Memoirs 10 (London: Dulau, 1910), 3.
43. The insurance handbook was James Pollock and James Chisholm, *Medical Handbook of Life Assurance*. In the fourth edition (London: Cassell, 1897), 49, we find, for insanity: "The disease generally increases in three or four generations," culminating often in death or sterility.
44. F. W. Mott, "Heredity and Eugenics in Relation to Statistics," from a lecture to the First International Eugenics Congress and to the London County Council, *British Medical Journal* (11 May 1912), 1053–1060 (also printed by the Eugenics Education Society in *Problems in Eugenics*). Pearson's response is "On an Apparent Fallacy in the Statistical Treating of 'Antedating' in the Inheritance of Pathological Conditions," *Nature* 90, no. 2247 (21 Nov. 1912), 334–335; see David Heron, "An Examination of Some Recent Studies of the Inheritance of Insanity," *Biometrika* 10 (1914–15), 356–383. Julia Bell took up this topic in relation to several diseases for the *Treasury of Human Inheritance*; see Curt Stern, *Principles of Human Genetics* (San Francisco: W. H. Freeman, 1949), 296–300.
45. Karl Pearson, *Nature and Nurture: The Problem of the Future. A Presidential Address . . . at the Annual Meeting of the Social and Political League, April 28, 1910*, 2nd ed. (London: Dulau, 1913), 10.
46. See Porter, *Karl Pearson*, 136, 147–148.
47. Karl Pearson Papers, 7/21, UCL. See also G. B. Griffiths, "Measurements of One Hundred and Thirty Criminals," with introductory note by Donkin, *Biometrika* 3 (1903), 60–62 plus tables.
48. Griffiths, "Measurements"; Charles Goring, *The English Convict. A Statistical Study* (London: HMSO, 1913), preface by E. Ruggles Bries, 6–9, and Goring's conclusions, 370–374; Karl Pearson, "Charles B. Goring," *Biometrika* 12 (1919), 297–299.
49. Horatio Bryan Donkin, *On the Inheritance of Mental Characters: The Harveian Oration for 1910* (London: Adlard and Son, Bartholomew Press, 1910), quotes on 20–22; William Bateson, *Mendel's Principles of Heredity* (Cambridge: Cambridge University Press, 1909), 305–306.
50. Donkin, *Inheritance of Mental Characters*, 10, 34.
51. H. B. Donkin, introductory note to E. B. Sherlock, *The Feeble-Minded. A Guide to Study and Practice* (London: Macmillan, 1911), xvii–xx. See also Thomson, *Problem of Mental Deficiency*, chap. 1.
52. Royal Commission on the Care and Control of the Feeble-Minded, *Report*, 8 vols. (London: HMSO, 1908), 8: 179–185. For an early discussion of Mendelism and disease, see R. C. Punnett, "Mendelism in Relation to Disease," *Proceedings of the Royal Society of Medicine* 1 (Sect. Epidemiol. State Med.) (1908), 135–168, held on 28 February, including comments by statisticians Major Greenwood (162–164) and George Udny Yule (164–166).
53. *Proceedings of the Royal Society of Medicine*, vol. 2, *General Reports. Comprising the Report of the Proceedings for the Session 1908–09* (London: Longmans, Green, 1909), session of 11 Nov. 1908; introductory address by William S. Church, 9–15, 14. On these debates see also Alan G. Cock and Donald R. Forsdyke, *Treasure Your Exceptions: The Science and Life of William Bateson* (New York: Springer, 2008), 314–317.
54. *Proceedings RSM* 2, *General Reports*, 18 Nov. 1908: Sir William Gowers, 15–22, 15, 21.
55. *Proceedings RSM* 2, *General Reports*, 18 Nov. 1908: Bateson, 22–30. On Nettleship's collaborations with Bateson and Pearson in the investigation of eye traits and of albinism, see Alan R. Rushton, "Nettleship, Pearson, and Bateson: The Biometric-

Mendelian Debate in a Medical Context," *Journal of the History of Medicine and Allied Sciences* 55 (2000), 134–157.

56. *Proceedings RSM* 2, *General Reports*, 18 Nov. 1908: Mercier, 40–45; Latham, 45–54; Pearson, 54–55.
57. *Proceedings RSM* 2, *General Reports*, 18 Nov. 1908: Pearson, 56–60.
58. *Proceedings RSM* 2, *General Reports*, 2 Dec. 1908: Mudge, 108–130
59. *Proceedings RSM* 2, *General Reports*, 2 Dec. 1908: Mudge; letter by William Bateson to the secretary of the Sociological Society, 4 May 1904, William Bateson Papers, John Innes Centre, bound papers, vol. 8 (typescripts).
60. *Proceedings RSM* 2, *General Reports*, 2 Dec. 1908: Pearson, 131–134; W.F.R. Weldon, "On the Ambiguities of Mendel's Categories," *Biometrika* 2 (1902), 44–55. Pearson's critique of the practices that give Mendelian ratios referred to George Percival Mudge, "On Some Features in the Hereditary Transmission of the Albino Character and the Black Piebald Coat in Rats—Paper II," *Proceedings of the Royal Society of London B* 80 (1908), 392. H. Crampe, "Die Gesetze der Vererbung der Farbe: Zuchtversuche mit zahmen Wanderratten," *Landwirtschaftliche Jahrbücher* 14 (1885), 379–399 and 539–619, had become well-known in English-language genetics.
61. *Proceedings RSM* 2, *General Reports*, 2 Dec. 1908: Pearson, esp. 133 on skin color; Pearson, "Note on the Skin-Colour of Crosses between Negro and White," *Biometrika* 6 (1908–9), 348. On race, skin color, and blending, see Theodore M. Porter, "The Curious Case of Blending Inheritance," *Studies in History and Philosophy of Biological and Biomedical Sciences* 46 (2014), 125–132.
62. *Proceedings RSM* 2, *General Reports*, 2 Dec. 1908: Pearson, 131–134; Pauline M. H. Mazumdar, *Eugenics, Human Genetics and Human Failings: The Eugenics Society, Its Sources and Its Critics in Britain* (London: Routledge, 1992), 58–59.

Chapter 11. Genetic Ratios and Medical Numbers Give Rise to Big Data Ambitions in America, 1902–1920

Epigraphs: Charles B. Davenport to David Starr Jordan, 18 Mar. 1909, B: D27, APS; Willet M. Hays, "Constructive Eugenics," *American Breeders Magazine* 3 (1912), 5–11 and 113–119, 116. This chapter and the next draw from a paper I finished in 2010, "Asylums of Hereditary Research in the Efficient Modern State" in Staffan Müller-Wille and Christina Brandt, eds., *Heredity Explored: Between Public Domain and Experimental Science* (MIT Press, 2016), 81-109.

1. Porter, "Blending Inheritance"; Nathaniel Comfort, *The Science of Human Perfection* (New Haven, CT: Yale University Press, 2013), 23–24, 72–75.
2. APS, B:J44, Herbert Spencer Jennings papers, Pearl to Jennings, 8 Oct. 1905.
3. The book is Charles B. Davenport, *Statistical Methods: With Special Reference to Biological Variation* (New York: John Wiley & Sons, 1899). His autobiographical sketch is in the letter to Helen M. Walker of 20 Nov. 1926, in APS, B:D27, Charles B. Davenport Papers (CBDP). See also Helen M. Walker, *Studies in the History of Statistical Method* (Baltimore: Williams & Wilkins, 1929; reprinted 1931), 158–160.
4. See APS, B:D27, CBDP, Hays to Davenport (quoting Davenport's words back to him), 10 Nov. 1902; Davenport to Hays, 20 Feb. 1904, B:D27.
5. APS, B:P312 Raymond B. Pearl Papers: Pearson to Carnegie Inst. 28 Oct. 1905; Pearson to Pearl, 28 Feb. 1910. For Galton's and Pearson's negative assessments of Davenport see Kevles, *Name of Eugenics*, 48.
6. William Castle, *Genetics and Eugenics: A Text-Book for Students of Biology and a Reference Book for Animal and Plant Breeders* (Cambridge, MA: Harvard University Press, 1916), 241–242; APS, B:D27, CBDP, Davenport to Walker, 20 Nov. 1926; APS,

B:P312, Pearl Papers, Pearl to Major Greenwood, 26 Oct. 1921. By the third edition of 1925, Castle had excised his critical remarks.

7. APS, B:D27, CBDP, Davenport to Bell, 7 Mar. and 12 Mar. 1904. See also his efforts in March–April 1904 to procure Naples goats with pendants behind their ears from Herbert Spencer Jennings, then in residence at the Naples Zoological Station: APS, B:J44, Jennings Papers.
8. APS, B:D27, CBDP, Bell to Davenport, 11 July 1906.
9. APS, B:D27, CBDP, Davenport to Bell, 19 July 1906; Bell to Davenport, 4 Aug. 1906.
10. APS, B:D27, CBDP, Bell to Davenport, 14 Sept. 1906; Davenport to Bell, 29 Jan. 1907.
11. APS, B:D27, CBDP, correspondence with Boas from 1899 and Thorndike from 1901.
12. The other sections were devoted to animal and to plant breeding, each divided into committees according to type: draft horses, pet stock, wild birds, carnations, sugar crops, and so on. See Barbara A. Kimmelman, "The American Breeders' Association: Genetics and Eugenics in an Agricultural Context, 1903–13," *Social Studies of Science* 13 (1983), 163–204; Diane B. Paul and Barbara Kimmelman, "Mendel in America: Theory and Practice, 1900–1919," in Ronald Rainger, Keith R. Benson, and Jane Maienschein, eds., *The American Development of Biology* (New Brunswick, NJ: Rutgers University Press, 1988), 281–310.
13. Charles B. Davenport, "Eugenics, a Subject for Investigation rather than Instruction," *American Breeders Magazine* 1 (1910), 68–69.
14. Charles B. Davenport, "Report of the Committee on Eugenics," *American Breeders Magazine* 1 (1910), 126.
15. Davenport, "Report of the Committee," 129.
16. Maternal impression refers to the possible effect of the things seen or imagined by the mother on the body of her unborn child (possibly at the moment of conception), quite an old-fashioned idea by 1907.
17. APS, B:D27, CBDP, correspondence with Willet Hays, 9–12 Mar. 1907.
18. APS, B:D27, CBDP, Hays to Davenport, 30 Jan. 1909, and Davenport to Laughlin, 5 Feb. 1909.
19. APS, B:D27, CBDP, Davenport to Hays, 12 Feb. 1912; undated resolutions of Eugenics Section under "American Breeders' Association" (ABA); Davenport to Southard, 18 Jan. 1912.
20. APS, B:D27, CBDP, file for ABA; see also APS, B:D27 CDBP, Southard to Davenport, 13 Nov. 1911; on the role of Indian and African blood, Davenport to Jordan, 6 Feb. 1913.
21. APS, B:D27, CBDP, Davenport to Samuel L. Rogers, 21 Sept. and 4 Oct. 1917; Rogers to Davenport, 27 Sept. 1917.
22. APS, B:D27, CBDP, Davenport to Jordan, 18 Mar. 1909. Quote continues in first epigraph.
23. APS, B:D27, CBDP, ABA, folder 2. They had already discussed the composition of some of these subcommittees: for example, APS, B:D27, CBDP, Jordan to Davenport, 26 Apr. 1909, on feeble-mindedness.
24. APS, B:D27, CBDP, Davenport to Bell, 9 Jan. 1913.
25. Davenport, "Report on the Committee," 127–128.
26. Dahl is mentioned in APS, B:D27 CBDP, Southard to Davenport, 23 Nov. 1910.
27. Garland E. Allen, "The Eugenics Record Office at Cold Spring Harbor, 1910–1940: An Essay in Institutional History," *Osiris* 2 (1986), 225–264; Amy Sue Bix, "Experiences and Voices of Eugenics Field-Workers: 'Women's Work' in Biology," *Social Studies of Science* 27 (1997), 625–668; Leila Zenderland, *Measuring Minds: Henry Herbert Goddard and the Origins of American Intelligence Testing* (Cambridge: Cambridge University Press, 1998), 159–163. See also Charles Davenport, *The Family-History*

Book, Eugenics Record Office Bulletin No. 7 (Cold Spring Harbor, NY, 1912) and *The Trait Book*, Eugenics Record Office Bulletin No. 6 (Cold Spring Harbor, NY, 1912).

28. James Wilson, "Presidential Address: Ninth Annual Meeting," *American Breeders Magazine* 4 (1913): 55.
29. For example, Davenport, *Heredity in Relation to Eugenics* (New York: Henry Holt, 1911), 111–112, on hereditary cataract. Rushton, *Genetics and Medicine in the United States*, chap. 5, shows how this book guided medical-eugenic advice. D. F. Jones and S. L. Mason, "Inheritance of Congenital Cataract," *American Naturalist* 50 (1916), 119–126 and 751–757, challenged Davenport's idea that cataract was a Mendelian dominant. C. H. Danforth, "Inheritance of Congenital Cataract," *American Naturalist* 50 (1916), 442–448 gave a partial defense. Only Danforth, however, expressed doubts that this cataract could be reduced to any kind of Mendelian factor.
30. APS, B:D27, CBDB, Davenport to Bell, 8 Dec. 1909, and correspondence involving Marsh and Arthur Hunter from April 1913 to March 1915; E. J. Marsh, *Value of Family History and Personal Condition in Estimating a Liability to Consumption* (New York: Mutual Life Insurance Company, 1895).
31. APS, B:D27, CBDP, correspondence with Max Schlapp, 3 Feb., 11 May, and 8–10 Nov. 1910. See also Davenport to Henderson, 11 May 1910, explaining that his chairmanship was to recognize past service to the field, and it would be enough if he lent his name to the endeavor.
32. Ira Van Gieson, "The Pathological Institute of the New York State Hospitals," in State of New York, *State Commission in Lunacy, Ninth Annual Report*, 1896–97 (Albany, 1897), 73–240, 93–95; also *Ninth Annual Report*, 11.
33. Aldon G. Blumer, "Report of the Superintendent," "Fifty-Fourth Annual Report of the Managers of the Utica State Hospital," in State of New York, State Commission in Lunacy, *Eighth Annual Report*, 1895–1896 (Albany, NY, 1897), 283–293, 288.
34. Van Gieson, "Pathological Institute," 96, 100, 126, 131, 189, 193–195, 212; Van Gieson, "Correlation of Sciences in the Investigation of Nervous and Mental Diseases," *Archives of Neurology and Psychopathology* 1 (1898), 25–262. On pathological institutes and scientific aspirations see Grob, *Mental Illness*, 127–135.
35. Adolf Meyer, "Report of the State Pathological Institute for the Year Ending September 30, 1904," in State of New York, State Commission in Lunacy, *Sixteenth Annual Report*, 1903–4 (Albany, NY, 1905), 31–48, 33–35, 42; compare the statistics in the *Fourteenth Annual Report* for 1901–2. See also Lamb, *Pathologist of the Mind*, 55–58.
36. Adolf Meyer, "Insanity: General Pathology," in Alfred H. Buck, ed., *Reference Handbook of the Medical Sciences*, 8 vols. (New York: William Wood, 1900–1908), 5 (1902): 36–43, 36–37. He titled this section "The Data of Heredity." See Ian Robert Dowbiggin, *Keeping America Sane: Psychiatry and Eugenics in the United States and Canada, 1880–1940* (Ithaca, NY: Cornell University Press, 1997), 100–104.
37. Quote from APS, B:D27, CBDP, Davenport to Southard, 10 Nov. 1909. See also Southard to Davenport, 5 Oct., 2 Nov., and 12 Nov. 1909, and Davenport to Southard, 17 Nov. 1909. Southard's letter accepting the appointment is dated 20 Nov. 1909.
38. APS, B:D27, CBDP, Davenport to Southard, 9 Mar. 1910 and 29 Oct. 1910, and Southard to Davenport, 31 Oct. 1910.
39. *Thirteenth Annual Report of the Kings Park State Hospital at Kings Park to the State Commission in Lunacy for the Year Ending September 1908*, 52–60 and patient list, 92–162; *Fourteenth Annual Report . . . 1909*, 11.

40. APS, B:D27, CBDP, files for ABA, folder 1; Rosanoff to Davenport, 5 Oct. 1910; Davenport to Southard, 18 Nov. 1910 and 1 Nov. 1910; Southard to Davenport, 2 Dec. 1910; Davenport to Southard, 5 Dec. 1910.
41. APS, B:D27, CBDP, Davenport to Southard 10 Nov. 1910; Gertrude L. Cannon and A. J. Rosanoff, "Preliminary Report of a Study of Heredity in the Light of the Mendelian Laws," *Journal of Nervous and Mental Disease* 38 (1911), 272–279.
42. Davenport comment in reprint of Cannon and Rosanoff, "Preliminary Report," as Eugenics Record Office Bulletin No. 3 (Cold Spring Harbor, NY, May 1911), 10; APS, B:D27, CBDP, Rosanoff to Davenport 20 Sept. 1910.
43. Charles B. Davenport, "Euthenics and Eugenics," *Popular Science Monthly* (January 1911), 16–20; APS, B:D27, CBDP, Jelliffe to Davenport, 26 Dec. 1910. Kevles, *Name of Eugenics*, 49, quotes from this letter.
44. APS, B:D27, CBDP, Davenport to Jelliffe, 27 Dec. 1910.
45. Smith Ely Jelliffe, "Predementia Praecox: The Hereditary and Constitutional Features of the Dementia Praecox Make Up," *Journal of Nervous and Mental Disease* 38 (1911), 4–5. Meyer argued similarly in "The Pathology of Mental Disease," *Reference Handbook of the Mental Sciences*, 3rd ed., 1916, reprinted in Winters, ed., *Collected Papers of Adolf Meyer*, 1: 289–310, 292–293.
46. Noll, *American Madness*; A. J. Rosanoff, "The Inheritance of the Neuropathic Constitution," *Journal of the American Medical Association* 58, no. 17 (April 1912), 1266–1269. Reprinted from a paper presented on 15 Feb. 1912 to the quarterly conference of New York State Hospitals, and first printed in *New York State Hospital Bulletin* 5, no. 2 (August 1912), 278–291.
47. A. J. Rosanoff and Florence I. Orr, "A Study of Heredity of Insanity in the Light of the Mendelian Theory," *AJI* 68 (1911), 221–261; Rosanoff, "Inheritance of the Neuropathic Constitution," 1268.
48. Jean-Paul Gaudillière and Ilana Löwy, "The Hereditary Transmission of Human Pathologies between 1900 and 1940: The Good Reasons Not to Become, 'Mendelian,'" in Müller-Wille and Brandt, *Heredity Explored*, 311–335.
49. Jelliffe, "Predementia Praecox," 5; C. B. Davenport and David Weeks, "A First Study of Inheritance in Epilepsy," *Journal of Nervous and Mental Disease* 38 (1911), 641–670, reprinted as Eugenics Record Office Bulletin No. 4.
50. A. C. Rogers, "Report of Committee on Heredity of Feeble-Mindedness," *American Breeders Magazine* 2 (1911), 269–270; Henry H. Goddard, "Heredity of Feeble-Mindedness," *American Breeders Magazine* 1 (1910), 165–166; Rogers, "Report of the Committee on the Heredity of Feeblemindedness," *American Breeders Magazine* 3 (1912), 134–136. Compare the five-part query sent out by the Hanwell Asylum in England, including several questions on mental disease in the family, with the remark that while a response was entirely optional, a full response would provide the "greatest assistance in the treatment of the case." LMA, Hanwell Case Book H11/HLL/B20/30, from 1901 to 1902.
51. APS, B:D27, CBDP, Johnstone to Davenport 23 July 1910; Davenport to Johnstone, 22 July 1912.
52. Zenderland, *Measuring Minds*, 153–154; APS, B:D27, CBDP, Goddard to Davenport, 15 Mar. 1909.
53. APS, B:D27, CBDP, Davenport to Goddard, 18 Mar. and 19 Apr. 1909.
54. APS, B:D27, CBDP, Davenport to Jordan, 27 Jan. 1909; Jordan to Davenport, 1 Feb. and 3 Feb. 1909; Davenport to Jordan, 18 Mar. 1909. On the census, see also Davenport to Laughlin, 5 Feb. 1909.
55. APS, B:D27, CBDP, Davenport to Jordan, 29 Mar. 1909; Jordan to Davenport, 8 Apr. 1909; Davenport to Jordan, 14 Apr. 1909; Jordan to Davenport, 1 Nov. 1909.

56. APS, B:D27, CBDP, Davenport to Meyer, 1 May 1909; Davenport to Irving Fisher, 27 Jan. 1910.
57. APS, B:D27, CBDP, Laughlin to Davenport, 30 Mar. 1908; Davenport to Laughlin, 1 Apr. 1908; Laughlin to Davenport, 24 Feb. 1909; Davenport to Laughlin, 9 Mar. 1909.
58. Zenderland, *Measuring Minds*, 156–158; Henry H. Goddard, *Feeble-Mindedness: Its Causes and Consequences* (New York: Macmillan, 1914), 556.
59. A. C. Rogers and Maud A. Merrill, *Dwellers in the Vale of Siddem: A True Story of the Social Aspect of Feeble-Mindedness* (Boston: Gorham Press, 1919), 12–14.
60. Goddard, *Feeble-Mindedness*, viii, 436–439, 548–556; David Barker, "The Biology of Stupidity: Genetics, Eugenics and Mental Deficiency in the Inter-War Years," *British Journal for the History of Science* 22 (1989), 347–375.
61. Zenderland, *Measuring Minds*, 158–163, 169–185; Henry H. Goddard, *The Kallikak Family: A Study in the Heredity of Feeble-Mindedness* (New York: Macmillan, 1919), 111.
62. Rosanoff and Orr, "Heredity of Insanity," 225, 228–229.
63. APS, B:D27, CBDP, Davenport to Woods, 15 Mar. 1910.
64. See Davenport's review in *Science* 42 (1915), 837–838. The 1:1 ratio applied for dominant traits displayed by just one parent or recessive ones for which one parent was homozygous and the other heterozygous.
65. Christophe Bonneuil, "Pure Lines as Industrial Simulacra: A Cultural History of Genetics from Darwin to Johannsen," in Müller-Wille and Brandt, *Heredity Explored*, 213–242.
66. Alice Wexler, *The Woman Who Walked into the Sea: Huntington's and the Making of a Genetic Disease* (New Haven, CT: Yale University Press, 2008), 60.
67. APS, B:D27, CBDP, correspondence with Melvil Dewey, 21–24 Feb. 1908 and 28 Mar. 1914; Markus Krajewski, "Zettelwirtschaft: Die Geburt der Kartei aus dem Geiste der Bibliothek (Berlin: Kulturverlag Kadmos, 2002), translated by Peter Krapp as *Paper Machines. About Cards & Catalogs 1548-1929* (Cambridge, MA: MIT Press, 2011); Porter, "Asylums of Hereditary Research."
68. Charles B. Davenport, *The Feebly Inhibited. Nomadism, or the Wandering Impulse, with Special Reference to Heredity* (Washington, DC: Carnegie Institution of Washington, 1915). Files containing Single Trait Sheets are at the APS. For example, insanity is SPR A: 316. I examined these documents in March 2015 at the exhibition "Haunted Files: The Eugenics Record Office," Asian/Pacific/American Institute, New York University.
69. Karl Pearson, "On the Ancestral Gametic Correlations of a Mendelian Population Mating at Random," *Proceedings of the Royal Society of London B* 81 (5 June 1909), 225–229, 229.
70. Todd M. Endelman, "Anglo-Jewish Scientists and the Science of Race," *Jewish Social Studies* 11 (2004), 60–61; Raphael Falk, "Three Zionist Men of Science: Between Nature and Nurture," in Ulrich Charpa and Ute Deichmann, eds., *Jews and Sciences in German Contexts* (Tübingen: Mohr Siebeck, 2007), 129–154; H. J. Laski, "A Mendelian View of Racial Heredity," *Biometrika* 9 (1911–1912), 424.
71. APS, B:D27, CBDP, Hrdlička to Davenport, 19 May 1915; also letters of 13 Feb. 1909 and 3 May 1915.
72. David Heron, *Mendelism and the Problem of Mental Defect. I. A Criticism of Recent American Work*, vol. 7 of Questions of the Day and the Fray (London: Cambridge University Press, 1913). Pearson wished he had access to field-workers; see Pearson, "On the Inheritance of Mental Disease," *Annals of Eugenics* 4 (1930–31), 362.
73. Karl Pearson and Gustav A. Jaederholm, *Mendelism and the Problem of Mental Defect, II On the Continuity of Mental Defect*, vol. 8 of Questions of the Day and Fray (London: Dualu, 1914); Karl Pearson, *Mendelism and Problem of Mental Defect. III. On the*

Graduated Character of Mental Defect, and on the Need for Standardizing Judgments as to the Grade of Social Inefficiency Which Shall Involve Segregation, vol. 9 of Questions of the Day and the Fray (London: Cambridge University Press, 1914), 18; Heron, *Criticism of Recent American Work*, 38.

74. Pearson, *On the Graduated Character*, 50; Pearson and Jaederholm, *On the Continuity of Mental Defect*, 28–29, 37.
75. A. J. Rosanoff, "Mendelism and Neuropathic Heredity: A Reply to some of Dr. David Heron's Criticisms of Recent American Work," *AJI* 70, no. 3 (January 1914), 571–587. See also APS, B:D27, CBDP, Rosanoff to Davenport, 25 Oct. 1913, and Davenport to Rosanoff, 27 Oct. and 7 Nov. 1913; Rosanoff to Davenport, 14 Dec. 1913.
76. Barker, "Biology of Stupidity"; William Bateson, "Address on Heredity. Delivered at the Seventeenth International Congress of Medicine," *British Medical Journal* (16 Aug. 1913), 359–360; Bateson, *Biological Fact and the Structure of Society* (Oxford: Clarendon Press, 1912), 13; Bateson, "Common Sense in Racial Problems," *Eugenics Review* 13 (1921), 327, 329. On Bateson's uncertainties ca. 1907, see Rushton, "Nettleship."
77. Hamish G. Spencer and Diane B. Paul, "The Failure of a Scientific Critique: David Heron, Karl Pearson, and Mendelian Eugenics," *British Journal for the History of Science* 31 (1998), 441–452.
78. See Fisher's student paper, presented 10 Nov. 1911, in Bernard Norton and E. S. Pearson, "A Note on the background to, and Refereeing of, R. A. Fisher's 1918 Paper 'On the Correlation Between Relatives on the Supposition of Mendelian Inheritance,'" *Notes and Records of the Royal Society of London* 31 (1976), 151–162; R. A. Fisher, "The Elimination of Mental Defect," *Eugenics Review* 16 (1924), 114–116; Thomson, *Problem of Mental Deficiency*, 185.
79. Lionel S. Penrose, *Mental Defect* (London: Sidgwick & Jackson, 1933), 6–7, 77, 145.

Chapter 12. German Doctors Link Genetics to Rigorous Disease Categories Then Settle for Statistics, 1895–1920

Epigraphs: Julius Wagner-Jauregg, "Einiges über erbliche Belastung," *Wiener klinische Wochenschrift* 19, no. 1 (4 Jan 1906), 2; Ernst Wittermann, "Psychiatrische Familienforschungen," *Zeitschrift für die gesamte Neurologie und Psychiatrie* (*ZgNP*) 20 (1913), 153.

1. Emanuel Mendel, "Hereditäre Anlage und progressive Paralyse der Irren," *Archiv für Psychiatrie und Nervenkrankheiten* 10 (1880), 780–787. On marriage, Mendel, "Geisteskrankheiten und Ehe," in H. Senator and S. Kaminer, eds., *Krankheiten und Ehe* (Munich: J. E. Lehmanns Verlag; New York: Rebman, 1904), 808–809, 814–815, 822. See Uta Fleckner, "Emanuel Mendel (1839–1907). Leben und Werk eines Psychiaters im Deutschland der Jahrhundertwende," medical diss., Freie Universität Berlin, 1994, 67, 75, 79.
2. Ernst Rüdin, "Einige Wege und Ziele der Familienforschung mit Rücksicht auf die Psychiatrie," *ZgNP* 7 (1911), 487.
3. Emil Oberholzer, "Erblichkeitsverhältnisse und Erbgang bei Dementia praecox," in *Verhandlungen der Gesellschaft deutscher Naturforscher und Ärzte* 85, no. 2 (1913), 636; an expanded paper with the same title was published in *Zur 50. Jahresversammlung des Verein Schweizerischer Irrenaerzte* (1–2 June 1914) (Geneva: Buchdruckerei Albert Kündig, 1914), 8–27. He also described in detail the various symptoms of heterozygosity, inspiring F. Riklin in discussion to speculate that Carl Jung's introversion type might be such a heterozygote.

4. APS, B:D27, CBDP, Rüdin to Davenport, 31 Dec. 1910; Rüdin, "Einige Wege," 519. The citation is to Heron, *First Study of the Statistics of Insanity*, 17–18.
5. Rüdin, "Einige Wege," 506.
6. Ernst Rüdin, *Studien über Vererbung und Entstehung geistiger Störungen, I. Zur Vererbung und Neuentstehung der Dementia Praecox* (Berlin: Julius Springer, 1916), 51–55. The manuscript, he said, was completed on 14 May 1914 and delayed by the outbreak of the war.
7. Heinrich Neumann, *Die Irrenanstalt zu Pöpelwitz bei Breslau im ersten Decennium ihrer Wirksamkeit* (Erlangen: Ferdinand Enke, 1862), 49–50; John S. Billings, *Report on the Insane, Feeble-Minded, Deaf and Dumb, and Blind in the United States at the Eleventh Census, 1890* (Washington DC: Government Printing Office, 1895), 35.
8. According to the *Promotionsakten* of Zürich University U106g.9.1, she took her first degree there on 15 Dec. 1892, and her medical degree on 21 Mar. 1895. In 1900 she married Heinrich Thomann, head of the Zurich Statistical Office, and in 1901, Koller was among the founding physicians of the Schweizerische Pflegerinnenschule, a hospital for women, where she specialized in internal medicine. On Koller's background and the reception of her study, see Heidi Thomann Tewarson, *Die ersten Zürcher Ärztinnen: Humanitäres Engagement und wissenschaftliche Arbeit zur Zeit der Eugenik* (Basel: Schwabe Verlag, 2018).
9. Ritter, *Psychiatrie und Eugenik*, 23; Auguste Forel, "Zum Entwurf eines schweizerischen Irrengesetzes," *Zeitschrift für Schweizer Strafrecht* 6 (1893), 313–331; *Die Ergebnisse der Irrenzählung vom 1. Dezember 1888*, in *Statistische Mittheilungen betreffend den Kanton Zürich* (Zurich, 1890), 231 and 187–188. On this census, see Ritter, "Von den Irrenanstalten," 64.
10. Jenny Koller, "Beitrag zur Erblichkeitsstatistik der Geisteskranken im Canton Zürich: Vergleichung derselben mit der erblichen Belastung gesunder Menschen durch Geistesstörungen u. dergl.," *Archiv für Psychiatrie und Nervenkrankheiten* 27 (1895), 269–270; see also Tewarson, *Ersten Zürcher Ärztinnen*.
11. Koller, "Beitrag," 276–277; Otto Diem, "Die psycho-neurotische erbliche Belastung der Geistesgesunden und der Geisteskranken: Eine statistisch-kritische Untersuchung auf Grund eigener Beobachtungen," *Archiv für Rassen- und Gesellschaftsbiologie* (*ARGB*) 2 (1905), 337.
12. *Rechenschaftsbericht über die zürcherische kantonale Irrenheilanstalt Burghölzli für das Jahr 1880* (Zurich, 1881), 4–5, from Staatsarchiv des Kantons Zürich, Irrenanstalt Burghölzli S 323, Jahresberichte and other reports.
13. Sabine Richebächer, *Sabina Spielrein: "Eine fast grausame Liebe zur Wissenschaft"* (Zurich: Dörlemann, 2005), 79; Burghölzli report for 1904 in Zürich Staatsarchiv S 323. On psychiatry and eugenics at Burghölzli, see Marietta Meier, Brigitta Bernet, Roswitha Dubach, and Urs Germann, *Zwang zur Ordnung: Psychiatrie im Kanton Zürich, 1870–1970* (Zürich: Chronos, 2007), esp. 66–67.
14. Koller, "Beitrag," 278–279.
15. Koller, "Beitrag," 279–284.
16. Sigmund Freud, "L'Hérédité et l'étiologie des névroses," *Revue neurologique* 4, no. 6 (30 Mar. 1896), 164, 162.
17. For example, A. Cramer, *Über die ausserhalb der Schule liegenden Ursachen der Nervosität der Kinder* (Berlin: Reuther & Reichard, 1899), 9.
18. Julius Wagner-Jauregg, "Ueber erbliche Belastung," *Wiener klinische Wochenschrift* 15 (30 Oct. 1902), 1154–1155; Magda Whitrow, *Julius Wagner-Jauregg (1857–1940)* (London: Smith-Gordon, 1993), 186–187.
19. Diem, "Psycho-neurotische erbliche Belastung," 224–225, 227–228, 232, 240.

20. Wagner-Jauregg, "Ueber erbliche Belastung," 1155–1157, 1159; Wagner-Jauregg, "Einiges über erbliche Belastung," 2; Diem, "Psycho-neurotische erbliche Belastung," 355, 359. However, in "Ueber erbliche Belastung" (1159), Wagner-Jauregg warned against inbreeding (*Inzucht*).
21. Diem, "Psycho-neurotische erbliche Belastung," 358; Ernst Rüdin, "Ergänzenden Bemerkungen zu Otto Diems Artikel," *ARGB* 2 (1905), 470–473; Wagner-Jauregg, "Ueber erbliche Belastung," 1157; Wagner-Jauregg, "Einiges über erbliche Belastung," 4.
22. Robert Sommer, *Familienforschung und Vererbungslehre* (Leipzig: Johann Ambrosius Barth, 1907), 208–210. He also described Thomas Mann's *Buddenbrooks* as a novel of degeneration. Wilhelm Weinberg, "Aufgabe und Methode der Familienstatistik bei medizinisch-biologischen Problemen," *Zeitschrift für soziale Medizin* 3–4 (1907), 13, called attention to Sommer's description of Dr. Pascal as an exemplary *Familienforscher*. See also review of Zola's *Docteur Pascal* in *JMS* 40 (1894), 274–277.
23. Paul Julius Möbius, *Die Nervosität*, 2nd ed. (Leipzig: Weber, 1885), 5; Emil Kraepelin, *Psychiatrie: Ein Lehrbuch für Studierende und Ärzte*, 8th ed., 4 vols. (Leipzig: Johann Ambrosius Barth, 1909), 1: 184–185; Paul Albrecht, "Gleichartige und ungleichartige Vererbung der Geisteskrankheiten," *ZgNP* 11 (December 1912), 542–545, 547, 577–578. On *Urschleim*, see Anja Haberer, *Zeitbilder: Krankheit und Gesellschaft in Theodor Fontanes Romanen* Cécile *(1886) und* Effi Briest *(1895)* (Würzburg: Königshausen & Neumann, 2012), 55.
24. Karl Grassmann, "Kritischer Überblick über die gegenwärtige Lehre von der Erblichkeit der Psychosen," *AZP* 52 (1896), 969, 1001–1002, 1008. Compare Joseph Jules Dejerine, *L'Hérédité dans les maladies du système nerveux* (Paris: Asselin et Houzeau, 1886), esp. 27–32.
25. Emil Kraepelin, "Ueber die Entartungsfrage," *Muenchener medizinische Wochenschrift* 28 (1908), 1512–1513, translated with introduction in Eric J. Engstrom, "'On the Question of Degeneration' by Emil Kraepelin (1908)," *History of Psychiatry* 18 (2007), 389–404. See also Kraepelin, "Wege und Ziele der klinischen Psychiatrie," *AZP* 53 (1897), 840–844, 843.
26. Wilhelm Weinberg, "Pathologische Vererbung und genealogische Statistik," *Deutsches Archiv für klinische Medizin* 78 (1903), 535, 527.
27. Ludwig Platte, *Vererbungslehre, mit besonderer Berücksichtigung des Menschen, für Studierende, Ärzte und Züchter* (Leipzig: Wilhelm Engelmann, 1913).
28. Paul Weindling, *Health, Race and German Politics between National Unification and Nazism, 1870–1945* (Cambridge: Cambridge University Press, 1989), 230–232; Bernd Gausemeier, "Auf der 'Brücke zwischen Natur- und Geschichtswissenschaft': Ottokar Lorenz und die Neuerfindung der Genealogie um 1900," in Florence Vienne and Christina Brandt, eds., *Wissensobjekt Mensch: Humanwissenschaftliche Praktiken im 20. Jahrhundert* (Berlin: Kulturverlag Kadmos, 2008), 137–164; Ottokar Lorenz, *Lehrbuch der gesammten wissenschaftlichen Genealogie* (Berlin: Wilhelm Hertz, 1898).
29. Wilhelm Strohmeyer, "Ueber die Bedeutung der Individualstatistik bei der Erblichkeitsfrage in der Neuro- und Psychopathologie," *Münchener medicinische Wochenschrift*, 48, no. 45 (5 Nov. 1901), 1786–1789 and no. 46 (12 Nov. 1901), 1824–1844.
30. Wilhelm Strohmeyer, "Ziele und Wege der Erblichkeitsforschung in der Neuro- und Psychopathologie," *AZP* 61 (1904), quote on 365.
31. Grassmann, "Kritischer Überblick"; Johannes Vorster, "Ueber die Vererbung endogener Psychosen in Beziehung zur Classifikation," *Monatsschrift für Psychiatrie und Neurologie* 9 (1901), 161–176, 301–315, 367–392, 164, 314, 392; W. Geiser, "Über

familiaere Geisteskrankheiten zur Beobachtung gelangt in den Jahren 1888–1903 in der Genf Irrenanstalt 'Bel-Air,'" Diss. für Doctorwürde, Med. Fak. Uni. Genf, 1903, Thèse no. 13 (Geneva: F. de Siebenthal & Cie., 1903), 7–10, 41. The classic authority on similar heredity was Emil Sioli (of the institution in Bunzlau/Bolesławiec, Silesia), "Ueber directe Vererbung von Geisteskrankheiten," *Archiv für Psychiatrie und Nervenkrankheiten* 16 (1885), 605. See Anne Cottebrune, "Zwischen Theorie und Deutung der Vererbung psychischer Störungen: Zur Übertragung des Mendelismus auf die Psychiatrie in Deutschland und in den USA, 1911–1930," *NTM* 17 (2009), 36–37.

32. Heron, *First Study of the Statistics of Insanity*, 5.
33. Robert Sommer, "Eine psychiatrische Abteilung der Reichsgesundheitsamts," *Psychiatrisch-Neurologische Wochenschrift* 12, no. 31 (29 Oct. 1910), 295–297; Sommer, "Zur Frage einer psychiatrischen Abteilung des Reichsgesundheitsamtes," *Psychiatrisch-Neurologische Wochenschrift* 13, no. 4 (22 Apr. 1911), 31–32; Alois Alzheimer, "Ist die Einrichtung einer psychiatrischen Abteilung im Reichsgesundheitsamt erstrebenewert?," *ZgNP* 6 (1911), 242–246.
34. Wilhelm Strohmeyer, "Die Bedeutung des Mendelismus für die klinische Vererbungslehre," *Die deutsche Klinik am Eingänge des zwanzigsten Jahrhunderts in akademischen Vorlesungen* 14 (1913), 331–339; Strohmeyer, "Ueber den Wert genealogischer Betrachtungsweise in der psychiatrischen Erblichkeitslehre," *Monatsschrift für Psychiatrie und Neurologie* 22 (1907), esp. 125; Max von Gruber und Ernst Rüdin, eds., *Fortpflanzung, Vererbung, Rassenhygiene: Katalog der Gruppe Rassenhygiene der Internationalen Hygiene-Ausstellung 1911 in Dresden* (Munich: J. F. Lehmanns Verlag, 1911), 59–90, esp. 73, 84, includes reprinted tables.
35. Strohmeyer, "Ziele und Wege," 369; Strohmeyer, "Bedeutung des Mendelismus," 346, 348–349.
36. Wittermann, "Psychiatrische Familienforschungen," 153–156, 164.
37. Wittermann, "Psychiatrische Familienforschungen," 158–159, 251, 262–264. See also Hermann Lundborg, *Medizinisch-biologische Familienforschungen innerhalb eines 2232 köpfigen Bauerngeschlechtes in Schweden (Provinz Blekinge)*, 2 vols. (Jena: Gustav Fischer, 1913), 1: 471–476; Matthias M. Weber, *Ernst Rüdin: Eine kritische Biographie* (Berlin: Springer Verlag, 1993), 98. Wittermann was equally satisfied to get 47% and then 44% when one parent had a single *Anlage* and the other, being sick, had two.
38. Wilhelm Weinberg, "Vererbungsforschung und Genealogie: Eine nachträgliche Kritik des Lorenzschens Lehrbuches," *ARGB* 8 (1911), 759. See also Bernd Gausemeier, "In Search of the Population: The Study of Human Heredity before and after the Mendelian Break," in Müller-Wille and Brandt, eds., *Heredity Explored*, 337–363.
39. On Weinberg's family and personal life, see Eugen Hübler, "Zum 100. Geburtstag von Wilhelm Weinberg," *Jahreshefte des Vereins für vaterländische Naturkunde in Württemberg* 118/119 (1964), 57–66. See also Hans Luxenburger, "Wilhelm Weinberg," *AZP* 107 (1938), 378–381; Franz J. Kallmann, "Obituary: Wilhelm Weinberg, M.D.," *Journal of Nervous and Mental Disease* 87 (February 1938), 263–264.
40. Hübler, "Zum 100. Geburtstag," 60; Curt Stern, "Wilhelm Weinberg," *Zeitschrift für menschliche Vererbungs- und Konstitutionslehre* 36 (1962), 374–382; Stern, "Wilhelm Weinberg, 1862–1937," *Genetics* 47 (1962), 1–6; Stern, "Weinberg, Wilhelm," *Complete Dictionary of Scientific Biography*, 16 vols. (1976; Detroit: Charles Scribner's Sons, 2008), 14: 230–231. On sterilization, see Wilhelm Weinberg, "Unfruchtbarkeit vom Standpunkte der Statistik," in Siegfried Placzek, ed., *Künstliche Fehlgeburt und künstliche Unfruchtbarkeit, ihre Indikationen, Technik und Rechtslage* (Leipzig: Georg Thieme, 1918), 437–448.

41. In the review of the revised edition of *Elemente der exakten Erblichkeitslehre*, by Wilhelm Johannsen, in *ARGB* 6 (1909), 555, Weinberg claimed that the principle was first shown by Pearson and himself and was falsely credited to Hardy. In "Weitere Beiträge zur Theorie der Vererbung," *ARGB* 7 (1910), 36, he wrote that he and Hardy had almost simultaneously extended Pearson's result. In *ARGB* 8 (1911), 400–401, a review of an article on Mendel's rules by van den Veld, Weinberg again mentioned only himself and Pearson. A.W.F. Edwards, "G. H. Hardy (1908) and Hardy-Weinberg Equilibrium," *Genetics* 179 (1908), 1143–1150, provides a chronology, from Udny Yule and Pearson to Hardy and beyond, while Alan Stark and Eugene Seneta, "Wilhelm Weinberg's Early Contribution to Segregation Analysis," *Genetics* 195 (September 2013), 1–6, analyze Weinberg's role. On the multiplication table, see G. H. Hardy, "Mendelian Proportions in a Mixed Population," *Science* 28, no. 706 (10 July 1908), 49–50, which also acknowledged Pearson for recognizing a special case of this equilibrium.
42. See, for example, W. Weinberg, ed., *Medizinisch-statistischer Jahresbericht über die Stadt Stuttgart im Jahre 1893. Einundzwanzigster Jahrgang* (Stuttgart: J. B. Metzlersche Buchhandlung, 1894). On 94, Weinberg is listed as *Armenarzt*, physician to the poor. See also Weinberg, "Sterblichkeit, Lebensdauer, und Todesursachen der württembergischen Aerzte von 1810–1895 und der Ärzte überhaupt," *Württembergische Jahrbücher für Statistik und Landeskunde* (1896), 104–170.
43. Hauptstaatsarchiv Stuttgart, E146 Bü 8052 Acten. Statistik über die Geisteskranken, 1832–33; Wilhelm Köstlin, *Zur Statistik der Geisteskrankheiten. Eine Inaugural-Dissertation welche zur Erlangung der Doctor-Würde in der Medicin und Chirurgie . . .* (Tubingen, 1840).
44. Staatsarchiv Ludwigsburg (SAL), E163 Bü 103, registration forms sent out by the Statistisch-Topographischen Bureau; P. Sick, "Statistik der Geisteskranken und der zu ihrer Pflege und Heilung bestehenden Anstalten im Königreich Württemberg," *Württembergische Jahrbücher für vaterländische Geschichte, Geographie, Statistik, und Topographie* (1855, no. 2), (Stuttgart: Eduard Hallberger, 1856). For 1864, see SAL, E163 Bü 104, "Statistische Aufnahmen der Geisteskranken in Württemberg vom Jahre 1864."
45. See *Bericht über die im Königreich Württemberg bestehenden Staats- und Privatanstalten für Irre, Schwachsinnige und Epileptische auf die Jahre 1892 und 1893* (Stuttgart, 1894).
46. A. Köhler, review of *Zur Statistik der Geisteskrankheiten*, by Julius Ludwig August Koch, in *AZP* 35 (1879), 657–670; Julius Ludwig August Koch, *Zur Statistik der Geisteskrankheiten in Württemberg und der Geisteskranken überhaupt* (Stuttgart, 1878), 152.
47. Koch, *Zur Statistik*, 46, 56–58, 87–88.
48. Koch, *Zur Statistik*, 33, 152–153. He referred to the *hereditär belastender Momente*, perhaps impulses of hereditary burden.
49. Koch, *Zur Statistik*, 158–159. He also noted that, since the insane are not registered by name, there will often be double counting of pairs of insane siblings. But this, he continued, will be balanced by the invisibility of those who have not yet manifested the insanity.
50. Paul Mayet, "Die Verwandtenehe und die Statistik," lecture on 29 Jan. 1902, printed in *Jahrbuch der internationalen Vereinigung der Vergleichende Rechtswissenschaft und Volkswirtschaftslehre zu Berlin*, vols. 6–7 (Berlin: Julius Springer, 1902), 193–210; Wilhelm Weinberg, "Verwandtenehe und Geisteskrankheit," *ARGB* 4 (1907), 471–475; Weinberg, "Pathologische Vererbung," 521–524.
51. Wilhelm Weinberg, "Über den Nachweis der Vererbung beim Menschen," *Jahreshefte des Vereins für vaterländische Naturkunde in Württemberg* 64 (1908), 369–382;

Weinberg, "Einige Tatsachen der experimenteller Vererbungslehre," in Robert Sommer, ed., *Bericht über den II. Kurs mit Kongreß für Familienforschung, Vererbungs- und Regenerationslehre in Gießen vom 9. bis 13. April 1912* (Halle: Carl Marhold Verlagsbuchhandlung, 1912), 71–80.

52. Wilhelm Weinberg, comp., "Die Tuberkulöse in Stuttgart, 1873–1901: Ergebnisse der Untersuchungen einer vom Stuttgarter ärztlichen Vereine eingesetzten Kommission," *Medizinisches Correspondenz-Blatt des Württembergischen ärztlichen Landesvereins* 76 (1906), offprint; Weinberg, "Die familiäre Belastung der Tuberkulösen und ihre Beziehung zu Infektionen und Vererbung," in Ludolph Brauer, ed., *Beiträge zur Klinik der Tuberkulose*, vol. 7 (Würzburg: A. Stuber's Verlag, 1907), 257–289.
53. Weinberg, "Einige Tatsachen."
54. Weinberg, "Über den Nachweis der Vererbung," 372; Weinberg, "Zur Bedeutung der Mehrlingsgeburten für die Frage der Bestimmung des Geschlechts," *ARGB* 6 (1909), 32; Weinberg, "Aufgabe und Methode," 8, 12.
55. Wilhelm Weinberg, review of *A First Study of the Statistics of Pulmonary Tuberculosis*, by Karl Pearson, *ARGB* 5 (1908), 572; Weinberg, "Vererbung und Soziologie," *Berliner klinische Wochenschrift* 49 (27 May 1912), 1081. See also Robert von Lendenfeld, "Karl Pearsons Untersuchungen über verwandtschaftliche Ähnlichkeit und Vererbung geistiger Eigenschaften," *ARGB* 1 (1904), 78–83, and the translation of Pearson's lecture in *ARGB* 5 (1908), 67–96. On temperance movements and the origins of this journal, see Weindling, *Health, Race*, 71–73, 127–131; Weber, *Ernst Rüdin*, 22–24, 54–59.
56. See Wilhelm Weinberg, "Die rassenhygienischen Bedeutung der Fruchtbarkeit," *ARGB* 7 (1910), 684–696, and 8 (1911), 25–32. On these matters, including Weinberg's involvement in insurance, see Bernd Gausemeier, "Borderlands of Heredity: The Debate about Hereditary Susceptibility to Tuberculosis, 1882–1945," in Gausemeier, Müller-Wille, and Ramsden, *Human Heredity*, 13–26.
57. On alcohol: Pearson's letter and Allers' response are in *ARGB* 8 (1911), 377–378 and 655–664; Weinberg's critical remarks are in "Die rassenhygienischen Bedeutung," 686–687; Weinberg, review of "The Fight against Tuberculosis and the Death Rate from Phthisis," by Pearson, *ARGB* 8 (1911), 813–815. See *ARGB* 9 (1912), 87, for Pearson's protest in German, and 221–222 for Weinberg's response. On the critique of Mott, see Judith E. Friedman, "The Disappearance of the Concept of Anticipation in the Post-War World," in Gausemeier, Müller-Wille, and Ramsden, *Human Heredity*, 154.
58. Ernest G. Pope and Karl Pearson, *A Second Study of the Statistics of Pulmonary Tuberculosis: Marital Infection*, vol. 3 of Draper's Company Research Memoirs (Cambridge: Cambridge University Press, 1908), 5–7. See Weinberg's two surviving letters to Pearson, both from 1909, in Karl Pearson Papers 11/22/35 and 3/13/81.
59. Wilhelm Weinberg, "Über Vererbungsgesetze beim Menschen," *Zeitschrift für inductive Abstammungs- und Vererbungslehre* 1 (1908), and 2 (1909), esp. 377–379, 382, 387. On Johannsen and genetic data, see Bonneuil, "Pure Lines."
60. *ARGB* 5 (1908), 572; *ARGB* 10 (1913), 710–717.
61. APS, B:D27, CBDP, Rüdin to Davenport, 23 Nov. 1912; Rüdin review of Goddard, *Heredity of Feeble-Mindedness*, *ARGB* 8 (1911), 531–534; Rüdin, review of Heron, *First Study of the Statistics of Insanity*, *ARGB* 5 (1908), 133–135; Cottebrune, "Zwischen Theorie und Deutung," 36–38.
62. See Allers review of Cannon and Rosanoff in *ARGB* 8 (1911), 527–528; Allers review of Pearson and Jaederholm, *Mendelism and Problem of Mental Defect*, in *ARGB* 11 (1914–15), 385–388, 385.

63. Wilhelm Weinberg, "Die neuere psychiatrische Vererbungsstatistik," *ARGB* 10 (1913), 303–312. See also, Bernd Gausemeier, "Pedigrees of Madness: The Study of Heredity in Nineteenth- and Early Twentieth-Century Psychiatry," *History and Philosophy of the Life Sciences* 36 (2015), 478–479.
64. Weinberg, "Neuere psychiatrische Vererbungsstatistik."
65. Weinberg's review of Davenport on skin color in white-negro crosses, *ARGB* 11 (1913–14), 519–522, quotes on 522, 521; see also Porter, "Blending Inheritance." Since this review is so sarcastic, I at first thought his conclusion was ironical, but he had already endorsed Davenport's work on discontinuity of skin color in Weinberg, "Über Methoden der Vererbungsforschung beim Menschen," *Berliner klinische Wochenschrift*, 49 (1912), 700; also Weinberg, "Einige Tatsachen," 71. Italicized words are English in the original.
66. Weinberg, "Neuere psychiatrische Vererbungsstatistik," 312; William Boven, "Similarité et Mendélisme dans l'heredité de la démence précoce et de la folie maniaque-dépressive," Thèse pour l'obtention du grade de Docteur en Médecine, Univ. de Lausanne, 26 Nov. 1915, 210; Weinberg review of Boven in *ARGB* 13 (1915–16), 214–216; Weber, *Ernst Rüdin*, 127.
67. Rüdin, *Studien über Vererbung*, iii. See also Bernd Gausemeier, "In Search of the Ideal Population, The Study of Human Heredity Before and After the Mendelian Break," in Müller-Wille and Brandt, *Heredity Explored*, 337–363.
68. Philipp Jolly, "Die Heredität der Psychosen," *Archiv für Psychiatrie und Nervenkrankheiten* 52 (1913), 379–381; Rudolf Allers, review of "Heredität der Psychosen" by Jolly and "Psychiatrische Familienforschung," by Wittermann *ARGB* 11 (1914–15), 111–115.
69. SAL, E163 Bü 105 Bl. 54–57; Wilhelm Weinberg, "Bemerkungen zur Reform der deutschen Bevölkerungs- und Gesundheitsstatistik," *Oeffentliche Gesundheitspflege* 4 (1919), 523.
70. SAL, E 163 Bü 105 Bl. 59–61; Pauline M. H. Mazumdar, "Two Models for Human Genetics: Blood Grouping and Psychiatry between the Wars," *Bulletin of the History of Medicine* 70 (1996), 645.
71. Wilhelm Weinberg, "Methodologische Gesichtspunkte für die statistische Untersuchung der Vererbung bei Dementia praecox," *ZgNP* 59 (1920), 39, 45, 49; Weinberg, "Vererbungsstatistik und Dementia praecox (Vorläufige Mitteilung)," *Münchener medizinischen Wochenschrift* 67, no. 3 (4 June 1920), 667–668.
72. Wilhelm Weinberg, "Zur Vererbung bei manisch-depressive Irresein," *Zeitschrift für angewandte Anatomie und Konstitutionslehre* 6 (1920), 383, 385–386.
73. Weinberg, "Methodologische Gesichtspunkte," 46, 43.
74. Weinberg, "Zur Vererbung," 380, 385; Weinberg, "Vererbungsstatistik"; Weinberg, "Methodologische Gesichtspunkte," 44.
75. Weinberg, "Methodologische Gesichtspunkte," 50.

Chapter 13. Psychiatric Geneticists Create Colossal Databases, Some with Horrifying Purposes, 1920–1939

Epigraph: E. B. Robson, "Lionel Penrose as Galton Professor," in *Penrose: Pioneer in Human Genetics : Report on a Symposium held to celebrate the Centenary of the Birth of Lionel Penrose* (London: Published by the Centre for Human Genetics at UCL, March 1998), 17.

1. *Speeches at a dinner held at University College in Honour of Professor Karl Pearson, 23 April, 1934* (Cambridge: Cambridge University Press, 1934), 22–23; Porter, *Rise of Statistical Thinking*, 302; Porter, *Karl Pearson*, 284–286.

2. Ewald Bodewig, "Mathematische Betrachtungen zur Rassenhygiene, insbesondere zur Sterilisation," *Annals of Eugenics* 5 (1932–33), 339–363; Karl Pearson, "Memorandum on Dr. Bodewig's Paper," *Annals of Eugenics* 5 (1932–33), 413. See also Weber, *Ernst Rüdin*, 189.
3. Fisher, "Elimination of Mental Defect"; see also Diane B. Paul and Hamish G. Spencer, "Did Eugenics Rest on an Elementary Mistake," in Paul, *Politics of Heredity*, 117–132.
4. Ewald Bodewig, "Entgegnung auf die Bemerkungen von Professor Karl Pearson," *Annals of Eugenics* 7 (1934–35), 128; Siegfried Koller, "Die deutsche Rassenhygienische Gesetzgebung in mathematischer Betrachtung, Antwort auf den Aufsatz von E. Bodewig . . . ," *Annals of Eugenics* 7 (1934–35), 129–131; Weber, *Ernst Rüdin*, 189.
5. Siegfried Koller, "Die Auslesevorgänge beim Kampf gegen die Erbkrankheiten," *Zeitschrift fur menschliche Vererbungs- und Konstitutionslehre*, 19 (1935), 253–322; Curt Stern, *Principles of Human Genetics* (San Francisco: Freeman 1949), 522–536. The catalogue of the Deutsche Nationalbibliothek places Bodewig in The Hague.
6. Adelaide University Special Collections, R. A. Fisher Digital Archive: Fisher to Weinberg, 10 May 1935, http://hdl.handle.net/2440/68104. This topic resumes below.
7. Karl Pearson and Ethel M. Elderton, "Foreword," *Annals of Eugenics* 1 (1925), 1–4, 1, 4; Porter, *Karl Pearson*, 269–270.
8. Mazumdar, *Eugenics*, 205. See also note 3 above.
9. C. P. Blacker, R. A. Fisher, R. A. Gibbons, R. Langdon-Down, J. A. Ryle, and C. Collyer, "Sterilisation of the Unfit" (letter to the editor), *Lancet* 216, no. 5577 (19 July 1930), 161. "One of us has shown" that a complete halt to reproduction by the mentally defective would reduce its rate by up to 17% in one generation, citing R. A. Fisher, "The Elimination of Mental Defect," *Eugenics Review* 16 (1927), 11. See also Fisher, *The Genetical Theory of Natural Selection* (London: Oxford University Press, 1930), 54.
10. [Brock Committee], *Report of the Departmental Committee on Sterilisation* (London: HMSO, 1934), 13, 26, 22.
11. Abraham Myerson, *The Inheritance of Mental Diseases* (Baltimore: Williams & Wilkins, 1925), 67, 64; [Brock Committee], *Report*, 13.
12. Kevles, *Name of Eugenics*, 99 and 355n18; see also Peter S. Harper, *A Short History of Medical Genetics* (Oxford: Oxford University Press, 2008), 67–68, 414. Diane B. Paul's classic paper, "Eugenics and the Left," in *Politics of Heredity*, 12–35, demonstrates how much of what people now condemn in eugenics arose from progressive and left-wing politics.
13. P. S. Harper, "Julia Bell and the Treasury of Human Inheritance," *Human Genetics* 116 (2005), 422–432.
14. Kevles, *Name of Eugenics*, esp. chap. 13; R. A. Fisher, "The Correlation between Relatives on the Supposition of Mendelian Inheritance," *Transactions of the Royal Society of Edinburgh* 52 (1918), 399–433; see also Mazumdar, *Eugenics*, 108–110.
15. Pearson, "On the Inheritance," 366, 379.
16. Penrose, *Mental Defect*, 152; Lionel S. Penrose, *A Clinical and Genetic Study of 1280 Cases of Mental Defect (The "Colchester Survey")*, Medical Research Council Special Report 229 (London: Medical Research Council, 1938), 8; Mazumdar, *Eugenics*, 161–169; Mazumdar, "Two Models."
17. Harry H. Laughlin, "An Account of the Work of the Eugenics Record Office," *American Breeders Magazine* 3 (1912), 119–123. On classification of biological traits at the ERO, see Philip Thurtle, *The Emergence of Genetic Rationality: Space, Time, and Information in American Biological Science, 1870–1920* (Seattle: University of Washington Press, 2007), 290–293.

18. Countway Library, Worcester State Hospital Records, box 7 vol. 40, 1871–74 female, 462; box 12, vol. 61, 414–415; box 12, vol. 45, 1879–1881, 464–465; Myrtelle M. Canavan and Louise Eisenhardt, *The Brains of Fifty Insane Criminals: Shapes and Patterns* (Boston: Warren Anatomical Museum, 1942), vii–viii, 172–173.
19. See Staatsarchiv Freiburg im Breisgau, B8212/2 Patientakten, Badische Heil- und Pflege-Anstalt Illenau, 20884, case record of a patient admitted 11 Jan. 1923.
20. Missing file from Generallandesarchiv Karlsruhe, Badische Heil- und Pflegeanstalt Pforzheim 567 Zugang 1990–44 656, admitted 1897; Staatsarchiv Sigmaringen, Wü 68/3 T3 Psychiatrisches Landeskrankenhaus Zwiefalten. Patientenakten No. 82, admitted 1905. On sterilization, see Robert N. Proctor, *Racial Hygiene: Medicine under the Nazis* (Cambridge, MA: Harvard University Press, 1988), 108.
21. ALVR, 7930 (Statistik der Irrenanstalten), 14717 (Verwaltungsberichte), 14850 (Erbbiologischen Arbeiten des Prof. Dr. Otto Löwenstein); Wolfgang Schaffer, "Erbbiologische Bestandsaufnahme im Rheinland—Das Institut fur Psychiatrisch-Neurologische Erbforschung in Bonn," in Erik Gieseking and Herbert Hörmig, eds., *Zum Ideologieproblem in der Geschichte* (Lauf an der Pegnitz: Europaforum-Verlag, 2006), 419–444; Ralf Forsbach, *Die Medizinische Fakultät der Universität Bonn in 'Dritten Reich'* (Munich: Oldenbourg, 2006), 201–206, 484–493.
22. Richard F. Wetzell, *Inventing the Criminal: A History of German Criminology, 1880–1945* (Chapel Hill: University of North Carolina Press, 2000), 48–49, 54–56, 100–103, 184.
23. Carl Rath, *Über die Vererbung von Dispositionen zum Verbrechen: Eine statistische und. Psychologische Untersuchung* (Stuttgart: Spemann, 1914), 37, 39, 53–58, 61, 72–73, table 2; Wetzell, *Inventing the Criminal*, 178.
24. Wetzell, *Inventing the Criminal*, 126–135, 159–165. For an introduction to genetic twin studies, see Susan Lindee, *Moments of Truth in Genetic Medicine* (Baltimore: Johns Hopkins University Press, 2005), 120–155.
25. On eugenics and public health: Alexandra Minna Stern, *Eugenic Nation: Faults and Frontiers of Better Breeding in Modern America* (Berkeley: University of California Press, 2005); Peter Weingart, Jürgen Kroll, and Kurt Bayertz, *Rasse, Blut, und Gene: Geschichte der Eugenik und Rassenhygiene in Deutschland* (Frankfurt am Main: Suhrkamp, 1988), 243–273.
26. Rüdin, "Einige Wege," 571–572; Weindling, *Health, Race*, esp. 80–81, 239, 331, 454; numbers from Michael Burleigh, *Death and Deliverance: Euthanasia in Germany, 1900-1945* (Cambridge: Cambridge University Press, 1994), 25, 29; Volker Roelcke, "Politische Zwänge und individuelle Handlungsspielräume: Karl Bonhoeffer und Maximinian de Crinis im Kontext der Psychiatrie im Nationalsozialismus," in Sabine Schleiermacher and Udo Schagen, eds., *Die Charité im Dritten Reich* (Paderborn: Ferdinand Schöningh, 2008), 68–84, 70.
27. Friedrich Stumpfl, *Erbanlagen und Verbrechen: Charakterologische und psychiatrische Sippenuntersuchungen* (Berlin: Julius Springer, 1935), part 2; Wetzell, *Inventing the Criminal*, 195–198, 282–283. On scientists and racial purity, see Anne Cottebrune, *Der planbare Mensch: Die Deutsche Forschungsgemeinschaft und die menschliche Vererbungswissenschaft, 1920–1970* (Stuttgart: Franz Steiner Verlag, 2008), 69–73.
28. Relevant works of scholarship include Benno Müller-Hill's pioneering *Murderous Science*, trans. George Fraser (Oxford: Oxford University Press, 1988). Especially illuminating is Hans-Walter Schmuhl, *The Kaiser Wilhelm Institute for Anthropology, Human Heredity, and Eugenics, 1927–1945*, trans. Sorcha O'Hagan (Dordrecht: Springer, 2008), which demonstrates the originality and scientific seriousness of this kind of genetics while integrating the scientific story with administrative and

political dimensions. Also Schmuhl, "Die Charité und die Forschungspolitik der Kaiser-Wilhelm-Gesellschaft und der Deutsche Forschungsgemeinschaft in der Zeit des Nationalsozialismus," in Sabine Schleiermacher and Udo Schagen, eds., *Die Charité im Dritten Reich* (Paderborn: Ferdinand Schöningh, 2008), 226–245. Jonathan Harwood, *Styles of Scientific Thought: The German Genetics Community, 1900–1933* (Chicago: University of Chicago Press, 1993), tracks this scientific tradition prior to the Nazi takeover. Sheila Faith Weiss, *The Nazi Symbiosis: Human Genetics and Politics in the Third Reich* (Chicago: University of Chicago Press, 2010) is a useful survey.

29. See Rüdin's lecture to the *Münchener Gesellschaft für Rassenhygiene*, "Über rassenhygienische Familienberatung," *ARGB* 16 (1924), 162–178, or, for a general introduction to the topic, a review essay (*Sammelreferat*) by Rainer Fetscher, "Der gegenwärtige Stand der Ehe- und Sexualberatung," *Zeitschrift für inductive Abstammungs- und Vererbungslehre* 48 (1928), 325–344.
30. Jürgen Simon, "Kriminalbiologie und Strafrecht von 1920 bis 1945," in Heidrun Kaupen-Haas and Christian Saller, eds., *Wissenschaftlicher Rassismus: Analysen einer Kontinuität in den Human- und Naturwissenschaften* (Frankfurt: Campus Verlag, 1999), 226–256, 235–237; Christiane Rothmaler, "Von 'haltlosen Psychopathinnen' und 'konstitutionellen Sittlichkeitsverbrechen': Die kriminalbiologische Untersuchungs- und Sammelstelle der Hamburgischen Gefangenenanstalten 1926 bis 1945," in Kauper-Haas and Saller, *Wissenschaftlicher Rassismus*, 257–303, 265; Weingart, Kroll, and Bayertz, *Rasse, Blut, und Gene*, 181–184.
31. Weber, *Ernst Rüdin*, chap. 7; Weiss, *Nazi Symbiosis*, 172–183.
32. Matthias M. Weber, "'Ein Forschungsinstitut für Psychiatrie . . . ' Die Entwicklung der Deutschen Forschungsanstalt für Psychiatrie in München zwischen 1917 und 1945," *Sudhoffs Archiv* 75 (1991), 73–89.
33. Ernst Rüdin, "Erblichkeit und Psychiatrie," *ZgNP* 93 (1924), 502–527; Alons Lokay, "Über die hereditären Beziehungen der Imbezillität," *ZgNP* 122 (1929), 90–143; Rüdin, "Erbbiologische-psychiatrische Streitfragen," *ZgNP* 108 (1927), 549–563.
34. A few notable works: Eugen Kahn, "Erbbiologisch-psychiastrische Übersicht," *Zeitschrift für die induktive Abstammungs- und Vererbungslehre* 38 (1925), 75–83; Bruno Schulz, "Zum Problem der Erbprognose-Bestimmung. Die Erkrankungsaussichten der Nefen und Nichten von Schizophrenen," *ZgNP* 102 (1926), 1–37; Hans Luxenburger, "Zur Methodik der empirischen Erbprognose in der Psychiatrie," *ZgNP* 117 (1928), 543–552.
35. Weber, *Ernst Rüdin*, 153, 172; Cottebrune, *Planbare Mensch*, 82–83.
36. Weingart, Kroll, and Bayertz, *Rasse, Blut, und Gene*, 494; Mazumdar, *Eugenics*, 205–206; Weber, *Ernst Rüdin*, 242; Rüdin, "Ueber die Vorhersage von Geistesstörung in der Nachkommenschaft," *ARGB* 20 (1928), 394–407, 401. For recent times, see Aaron Panofsky, "From Behavior Genetics to Postgenomics," in Sarah S. Richardson and Hallam Stevens, eds., *Postgenomics: Perspectives on Biology after the Genome* (Durham: Duke University Press, 2015), 150–173.
37. Ernst Rüdin, "Psychiatrische Indikation zur Sterilisierung," *Das kommende Geschlecht* 5 (1929), 1–19. *Das kommende Geschlecht* was previously used to translate the title of Bulwer-Lytton's 1871 novel, *The Coming Race*.
38. Mazumdar, *Eugenics*, 204–208; Ernst Rüdin, *Psychiatric Indications for Sterilisation* (London: Eugenics Society, 1930), 3–5. See also the summary of this pamphlet in *Lancet* (13 June 1931), 1306–1307.
39. Rüdin, *Psychiatric Indications*, 4–6; Rüdin, "Kraepelins sozialpsychiatrische Grundgedanken," *Archiv für Psychiatrie und Nervenkrankheiten* 87 (1929), 75–86; Rüdin, "Empirische Erbprognose," in Rüdin, ed., *Erblehre und Rassenhygiene im*

völkischen Staat (Munich: J. F. Lehmanns Verlag, 1934), 136–142, 139; Galton, *English Men of Science*, 23. Even Penrose accepted a much milder form of this analysis in *Mental Defect*, 138.

40. Weber, *Ernst Rüdin*, 211–212.
41. Diane Paul, "The Rockefeller Foundation and the Origins of Behavior Genetics," in Paul, *Politics of Heredity*, 53–89; Florian Mildenberger, "Auf der Spur des 'scientific pursuit,' Franz Joseph Kallmann (1897–1965) und die rassenhygienische Forschung," *Medizinhistorische Journal* 37 (2002), 183–200; Volker Roelcke, "Programm und Praxis der psychiatrischen Genetik an der Deutschen Forschungsanstalt für Psychiatrie unter Ernst Rüdin," in Hans-Walter Schmuhl, ed., *Rassenforschung an Kaiser-Wilhelm-Instituten vor und nach 1933* (Göttingen: Wallstein Verlag, 2003), 38–67; Roelcke, "Die Etablierung der psychiatrischen Genetik in Deutschland, Großbritannien und den USA, ca. 1910–1960: Zur untrennbaren Geschichten von Eugenik und Humangenetik," *Acta historica Leopoldina* 48 (2007), 173–190; Roelcke, "Psychiatry in Munich and Yale, ca. 1920–1935: Mutual Perceptions and Relations, and the Case of Eugen Kahn (1887–1973)," in Volker Roelcke, Paul J. Weindling, and Louise Westwood, eds., *International Relations in Psychiatry: Britain, Germany, and the United States to World War II* (Rochester, NY: University of Rochester Press, 2010), 156–178; Cottebrune, "Franz-Josef Kallmann (1897–1965) und der Transfer psychiatrisch-genetischer Wissenschaftskonzepte vom NS-Deutschland in die USA," *Medizinhistorisches Journal* 44 (2009), 269–324.
42. Weber, *Ernst Rudin*, 296; John D. Rainer, "Perspectives on the Genetics of Schizophrenia: A Re-Evaluation of Kallmann's Contribution—Its Influence and Current Relevance," *Psychiatric Quarterly* 46 (1972), 356–362; Elliott S. Gershon, "The Historical Context of Franz Kallmann and Psychiatric Genetics," *Archiv für Psychiatrie und Nervenkrankheiten* 229 (1981), 273–276, quotes 274–275. The charge of Lysenkoism referred to the XYY gene debate. A recent medical author takes the Nazi connection more seriously: Carlos A. Benbassat, "Kallmann Syndrome: Eugenics and the Man behind the Eponym," *Rambam Maimonides Medical Journal* 7 (2) (2016), doi:10.5041/RMMJ.10242.
43. Rüdin, "Psychiatrische Indikation," 16; Arthur Gütt, Ernst Rüdin, and Falk Ruttke, comp., *Gesetz zur Verhütung erbkranken Nachwuchses 1934*, 2nd rev. ed. (Munich: J. F. Lehmanns Verlag, 1936), 15–58, 46 for *Momente*.
44. Franz-Josef Kallmann, "Die Fruchtbarkeit der Schizophrenen," in Hans Harmsen and Franz Lohse, eds., *Bevölkerungsfragen: Bericht der Internationalen Kongresses für Bevölkerungswissenschaft* (Munich: J. F. Lehmanns Verlag, 1936), 725–729. On the politics of this congress see Paul-André Rosental, *Destins de l'Eugenisme* (Paris: Seuil, 2016), chap. 19, esp. 383–385.
45. Adelaide University Special Collections, Weinberg to Fisher, 5 May 1935 (not digitized).
46. Bruno Schulz, *Methodik der medizinischen Erbforschung* (Leipzig: Georg Thieme Verlag, 1936), 9–12, 159.
47. "Wie Perlen auf einer Kette," Harald Geppert and Siegfried Koller, *Erbmathematik: Theorie der Vererbung in Bevölkerung und Sippe* (Leipzig: Quelle & Meyer, 1938), 170, 5; Amir Teicher, "Social Mendelism: Genetics and the Politics of Racial Anxiety in Germany, 1900–1948" (unpublished manuscript, 2017) is excellent on Nazi appropriations of Mendelian arguments.
48. [Brock Committee], *Report*, 24–25; Roelcke, "Programm und Praxis." A few among many relevant papers: Aubrey J. Lewis, "Inheritance of Mental Disorders," *Annals of Eugenics* 25 (1933), 79–84; Eliot Slater, "The Inheritance of Manic-Depressive Insanity," in James Shields and Irving I. Gottesman, eds., *Man, Mind, and Heredity:*

Selected Papers of Eliot Slater on Psychiatry and Genetics (Baltimore: Johns Hopkins University Press, 1971), 43–53; Irving Gottesman and Peter McGuffin, "Eliot Slater and the Birth of Psychiatric Genetics in Great Britain," in Hugh Freeman and German Barrios, eds., *150 Years of British Psychiatry, Volume II: The Aftermath* (London: Athlone, 1996), 537–548.

49. Leslie Hearnshaw, *Cyril Burt, Psychologist* (Ithaca, NY: Cornell University Press, 1979); Edmund Ramsden, "Remodelling the Boundaries of Normality: Lionel S. Penrose and Population Surveys of Mental Ability," in Gausemeier, Müller-Wille, and Ramsden, *Human Heredity*, 39–54; Mazumdar, *Eugenics*, 227; Penrose, *Clinical and Genetic Study*, 36; Carl-Gottlieb Bennholdt-Thomsen, "Über den Mongolismus und andere angeborene Abartungen in ihrer Beziehung zu hohem Alter der Mütter," *Zeitschrift für Kinderheilkunde* 53 (1932), 427–454; Lionel S. Penrose, "A Method of Separating the Relative Aetiological Effects of Birth Order and Maternal Age, with Special Reference to Mongolian Imbecility," *Annals of Eugenics* 6 (1934), 108–122.
50. Penrose, *Clinical and Genetic Study*, 8–16; Lionel S. Penrose, "Intelligence Test Scores of Mentally Defective Patients and Their Relatives," *British Journal of Psychology* 30 (1939), 1–18.
51. Carl Brugger, "Genealogische Untersuchungen an Schwachsinnigen," *ZgNP* 130 (1930), 66–103; Lionel Penrose, "Eugenic Prognosis with Respect to Mental Deficiency," *Eugenics Review* 31 (1939), 35. On Rüdin and Brugger, see Hans Jakob Ritter and Volker Roelcke, "Psychiatric Genetics in Munich and Basel between 1925 and 1945: Programs—Practices—Cooperative Arrangements," *Osiris* 20 (2005), 263–288; Mazumdar, *Eugenics*, 228.

Aftermath: Data Science, Human Genetics, and History

Epigraph from Kevles, *Name of Eugenics*, 225.

1. Susan Lindee, *Suffering Made Real: American Science and the Survivors at Hiroshima* (Chicago: University of Chicago Press, 1994); Jenny Bangham, "Blood Groups and Human Groups: Collecting and Calibrating Genetic Data after World War II," *Studies in the History and Philosophy of Biological and Biomedical Science* 47A (2014), 74–86.
2. James V. Neel, "The Detection of the Genetic Carriers of Hereditary Disease," *American Journal of Human Genetics* 26 (1949), 19–36.
3. Kevles, *Name of Eugenics*, 205, 225; Mazumdar, *Eugenics*, 169; Stephen Pemberton, "'The Most Hereditary of all Diseases': Haemophilia and the Utility of Genetics for Haematology, 1930–70," in Gausemeier, Müller-Wille, and Ramsden, *Human Heredity*, 165–178.
4. Comfort, *Science of Human Perfection*, chap. 6; Stern, *Human Genetics*, 487–494. Another prominent practitioner of twin studies was Aaron Rosanoff.
5. Soraya de Chadarevian, "Putting Human Genetics on a Solid Basis: Human Chromosome Research, 1950s-1970s," in Gausemeier, Müller-Wille, and Ramsden, *Human Heredity*, 141–152; Chadarevian, "Chromosome Surveys of Human Populations: Between Epidemiology and Anthropology," *Studies in the History and Philosophy of Biological and Biomedical Sciences* 47A (2014), 90–91. I have benefited greatly from reading her book manuscript, *Heredity under the Microscope*.
6. Alexandra Minna Stern, *Telling Genes: The Story of Genetic Counseling in America* (Baltimore: Johns Hopkins University Press, 2012), 40–43; J. A. Böök and S. Reed, "Empiric Risk Figures in Mongolism," *Journal of the American Medical Association* 143, no. 8 (24 June 1950), 730–732; Hans Olof Åkesson, "Empiric Risk Figures in Mental Deficiency," *Acta Genetica* 12 (1962), 28–32; Penrose, "Intelligence Test

Scores"; Brugger, "Genealogische Untersuchungen." Curt Stern emphasized Swedish and Danish "heredity clinics" in *Human Genetics*, 113.

7. Aaron Panofsky, *Misbehaving Science: Controversy and the Development of Behavior Genetics* (Chicago: University of Chicago Press, 2014), chap. 6, quote on 168; Wilson, "Presidential Address."
8. Mike Fortun, *Promising Genomics: Iceland and deCODE Genetics in a World of Speculation* (Berkeley: University of California Press, 2008), 43.
9. See Staffan Müller-Wille and Hans-Jörg Rheinberger, *A Cultural History of Heredity* (Chicago: University of Chicago Press, 2012); Müller-Wille and Rheinberger, *Heredity Produced*; Müller-Wille and Brandt, *Heredity Explored*; Gausemeier, Müller-Wille, and Ramsden, *Human Heredity*.
10. Richardson and Stevens, *Postgenomics*, especially Evelyn Fox Keller, "The Postgenomic Genome," 9–31; Theodore M. Porter, "Thin Description: Surface and Depth in Science and Science Studies," *Osiris* 27 (2012), 209–226.
11. Quoted in Carl Zimmer, "Agriculture Linked to DNA Changes in Ancient Europe," *New York Times*, 23 Nov. 2015.

BIBLIOGRAPHY

Archival Sources

Australia

Adelaide University Special Collections, R. A. Fisher papers, including R. A. Fisher Digital Archive: https://digital.library.adelaide.edu.au/dspace/handle/2440/3860

Canada

Archives of Ontario at York University, Toronto: Queen St. Mental Health Centre

France

AN: Archives Nationales (Paris). F20 Statistique
Archives de la Ville de Paris

Italy

San Servolo Archives, Venice

Germany

ABO: Archiv des Bezirks Oberbayern (Munich)
ALVR: Archiv des Landesverbandes Rheinland, Pulheim-Brauweiler: Archives of the Irren-Heil Anstalt zu Siegburg.
Bayerisches Hauptstaatsarchiv, Mininsterium des Innern (MInn)
Geheimes Staatsarchiv Preussischer Kulturbesitz
Generallandesarchiv Karlsruhe
Hauptstaatsarchiv Stuttgart
Heimatmuseum, Achern, records from Illenau
Klinikum am Europakanal, Erlangen, draft reports from Kreisirrenanstalt zu Erlangen
Landesarchiv Karlsruhe
Landeshauptarchiv Sachsen-Anhalt, Abteilung Magdeburg
Staatsarchiv Freiburg im Breisgau
SAL: Staatsarchiv Ludwigsburg
Staatsarchiv München
Staatsarchiv Nürnberg
Staatsarchiv Nürnberg, Außenstelle Lichtenau (Patientenakten, Kreisirrenanstalt zu Erlangen)
Staatsarchiv Sigmaringen
Stadtarchiv Erlangen
Universität Erlangen Handschriftabteilung, Nachlaß F. W. Hagen
Württembergische Landesbibliothek

BIBLIOGRAPHY

Switzerland

Staatsarchiv des Kantons Zürich, Irrenanstalt Burghölzli
Universität Zürich (UZH) Archiv, Promotionsakten

United Kingdom

Bethlem Hospital Archives
BIA: Borthwick Institute for Archives, York University, Archives of York Retreat. Permission to reproduce documents in the custody of the Borthwick Institute, University of York, has been granted.
Darwin Correspondence Project (online): https://www.darwinproject.ac.uk/
John Innes Centre, William Bateson Papers
LMA: London Metropolitan Archives
Surrey History Centre, Royal Earlswood Asylum
University College, London (UCL)
  Francis Galton Papers
  Karl Pearson Papers
Wellcome Institute Library, London

United States

APS: American Philosophical Society, Philadelphia
  CBDP: Charles B. Davenport Papers
  Herbert Spencer Jennings Papers
  RPP: Raymond Pearl Papers
Harvard Countway Library, Center for History of Medicine, records of the Worcester State Hospital / Worcester Lunatic Hospital
Medical Center of New York-Presbyterian / Weill Cornell, New York, Archives of Bloomingdale Asylum
National Library of Medicine, Bethesda, Maryland
UCLA Louise Darling Biomedical Library, Special Collections
Washington State Archives, Olympia, Western State Hospital

Runs of Institutional Reports Consulted, Mainly Annual
Asylum Reports (Names Vary Somewhat)

Canada: Lower Canada Lunatic Asylum (Quebec), Toronto Provincial Lunatic Asylum
France: Administration générale de l'Assistance Publique de Paris
Germany: Irren-Heil-Anstalt zu Siegburg, *Preußisches Statistik*: Die Irrenanstalten im preußischen Staate
India: Report on the Lunatic Asylums in the Central Provinces
Norway: Generalberetning fra Gaustad Sindssygeasyl
Switzerland: Irrenanstalt Burghölzli
United Kingdom: Earlswood Asylum, Hanwell Asylum, Royal Hospital of Bethlem, St. Luke's Hospital (London)
United States: Asylum for the Relief of Persons Deprived of the Use of their Reason (Philadelphia), Connecticut Retreat for the Insane, Worcester Lunatic Hospital (Massachusetts), New York State Lunatic Asylum (Utica), Ohio Lunatic Asylum, Stockton State Hospital (California), Long Island State Hospital / Kings Park State Hospital, Annual Report of the State Commissioner in Lunacy to the State Board of Charities (New York)

BIBLIOGRAPHY

Selected Primary Sources

Åkesson, Hans Olof. "Empiric Risk Figures in Mental Deficiency." *Acta Genetica* 12 (1962), 28–32.

Albrecht, Paul. "Gleichartige und ungleichartige Vererbung der Geisteskrankheiten." *Zeitschrift für die gesamte Neurologie und Psychiatrie* 11 (December 1912), 541–580.

Alzheimer, Alois. "Ist die Einrichtung einer psychiatrischen Abteilung im Reichsgesundheitsamt erstrebenewert?" *Zeitschrift für die gesamte Neurologie und Psychiatrie* 6 (1911), 242–246.

Baillarger, Jules. "De la statistique appliquée à l'étude des maladies mentales." *Annales médico-psychologiques* 7 (1846), 163–168.

[Baillarger, Jules]. "Introduction." *Annales médico-psychologiques* 1 (1843), i–xxvii.

Baillarger, Jules. "Recherches statistiques sur l'hérédité de la folie." *Annales médico-psychologiques* 3 (1844), 328–339.

Bateson, William. "Address on Heredity. Delivered at the Seventeenth International Congress of Medicine." *British Medical Journal* 2 (16 Aug. 1913), 359–362.

Bateson, William. *Biological Fact and the Structure of Society* (Oxford: Clarendon Press, 1912).

Bateson, William. "Commonsense in Racial Problems." *Eugenics Review* 13 (1921), 325–338.

Bateson, William. *Mendel's Principles of Heredity.* Cambridge: Cambridge University Press, 1909.

Beck, Theodric Romeyn. "Statistical Notices of Some of the Lunatic Asylums of the United States." *Transactions of the Albany Institute* 1, part 1 (1830), 60–83.

Bennholdt-Thomsen, Carl-Gottlieb. "Über den Mongolismus und andere angeborene Abartungen in ihrer Beziehung zu hohem Alter der Mütter." *Zeitschrift für Kinderheilkunde* 53 (1932), 427–454.

Bernhardi, Dr. "Irrenstatistische Bemerkungen zu dem Vorschläge eines Normalschemas tabellarische Uebersichten." *Allgemeine Zeitschrift für Psychiatrie* 2 (1845), 264–295.

Billings, John S. *Report on the Insane, Feeble-Minded, Deaf and Dumb, and Blind in the United States at the Eleventh Census, 1890.* Washington DC: Government Printing Office, 1895.

Black, William. *An Arithmetical and Medical Analysis of the Diseases and Mortality of the Human Species.* 2nd ed. London: C. Dilly, 1789.

Black, William. *A Comparative View of the Mortality of the Human Species at All Ages; And of the Diseases and Casualties by Which They Are Destroyed or Annoyed.* London: C. Dilly, 1788.

Black, William. *Observations Medical and Political, on the Small-pox, and the Advantages and Disadvantages of General Inoculation, Especially in Cities, and on the Mortality of Mankind at Every Age in City and Country.* London: J. Johnson, 1781.

Bodewig, Ewald, "Mathematische Betrachtungen zur Rassenhygiene, insbesondere zur Sterilisation." *Annals of Eugenics* 5 (1932–33), 339–363.

Böök, J. A., and S. Reed. "Empiric Risk Figures in Mongolism." *Journal of the American Medical Association* 143, no. 8 (24 June 1950), 730–732.

Boven, William. "Similarité et Mendélisme dans l'heredité de la démence précoce et de la folie maniaque-dépressive." Thèse pour l'obtention du grade de Docteur en Médecine, Univ. de Lausanne, 26 Nov. 1915.

Brierre de Boismont, Alexandre Jacques François, "Sur l'aliénation mentale et les asiles d'aliénés en Norwège." *Annales médico-psychologiques*, 3rd ser., 5 (1858), 441–445.

[Brigham, Amariah]. "Statistics of Insanity." *American Journal of Insanity* 6 (1849–50), 141–145.

[Brock Committee]. *Report of the Departmental Committee on Sterilisation.* London: HMSO, 1934.

Brugger, Carl. "Genealogische Untersuchungen an Schwachsinnigen." *Zeitschrift für die gesamte Neurologie und Psychiatrie* 130 (1930), 66–103.

Buckle, Henry Thomas. *History of Civilization in England.* New York: D. Appleton, 1858. First published in 1857 by J. W. Walker, London.

Busch, Gerhard von dem. Review of *Bidrag til Kundskab om de Sindssyge i Norge*, by Ludvig Dahl. *Allgemeine Zeitschrift für Psychiatrie* 18 (1861), 474–518.

Busch, Gerhard von dem. Review of *Fortsatte Bidrag til Kundskab om de Sindssyge i Norge*," by Ludvig Dahl. *Allgemeine Zeitschrift für Psychiatrie* 21 (1864), 283–306.

Buxton, David. "On the Marriage and Intermarriage of the Deaf and Dumb." *Liverpool Medico-Chirurgical Journal* 1 (1857), 167–180.

Canavan, Myrtelle M., and Louise Eisenhardt. *The Brains of Fifty Insane Criminals: Shapes and Patterns.* Boston: Warren Anatomical Museum, 1942.

Cannon, Gertrude L., and A. J. Rosanoff. "Preliminary Report of a Study of Heredity in the Light of the Mendelian Laws." *Journal of Nervous and Mental Disease* 38 (1911), 272–279.

Castle, William. *Genetics and Eugenics: A Text-Book for Students of Biology and a Reference Book for Animal and Plant Breeders.* Cambridge, MA: Harvard University Press, 1916.

Committee on Madhouses. *Report from the Committee on Madhouses in England.* 1815.

Coxe, James. "On the Causes of Insanity, and the Means of Checking its Growth; Being the Presidential Address." *Journal of Mental Science* 18 (1872), 311–333.

Dahl, Ludvig. "Beretning om en med Kgl. Stipendium foretagen Reise i Danmark, Holland, Belgien og Storbritannien." *Norsk Magazin for Lægevidenskaben* 11 (1857), 387–411, and 12 (1858), 1–19, 81–111.

Dahl, Ludvig, *Bidrag til Kundskab om de Sindssyge i Norge.* Christiania: Steenske Bogtrykkeri, 1859.

Dahl, Ludvig. "De Sindssvage i Norge den 31te December 1865." *Norsk Magazin for Lægevidenskaben* 23 (1869), 705–724.

Dahl, Ludvig. "Fortsatte Bidrag til Kundskab om de Sindssyge i Norge." *Norsk Magazin for Lægevidenskaben* 16 (1862), 521–541 and 601–624.

Dahl, Ludvig. "Om Tilveiebringelse af en fælles Sindssygestatistik for Sverige, Danmark og Norge." *Norsk Magazin for Lægevidenskaben* 17 (1863), 448–461.

Dahl, Ludvig. "Ueber einige Resultate der Zählung der Geisteskranken in Norwegen, den 31. December 1865." *Allgemeine Zeitschrift für Psychiatrie* 25 (1868), 839–846.

Damerow, Heinrich. "Die Zeitschrift. Ein Blick rückwärts und vorwärts." *Allgemeine Zeitschrift für Psychiatrie* 3 (1846), 1–32.

Damerow, Heinrich. "Einleitung." *Allgemeine Zeitschrift für Psychiatrie* 1 (1844), i–xlviii.

Damerow, Heinrich. "Kritisches zur Irrenstatistik aus der Anstalt bei Halle." *Allgemeine Zeitschrift für Psychiatrie* 12 (1855), 440–467.

Damerow, Heinrich. Review of "Statistique de la France. Section III. Aliénés." *Allgemeine Zeitschrift für Psychiatrie* 2 (1845), 723–730.

Damerow, Heinrich. "Zur Statistik der Provinzial-Irren-Heil- und Pflege-Anstalt bei Halle vom 1 November 1844 bis Ende Dezember 1863 nebst besonderen Mittheilungen und Ansichten über Selbsttödtungen . . . ." *Allgemeine Zeitschrift für Psychiatrie* 22 (1865), 219–251.

Danforth, C. H. "Inheritance of Congenital Cataract." *American Naturalist* 50 (1916), 442–448.

Danielssen, David Cornelius, and Wilhelm Boeck. *Traité de la Spédalskhed ou Éléphantiasis des Grecs.* Translated by L. A. Cosson. Paris: J.-B. Baillière: Librairie de l'Académie Royale de Médecine, 1848. First published in 1847 by J.-B. Baillière, Paris.

Darwin, George. "Marriages between First Cousins in England and their Effects." *Journal of the Statistical Society of London* 38 (1875), 153–184 and 344–348.

Darwin, George. "On Beneficial Restrictions to Liberty of Marriage." *Contemporary Review* 22 (1873), 412–426.

Davenport, Charles B. *The Feebly Inhibited. Nomadism, or the Wandering Impulse, with Special Reference to Heredity*. Washington, DC: Carnegie Institution, 1915.

Davenport, Charles B. *Heredity in Relation to Eugenics*. New York: Henry Holt, 1911.

Davenport, Charles B. "Report of the Committee on Eugenics." *American Breeders Magazine* 1 (1910), 126–129.

Davenport, Charles B. *Statistical Methods: With Special Reference to Biological Variation*. New York: John Wiley & Sons, 1899.

Davenport, Charles B., and David Weeks. "A First Study of Inheritance in Epilepsy." *Journal of Nervous and Mental Disease* 38 (1911), 641–670.

Deboutteville, L., and Maximien Parchappe. *Notice statistique sur l'Asile des Aliénées de la Seine-Inférieure (Maison de Saint Yon de Rouen), pour la période comprise entre le 11 Juillet 1825 et le 31 Décembre 1843*. Rouen: Imprimé chez Alfred Péron, 1845.

Die Ergebnisse der Irrenzählung vom 1. Dezember 1888. In *Statistische Mittheilungen betreffend den Kanton Zürich*, 177–235. Zurich, 1890.

Diem, Otto. "Die psycho-neurotische erbliche Belastung der Geistesgesunden und der Geisteskranken: Eine statistisch-kritische Untersuchung auf Grund eigener Beobachtungen." *Archiv für Rassen- und Gesellschaftsbiologie* 2 (1905), 215–252, 336–368.

Donkin, Horatio Bryan. Introductory note to E. B. Sherlock, *The Feeble-Minded. A Guide to Study and Practice*, xvii-xx. London: Macmillan and Co., 1911.

Donkin, Horatio Bryan. *On the Inheritance of Mental Characters: The Harveian Oration for 1910*. London: Adlard and Son, Bartholomew Press, 1910.

Doutrebente, Gabriel. "Etude généalogique sur les aliénés héréditaires." *Annales médico-psychologiques*, 5th. ser., 2 (1869): 197–237.

Earle, Pliny. *The Curability of Insanity. A Series of Studies*. Philadelphia: J. B. Lippincott Company, 1887.

Earle, Pliny. *Institutions for the Insane in Prussia, Austria and Germany*. New York: Samuel S. and William Wood, 1854.

Earle, Pliny. "On the Causes of Insanity, as exhibited by the Records of the Bloomingdale Asylum from June 16th, 1821, to December 31st, 1844." *American Journal of Insanity* 4 (1847–48), 185–211.

Earle, Pliny. "On the Inability to Distinguish Colors." *American Journal of the Medical Sciences* 9 (April 1845), 346–354.

Earle, Pliny. "Researches in Reference to the Causes, Duration, Termination, and Moral Treatment of Insanity." *American Journal of Medical Sciences* 22 (1838), 339–356.

Earle, Pliny. *A Visit to Thirteen Asylums for the Insane in Europe; to Which are Added a Brief Notice of Similar Institutions in Transatlantic Countries and in the United States*. Philadelphia: J. Dobson, 1841.

Elderton, Ethel M, with the assistance of Karl Pearson. *A First Study of the Influence of Parental Alcoholism on the Physique and Ability of Offspring*. Eugenics Laboratory Memoirs X. London: Dulau, 1910.

Engel, Ernst. "Die Methoden der Volkszählung, mit besonderer Berücksichtigung der im preussischen Staate angewandten." *Zeitschrift des Königlichen Preussischen Statistischen Bureaus* no. 7 (1861), 149–212.

Esquirol, J.E.D. *Des maladies mentales considérées sous les rapports médical, hygiénique et médico-légal*. Vol. 1. Paris: J.-B. Baillière, 1838.

Esquirol, J.E.D. "Folie." *Dictionnaire des sciences médicales*. Vol. 16. Paris: C. L. F. Panckoucke, 1816, 151–240.

Esquirol, J.E.D. "Mémoire historique et statistique sur la Maison Royale de Charenton." *Annales d'hygiène publique et de médecine légale* 13 (1835), 5–192.

Esquirol, J.E.D. "Remarques sur la statistique des aliénés, et sur le rapport du nombre des aliénés à la population. Analyse de la statistique des aliénés de la Norwège." *Annales d'hygiene publique et de médecine légale* 4 (1830), 332–359.

Eugenics Record Office, *The Family-History Book*. Bulletin No. 7. Cold Spring Harbor, NY, 1912.

Farr, William. *On the Statistics of English Lunatic Asylums and the Reform of their Public Management*. London: Sherwood, Gilbert, and Piper, [1840?].

Fay, Edward Allen. *Marriages of the Deaf in America*. Washington, DC: Gibson Bros., 1898.

Fetscher, Rainer. "Der gegenwärtige Stand der Ehe- und Sexualberatung." *Zeitschrift für inductive Abstammungs- und Vererbungslehre* 48 (1928), 325–344.

Fisher, R. A. "The Correlation between Relatives on the Supposition of Mendelian Inheritance." *Transactions of the Royal Society of Edinburgh* 52 (1918), 399–433.

Fisher, R. A. "The Elimination of Mental Defect." *Eugenics Review* 16 (1924), 114–116.

Fisher, R. A. *The Genetical Theory of Natural Selection*. London: Oxford University Press, 1930.

Flemming, C. F. "Aerztlicher Bericht über die Heilanstalt Sachsenberg aus dem 10jährigen Zeiträume von 1840–1849." *Allgemeine Zeitschrift für Psychiatrie* 9 (1852), 377–414.

Flemming, C. F. "Einladung an die Irrenstalts-Direktoren zur Benutzung gemeinschaftlicher Schemata zu den tabellarischen Uebersichten." *Allgemeine Zeitschrift für Psychiatrie* 1 (1844), 430–440.

Foville, Achille. "Rapport sur la proposition de M. Lunier relative à une réunion des médecins aliénistes de tous les pays." *Annales médico-psychologiques*, 4th ser., 9 (1867), 286–294.

Freud, Sigmund. "L'Hérédité et l'étiologie des névroses." *Revue neurologique* 4, no. 6 (30 mars 1896), 161–169.

Further Report of the Commissioners in Lunacy to the Lord Chancellor. Presented to Both Houses of Parliament by Command of Her Majesty. London: Shaw and Sons, 1847, 177–222.

Galton, Francis. *English Men of Science: Their Nature and Nurture*. London: Macmillan, 1874.

Galton, Francis. *Hereditary Genius*. London: Macmillan, 1869; 2nd ed. with a new "Prefatory Chapter." London: Macmillan, 1892.

Galton, Francis. "Hereditary Talent and Character." *Macmillan's Magazine* 12 (1865), 57–66 and 318–327.

Galton, Francis. "On the Anthropometric Laboratory at the late International Health Exhibition." *Journal of the Anthropological Institute of Great Britain and Ireland* 14 (1884–85), 205–219.

Garson, J. G. "The Metric System of Identification of Criminals, as Used in Great Britain and Ireland." *Journal of the Anthropological Institute of Great Britain and Ireland* 33 (1900), 161–198.

Gavarret, Jules. *Principes généraux de statistique medicale*. Paris: Bechet Jeune et Labé, 1840.

Geiser, W. "Über familiäre Geisteskrankheiten zur Beobachtung gelangt in den Jahren 1888–1903 in der Genf Irrenanstalt 'Bel-Air.'" Diss. für Doctorwürde, Med. Fak. Uni. Genf, 1903, Thèse no. 13, Genève: F. de Siebenthal, 1903.

Geppert, Harald, and Siegfried Koller. *Erbmathematik: Theorie der Vererbung in Bevölkerung und Sippe*. Leipzig: Quelle & Meyer, 1938.

Gershon, Elliott S. "The Historical Context of Franz Kallmann and Psychiatric Genetics." *Archiv für Psychiatrie und Nervenkrankheiten* 229 (1981), 273–276.

Goddard, Henry H. *Feeble-Mindedness: Its Causes and Consequences*. New York: Macmillan, 1914.

Goddard, Henry H. "Heredity of Feeble-Mindedness." *American Breeders Magazine* 1 (1910), 165–178.

Goddard, Henry H. *The Kallikak Family: A Study in the Heredity of Feeble-Mindedness.* New York: Macmillan, 1919.

Goring, Charles. *The English Convict. A Statistical Study.* London: HMSO, 1913.

Granville, J. Mortimer. *The Care and Cure of the Insane: Being the Reports of the Lancet Commission on Lunatic Asylums, 1875–76–77.* 2 vols. London; Hardwicke and Bogue, 1877.

Grassmann, Karl. "Kritischer Überblick über die gegenwärtige Lehre von der Erblichkeit der Psychosen." *Allgemeine Zeitschrift für Psychiatrie* 52 (1896), 960–1022.

[Gray, John P.]. Review of *History of Civilization in England*, by Henry Thomas Buckle. *American Journal of Insanity* 15 (1858–59), 233–238.

[Gray, John P.]. "Statistics of Insanity." *American Journal of Insanity* 18 (1861–62), 1–14.

Griesinger, Wilhelm. *Die Pathologie und Therapie der psychischen Krankheiten für Aerzte und Studirende.* Stuttgart: Adolph Krabbe, 1845; 3rd ed. Braunschweig: Friedrich Wreden, 1871.

Grüber, Max von, and Ernst Rüdin, eds. *Fortpflanzung, Vererbung, Rassenhygiene: Katalog der Gruppe Rassenhygiene der Internationalen Hygiene-Ausstellung 1911 in Dresden.* Munich: J. F. Lehmanns Verlag, 1911; 2nd, larger-format ed., also 1911.

Gütt, Arthur, Ernst Rüdin, and Falk Ruttke, comp. *Gesetz zur Verhütung erbkranken Nachwuchses 1934.* 2nd rev. ed. Munich: J. F. Lehmanns Verlag, 1936.

Hagen, Friedrich Wilhelm. "Aerztlicher Bericht aus der Kreis-Irrenanstalt Irsee." *Allgemeine Zeitschrift für Psychiatrie* 10 (1853), 1–72.

Hagen, Friedrich Wilhelm. *Statistische Untersuchungen über Geisteskrankheit, nach den Ergebnissen der Ersten Fünfundzwanzig Jahre der Kreis-Irrenanstalt Erlangen.* Erlangen: Eduard Besold, 1876.

Hagen, Friedrich Wilhelm. "Ueber Statistik der Irrenanstalten mit besonderer Beziehung auf das im Auftrage des internationalen Congresses vom Jahre 1867 vorgeschlagene Schema." *Allgemeine Zeitschrift für Psychiatrie* 27 (1871), 267–294.

Haslam, John. *Observations on Insanity: with Practical Remarks on the Disease, and an Account of the Morbid Appearances on Dissection.* London: F. and C. Rivington, 1798.

Haslam, John. *Observations on Madness and Melancholy: Including Practical Remarks on those Diseases, together with Cases: and an Account of the Morbid Appearances on Dissection.* 2nd ed. (of Haslam, *Observations on Insanity*). London: J. Callow, Medical Bookseller, 1809.

Hays, Willet M. "Constructive Eugenics." *American Breeders Magazine* 3 (1912), 5–11 and 113–119.

Heron, David. *Mendelism and the Problem of Mental Defect. I. A Criticism of Recent American Work.* Vol. 7 of Questions of the Day and the Fray. London: Cambridge University Press, 1913.

Heron, David. "An Examination of Some Recent Studies of the Inheritance of Insanity." *Biometrika* 10 (1914–15), 356–383.

Heron, David. *The Influence of Defective Physique and Unfavourable Home Environment on the Intelligence of School Children, Being a Statistical Examination of the London County Council Pioneer School Survey.* Eugenics Laboratory Memoirs 8. London: Dulau, 1910.

Hohnbaum, Carl. "Ueber Erblichkeit der Geisteskrankheiten." *Allgemeine Zeitschrift für Psychiatrie* 5 (1848), 540–568.

Holst, Frederik. *Beretning, Betankning og Indstilling fra en til at undersøge de Sindsvages Kaar i Norge og gjøre Forslag til deres Forbedring i Aaret 1825.* Christiania: Trykt hos Jacob Lehmanns Enke, 1828.

Holst, Frederik. "Ueber die Anzahl der Geisteskranken, Blinden und Taubstummen in Norwegen im Jahre 1835." *Allgemeine Zeitschrift für Psychiatrie* 4 (1847), 479–487.

Hood, Charles W. *Statistics of Insanity; Being a Decennial Report of Bethlem Hospital from 1846 to 1855 Inclusive*. London: David Batten, 1856.

Hood, Charles W. *Statistics of Insanity: Embracing a Report of Bethlem Hospital from 1846 to 1860 Inclusive*. London: David Batten, 1862.

Howe, Samuel G. *The Causes of Idiocy*. Edinburgh: Maclachlan and Stewart, 1848.

Humphreys, Noel A. "The Alleged Increase of Insanity." *Journal of the Royal Statistical Society* 70 (June 1907), 203–241.

Humphreys, Noel A. "Statistics of Insanity in England, with Special Reference to its Alleged Increasing Prevalence." *Journal of the Royal Statistical Society* 53 (June 1890), 201–252.

Ireland, William W. *On Idiocy and Imbecility*. London: J & A. Churchill, 1877.

Jacobi, Maximilian. *On the Construction and Management of Hospitals for the Insane: with a Particular Notice of the Institution at Siegburg*. Translated by John Kitching. London: John Churchill, 1841.

Jarvis, Edward. "On the Supposed Increase of Insanity." *American Journal of Insanity* 8 (1851–52), 331–362.

Jelliffe, Smith Ely. "Predementia Praecox: The Hereditary and Constitutional Features of the Dementia Praecox Make Up." *Journal of Nervous and Mental Disease* 38 (1911), 1–26.

Jessen, P. W. "Aerztliche Erfahrungen in der Irrenanstalt bei Schleswig." *Zeitschrift für die Beurtheilung und Heilung der krankhaften Seelenzustände* 1 (1838), 580–701.

Johnstone, John. *Medical Jurisprudence*. Birmingham: J. Belcher, 1800.

Jolly, Philipp. "Die Heredität der Psychosen." *Archiv für Psychiatrie und Nervenkrankheiten* 52 (1913), 377–437 and 492–715.

Jones, D. F., and S. L. Mason. "Inheritance of Congenital Cataract." *American Naturalist* 50 (1916), 119–126 and 751–757.

Julius, Nicolaus Heinrich. *Beiträge zur britischen Irrenheilkunde aus eignen Anschauungen im Jahre 1841*. Berlin: Theod. Chr. Fr. Enslin, 1844.

Jung, Wilhelm. "Noch einige Untersuchungen über die Erblichkeit des Seelenstörungen." *Allgemeine Zeitschrift für Psychiatrie* 23 (1866), 211–257.

Jung, Wilhelm. "Untersuchungen über die Erblichkeit der Seelenstörungen." *Allgemeine Zeitschrift für Psychiatrie* 21 (1864), 534–653.

Kahn, Eugen. "Erbbiologisch-psychiastrische Übersicht." *Zeitschrift für die induktive Abstammungs- und Vererbungslehre* 38 (1925), 75–83.

Kallmann, Franz-Josef. "Die Fruchtbarkeit der Schizophrenen." In Hans Harmsen and Franz Lohse, eds., *Bevölkerungsfragen: Bericht der Internationalen Kongresses für Bevölkerungswissenschaft*, 725–729. Munich: J. F. Lehmanns Verlag, 1936.

Koch, Julius Ludwig August. *Zur Statistik der Geisteskrankheiten in Württemberg und der Geisteskranken überhaupt*. Stuttgart, 1878.

Koller, Jenny. "Beitrag zur Erblichkeitsstatistik der Geisteskranken im Canton Zürich: Vergleichung derselben mit der erblichen Belastung gesunder Menschen durch Geistesstörungen u. dergl." *Archiv für Psychiatrie und Nervenkrankheiten* 27 (1895), 268–294.

Koller, Siegfried, "Die Auslesevorgänge beim Kampf gegen die Erbkrankheiten." *Zeitschrift fur menschliche Vererbungs- und Konstitutionslehre* 19 (1935), 253–322.

Koller, Siegfried, "Die deutsche Rassenhygienische Gesetzgebung in mathematischer Betrachtung, Antwort auf den Aufsatz von E. Bodewig . . . ." *Annals of Eugenics* 7 (1934–35), 129–131.

Koster, Friedrich, and Wilhelm (Guilelmus) Tigges. *Geschichte und Statistik der westfälischen Provinzial—Irrenanstalt Marsberg, mit Rücksicht auf die Statistik anderer*

Anstalten. Berlin: August Hirschwald, 1867. Supplemental issue to vol. 24 of *Allgemeine Zeitschrift für Psychiatrie*.

Köstlin, Wilhelm. "Zur Statistik der Geisteskrankheiten. Eine Inaugural-Dissertation welche zur Erlangung der Doctor-Würde in der Medicin und Chirurgie . . . ." Tubingen, 1840.

Kraepelin, Emil. *Die psychiatrischen Aufgaben des Staates*. Jena: Fischer, 1900. Translated by Stewart Paton, as "The Duty of the State in the Care of the Insane." *American Journal of Insanity* 57 (1900–1901), 235–280.

Kraepelin, Emil. *Psychiatrie. Ein Lehrbuch für Studirende und Aerzte*. 5th ed. Leipzig: Johann Ambrosius Barth, 1896.

Kraepelin, Emil. *Psychiatrie: Ein Lehrbuch für Studierende und Ärzte*. 8th ed. Leipzig: Johann Ambrosius Barth, 1909.

Kraepelin, Emil. "Ueber die Entartungsfrage." *Muenchener medizinische Wochenschrift* 28 (1908), 1512–1513. Translated with introduction by Eric J. Engstrom, as "'On the Question of Degeneration' by Emil Kraepelin (1908)." *History of Psychiatry* 18 (2007), 389–404.

Laski, H. J. "A Mendelian View of Racial Heredity." *Biometrika* 9 (1911–12), 424–430.

Laughlin, Harry H. "An Account of the Work of the Eugenics Record Office." *American Breeders Magazine* 3 (1912), 119–123.

Lefebvre, Ferdinand. "Des Bases d'une bonne statistique internationale des maladies mentales: Rapport." *Bulletin de la Société de Médecine Mentale de Belgique* 37 (1885), 55–60.

Legrand du Saulle, Henri. *La folie héréditaire: Leçons professées à l'École Pratique*. Paris: Adrien Delahaye, 1873.

Lendenfeld, Robert von. "Karl Pearsons Untersuchungen über verwandtschaftliche Ähnlichkeit und Vererbung geistiger Eigenschaften." *Archiv für Rassen- und Gesellschaftsbiologie* 1 (1904), 78–83.

Leubuscher, Rudolf. "Bemerkungen über die Erblichkeit des Wahnsinns." *Archiv für Anatomie und Physiologie und für klinische Medicin* 1 (1847), 72–93.

Lindsay, W. Lauder. "On Insanity and Lunatic Asylums in Norway." *Journal of Psychological Medicine and Mental Pathology* 11 (1858), 246–295.

Lokay, Alons. "Über die hereditären Beziehungen der Imbezillität." *Zeitschrift für die gesamte Neurologie und Psychiatrie* 122 (1929), 90–143.

Lorenz, Ottokar. *Lehrbuch der gesammten wissenschaftlichen Genealogie*. Berlin: Wilhelm Hertz, 1898.

"Lunatics." *Literary Panorama*. 2 (1807), 1255–1263.

Lundborg, Hermann. *Medizinisch-biologische Familienforschungen innerhalb eines 2232 köpfigen Bauerngeschlechtes in Schweden (Provinz Blekinge)*. 2 vols. Jena: Gustav Fischer, 1913.

Lunier, Ludger. "Asile départemental d'aliénés de Blois (Loir et Cher)." *Compte-rendu du Service Médical pour l'année 1863* (Blois: H. Giraud, 1864), 8–9.

Lunier, Ludger. "De l'augmentation progressive du chiffre des aliénés et de ses causes." *Annales médico-psychologiques*, 6th ser., 3 (1870), 20–34.

Lunier, Ludger. *De l'influence des grandes commotions politiques et sociales sur le développement des maladies mentales, Mouvement d'aliénation en France pendant les années 1869 à 1873*. Paris: F. Savy, 1874.

Lunier, Ludger. "Projet de statistique applicable à l'étude des maladies mentales arrêté par le Congrès Alieniste International de 1867. Rapport et exposé des motifs." *Annales médico-psychologiques*, 5th ser., 1 (1867), 32–59.

Lunier, Ludger. "Recherches statistiques sur les aliénés du Département des Deux-Sèvres." *Mémoires de la Société de Statistique du Département des Deux-Sèvres* 16 (1853), 27–53.

Luxenburger, Hans. "Zur Methodik der empirischen Erbprognose in der Psychiatrie." *Zeitschrift für die gesamte Neurologie und Psychiatrie* 117 (1928), 543–552.

Lyttleton, E. "More about Biometry." *Oxford and Cambridge Review* 2 (1907), 36–40.
Macdonell, W. R. "On Criminal Anthropometry and the Identification of Criminals." *Biometrika* 1 (1901–2), 172–227.
MacPherson, John. *Mental Affections: An Introduction to the Study of Insanity*. London: Macmillan, 1899.
Marsh, E. J. *Value of Family History and Personal Condition in Estimating a Liability to Consumption*. New York: Mutual Life Insurance Company, 1895.
Maudsley, Henry. "Considerations with Regard to Hereditary Influence." *Journal of Mental Science* 8 (January 1863), 482–512, and 9 (January 1864), 506–530.
Maudsley, Henry. "Insanity and its Treatment." *Journal of Mental Science* 17 (October 1871), 311–334.
Maudsley, Henry. "On the Causes of Insanity." *Journal of Mental Science* 12 (1867), 488–502.
Mayet, Paul. "Die Verwandtenehe und die Statistik." In *Jahrbuch der internationalen Vereinigung der Vergleichende Rechtswissenschaft und Volkswirtschaftslehre zu Berlin*. Vols. 6–7, 193–210. Berlin: Julius Springer, 1902.
Mayr, Georg, comp. *Die Verbreitung der Blindheit, der Taubstummheit, des Blödsinns und der Irrsinns in Bayern, nebst einer allgemeinen internationalen Statistik dieser vier Gebrechen*. Munich: Commissionsverlag von Adolf Ackermann, 1877.
Mendel, Emanuel. "Geisteskrankheiten und Ehe." In H. Senator and S. Kaminer, eds., *Krankheiten und Ehe*, 801–832. Munich: J. E. Lehmanns Verlag; New York: Rebman, 1904.
Mendel, Emanuel. "Hereditäre Anlage und progressive Paralyse der Irren." *Archiv für Psychiatrie und Nervenkrankheiten* 10 (1880), 780–787.
Meyer, Adolf. "A Review of the Signs of Degeneration and of Methods of Registration." *American Journal of Insanity* 52 (1895–96), 344–363.
Mitchell, Arthur. "On the Influence which Consanguinity in the Parentage exercises upon the Offspring." *Edinburgh Medical Journal* 10, no. 2 (1865), 781–794, 894–913, 1074–1085.
Moreau de Tours, Jacques-Joseph. *La psychologie morbide dans ses rapports avec la philosophie de l'histoire, ou de l'influence des névropathies sur le dynamisme intellectuel*. Paris: Librairie Victor Massson, 1859.
Morel, B. A. "Rapport médical sur l'Asile de Maréville (Meurthe)." *Annales médico-psychologiques*, 2nd. ser., 2 (1850), 353–392.
Morel, B. A. *Traité des dégénérescences physiques, intellectuelles et morales de l'espèce humaine*. Paris: J. B. Baillière, 1857.
Morel, B. A. *Traité des maladies mentales*. Paris: Librairie Victor Masson, 1860.
Mott, F. W. "Heredity and Eugenics in Relation to Statistics." *British Medical Journal* (11 May 1912), 1053–1060.
Mudge, George Percival. "On Some Features in the Hereditary Transmission of the Albino Character and the Black Piebald Coat in Rats—Paper II." *Proceedings of the Royal Society of London, B* 80 (1908), 388–393.
Myerson, Abraham. *The Inheritance of Mental Diseases*. Baltimore: Williams & Wilkins, 1925.
Myres, John L. "A Bureau of Biometry." *Oxford and Cambridge Review* 1 (1907), 131–144.
Neel, James V. "The Detection of the Genetic Carriers of Hereditary Disease." *American Journal of Human Genetics* 26 (1949), 19–36.
Newth, A. H. "Systematic Case-taking." *Journal of Mental Science* 46 (1900), 255–260.
Oberholzer, Emil. "Erblichkeitsverhältnisse und Erbgang bei Dementia praecox." *Verhandlungen der Gesellschaft deutscher Naturforscher und Ärzte* 85, no. 2 (1913), 635–640. Expanded version republished in *Zur. 50. Jahresversammlung des Vereins Schweizerischer Irrenaerzte*, 8–27. Geneva: Buchdruckerei Albert Kündig, 1914.

Parchappe, Maximien. *Recherches statistiques sur les causes de l'aliénation mentale.* Rouen, 1839.

Paris, John, and John S. M. Fonblanque. *Medical Jurisprudence.* Vol. 1. London: W. Phillips, 1823.

Pearson, Karl. "Mathematical Contributions to the Theory of Evolution.—III. Regression, Heredity, and Panmixia." *Philosophical Transactions of the Royal Society of London A* 187 (1896), 253–318.

Pearson, Karl. *Mendelism and the Problem of Mental Defect. III. On the Graduated Character of Mental Defect and on the Need for Standardizing Judgments as to the Grade of Social Inefficiency Which Shall Involve Segregation.* Vol. 9 of The Questions of the Day and the Fray. London: Cambridge University Press, 1914.

Pearson, Karl. *Nature and Nurture: The Problem of the Future. A Presidential Address . . . at the Annual Meeting of the Social and Political Education League, April 28, 1910.* 2nd ed. London: Dulau, 1913.

Pearson, Karl. "Note on the Skin-Colour of Crosses between Negro and White." *Biometrika* 7 (1908–9), 348–353.

Pearson, Karl. "On an Apparent Fallacy in the Statistical Treating of 'Antedating' in the Inheritance of Pathological Conditions." *Nature* 90, no. 2247 (1912), 334–335.

Pearson, Karl. "On the Ancestral Gametic Correlations of a Mendelian Population Mating at Random." *Proceedings of the Royal Society of London B* 81 (5 June 1909), 225–229.

Pearson, Karl. "On the Correlation of Intellectual Ability with the Size and Shape of the Head." *Proceedings of the Royal Society of London* 69 (1901), 333–342.

Pearson, Karl. "On the Inheritance of Mental Disease." *Annals of Eugenics* 4 (1930–31), 362–380.

Pearson, Karl. "On the Laws of Inheritance in Man: II. On the Inheritance of the Mental and Moral Characters of Man, and its Comparison with the Inheritance of the Physical Characters." *Biometrika* 3 (1904), 131–190.

Pearson, Karl. "On the Relationship of Intelligence to Size and Shape of the Head, and to Other Physical and Mental Characters." *Biometrika* 5 (1906–7), 105–146.

Pearson, Karl. "On the Value of Teachers' Opinion of the General Intelligence of School Children." *Biometrika* 7 (1910), 542–548.

Pearson, Karl. *Speeches at a Dinner Held at University College in Honour of Professor Karl Pearson, 23 April, 1934.* Cambridge: Cambridge University Press, 1934.

Pearson, Karl, ed. *The Treasury of Human Inheritance.* Vol. 1. London: Cambridge University Press, 1912.

Pearson, Karl, and Ethel M. Elderton. "Editorial." *Annals of Eugenics* 1 (1925), 1–4.

Pearson, Karl, and Ethel M. Elderton. *A Second Study of the Influence of Parental Alcoholism on the Physique of the Offspring. Being a Reply to Certain Medical Critics of the First Memoir and an Examination of the Rebutting Evidence Cited by Them.* Eugenics Laboratory Memoirs 13. London: Dulau, 1910.

Pearson, Karl, and Gustav A. Jaederholm. *Mendelism and the Problem of Mental Defect, II. On the Continuity of Mental Defect.* Vol. 8 of Questions of the Day and Fray. London: Dualu, 1914.

Penrose, Lionel S. *A Clinical and Genetic Study of 1280 Cases of Mental Defect* (The "Colchester Survey"). Medical Research Council Special Report 229. London: HMSO, 1938.

Penrose, Lionel S. "Eugenic Prognosis with Respect to Mental Deficiency." *Eugenics Review* 31 (1939), 35–37.

Penrose, Lionel S. "Intelligence Test Scores of Mentally Defective Patients and Their Relatives." *British Journal of Psychology* 30 (1939), 1–18.

Penrose, Lionel S. *Mental Defect.* London: Sidgwick & Jackson, 1933.

Penrose, Lionel S. "A Method of Separating the Relative Aetiological Effects of Birth Order and Maternal Age, with Special Reference to Mongolian Imbecility." *Annals of Eugenics* 6 (1934), 108–122.

Percival, Thomas. *Medical Ethics, or a Code of Institutes and Precepts, adapted to the Professional Conduct of Physicians and Surgeons.* Manchester: S. Russell, 1803.

Platte, Ludwig. *Vererbungslehre, mit besonderer Berücksichtigung des Menschen, für Studierende, Ärzte und Züchter.* Leipzig: Wilhelm Engelmann, 1913.

Pope, Ernest G., and Karl Pearson. *A Second Study of the Statistics of Pulmonary Tuberculosis: Marital Infection.* Vol. 3 of Draper's Company Research Memoirs. Cambridge: Cambridge University Press, 1908.

Prichard, James Cowles. *A Treatise on Insanity and Other Disorders Affecting the Mind.* London: Sherwood, Gilbert, and Piper, 1835.

Proceedings of the Royal Society of Medicine. Vol. 2. *General Reports. Comprising the Report of the Proceedings for the Session 1908–09.* London: Longmans, Green, 1909.

Punnett, R. C. "Mendelism in Relation to Disease." *Proceedings of the Royal Society of Medicine* 1 (Sect. Epidemiol. State Med.) (1908), 135–168.

Quetelet, Adolphe. *Sur l'homme et le développement de ses facultés, ou essai de physique sociale.* 2 vols. Paris: Bachelier, 1835.

Rainer, John D. "Perspectives on the Genetics of Schizophrenia: A Re-Evaluation of Kallmann's Contribution—Its Influence and Current Relevance." *Psychiatric Quarterly* 46 (1972), 356–362.

Rath, Carl. *Über die Vererbung von Dispositionen zum Verbrechen: Eine statistische und psychologische Untersuchung.* Stuttgart: Spemann, 1914.

Ray, Isaac. *Mental Hygiene.* Boston: Ticknor & Fields, 1863.

Ray, Isaac. "Observations on the Principal Hospitals for the Insane." *American Journal of Insanity* 2 (April 1846), 289–390.

Ray, Isaac. "The Statistics of Insane Hospitals." *American Journal of Insanity* 6 (1849), 23–52.

Renaudin, L.F.E. "De la statistique appliquée à l'étude des maladies mentales . . . , Lettre à M. Baillarger . . . ." *Annales médico-psychologiques* 7 (1846), 467–469.

Renaudin, L.F.E. *Notice statistique sur les aliénés du Département du Bas-Rhin, d'après les observations recueillies à l'Hospice de Stéphansfeld, pendant les années 1836, 1837, 1838, 1839.* Paris: J. B. Baillière, 1840.

Renaudin, L.F.E. "Observations sur les Recherches Statistiques relatives à l'aliénation mentale." *Annales Médico-Psychologiques*, 3rd ser., 2 (1856), 486–504.

Report by Her Majesty's Commissioners appointed to inquire into the state of Lunatic Asylums in Scotland. Edinburgh: HMSO, 1857.

Report on the Scientific Study of the Mental and Physical Conditions of Childhood . . . . London: Committee, Parkes Museum, 1895.

Reports and other Documents Relating to the State Lunatic Hospital at Worcester, Mass. Boston: Dutton and Wentworth, 1837.

Richter, Hermann Eberhard. "Zur Darwin'schen Lehre." [Schmidt's] *Jahrbücher der in- und ausländische Medicin* 126 (1865), 243–249.

Rogers, A. C. "Report of Committee on Heredity of Feeble-Mindedness." *American Breeders Magazine* 2 (1911), 269–272, and 3 (1912), 134–136.

Rolle, Friedrich. *Darwin's Lehre von der Entstehung der Arten im Pflanzen- und Thierreich in ihrer Anwendung auf die Schöpfungsgeschichte.* Frankfurt: Joh. Christ. Hermann, 1863.

Rosanoff, A. J. "The Inheritance of the Neuropathic Constitution." *Journal of the American Medical Association* 58, no. 17 (April 1912), 1266–1269.

Rosanoff, A. J. "Mendelism and Neuropathic Heredity: A Reply to some of Dr. David Heron's Criticisms of Recent American Work." *American Journal of Insanity* 70, no. 3 (January 1914), 571–587.

Rosanoff, A. J., and Florence I. Orr. "A Study of Heredity of Insanity in the Light of the Mendelian Theory." *American Journal of Insanity* 68 (1911), 221–261.

Royal Commission on the Care and Control of the Feeble-Minded. *Report*. 8 vols. London: HMSO, 1908.

Royer-Collard, Hippolyte-Louis. "Rapport: *Recherches statistiques sur l'hérédité de la folie* par M. Baillarger . . . ." *Bulletin de l'Académie Royale de Médecine* 12 (1846–47), 760–777.

Rüdin, Ernst. "Einige Wege und Ziele der Familienforschung mit Rücksicht auf die Psychiatrie." *Zeitschrift für die gesamte Neurologie und Psychiatrie* 7 (1911), 487–585.

Rüdin, Ernst. "Empirische Erbprognose." In Ernst Rüdin, ed., *Erblehre und Rassenhygiene im völkischen Staat*, 136–142. Munich: J. F. Lehmanns Verlag, 1934.

Rüdin, Ernst. "Erbbiologische-psychiatrische Streitfragen." *Zeitschrift für die gesamte Neurologie und Psychiatrie* 108 (1927), 549–563.

Rüdin, Ernst. "Erblichkeit und Psychiatrie." *Zeitschrift für die gesamte Neurologie und Psychiatrie* 93 (1924), 502–527.

Rüdin, Ernst. "Kraepelins sozialpsychiatrische Grundgedanken." *Archiv für Psychiatrie und Nervenkrankheiten* 87 (1929), 75–86.

Rüdin, Ernst. "Psychiatrische Indikation zur Sterilisierung." *Das kommende Geschlecht* 5 (1929), 1–19. Partially translated as *Psychiatric Indications for Sterilisation*. London: Eugenics Society, 1930.

Rüdin, Ernst. *Studien über Vererbung und Entstehung geistiger Störungen I. Zur Vererbung und Neuentstehung der Dementia Praecox*. Berlin: Julius Springer, 1916.

Rüdin, Ernst. "Über rassenhygienische Familienberatung." *Archiv für Rassen- und Gesellschaftsbiologie* 16 (1924), 162–178.

Rüdin, Ernst. "Ueber die Vorhersage von Geistesstörung in der Nachkommenschaft." *Archiv für Rassen- und Gesellschaftsbiologie* 20 (1928), 394–407.

Rush, Benjamin. *Medical Inquiries and Observations upon the Diseases of the Mind*. Philadelphia: Kimber & Richardson, 1812.

Sagra, Ramon de la. "Statistique des aliénés et des Sourds-Muets dans les États-Unis de l'Amérique." *Annales médico-psychologiques* 1 (1843), 281–288.

Schallmayer, Wilhelm. "Grundlinien der Vererbungslehre." In Siegfried Placzek, ed., *Künstliche Fehlgeburt und künstliche Unfruchtbarkeit, ihre Indikationen, Technik, und Rechtslage*, 1–48. Leipzig: Georg Thieme, 1918.

Schmidt, Oscar. *Descendenzlehre und Darwinismus*. Leipzig: F. A. Brockhaus, 1873.

Schulz, Bruno. *Methodik der medizinischen Erbforschung*. Leipzig: Georg Thieme Verlag, 1936.

Schulz, Bruno. "Zum Problem der Erbprognose-Bestimmung. Die Erkrankungsaussichten der Nefen und Nichten von Schizophrenen." *Zeitschrift für die gesamte Neurologie und Psychiatrie* 102 (1926), 1–37.

Schuster, Edgar. "Hereditary Deafness. A Discussion of the Data Collected by Dr. E. A. Fay in America." *Biometrika* 4 (1905–6), 465–482.

Schuster, Edgar. *The Promise of Youth and the Performance of Manhood: A Statistical Inquiry into the Question Whether Success in the Examination for the B. A. Degree at Oxford is Followed by Success in Professional Life*. Eugenic Laboratory Memoirs 3. London: Dulau, 1907.

Schuster, Edgar, and Ethel M. Elderton. *The Inheritance of Ability: Being a Statistical Study of the Oxford Class Lists and of the School Lists of Harrow and Charterhouse*. London: Dulau, 1907.

Schweig, Georg. "Auseinandersetzung der statistischen Methoden in besonderem Hinblick auf das medicinische Bedürfniss." *Archiv für physiologische Heilkunde* 13 (1854), 305–355.

Scottish Lunacy Commission. *Report by Her Majesty's Commissioners Appointed to Inquire into the State of Lunatic Asylums in Scotland.* Edinburgh: HMSO, 1857.

Shuttleworth, George Edward. *Mentally-Deficient Children: Their Treatment and Training.* London: H. K. Lewis, 1895.

Shuttleworth, George Edward, and Fletcher Beach. "Idiocy and Imbecility, Aetiology of." In Daniel Hack Tuke, ed., *Dictionary of Psychological Medicine*, vol. 1, 659–665. London: J. & A. Churchill, 1892.

Sick, P. "Statistik der Geisteskranken und der zu ihrer Pflege und Heilung bestehenden Anstalten im Königreich Württemberg." In *Württembergische Jahrbücher für vaterländische Geschichte, Geographie, Statistik, und Topographie* (1855). 2nd ed. Stuttgart: Eduard Hallberger, 1856.

Sioli, Emil. "Ueber directe Vererbung von Geisteskrankheiten." *Archiv für Psychiatrie und Nervenkrankheiten* 16 (1885), 599–638.

Smith, Samuel Hanbury. "Superintendent's Report." *Thirteenth Annual Report of the Directors and Superintendent of the Ohio Lunatic Asylum to the 50th General Assembly of the State of Ohio for the Year 1851.* Columbus, OH, 1852.

Sommer, Robert. "Eine psychiatrische Abteilung der Reichsgesundheitsamts." *Psychiatrisch-Neurologische Wochenschrift* 12, no. 31 (29 Oct. 1910), 295–297.

Sommer, Robert. *Familienforschung und Vererbungslehre.* Leipzig: Johann Ambrosius Barth, 1907.

Sommer, Robert. "Zur Frage einer psychiatrischen Abteilung des Reichsgesundheitsamtes." *Psychiatrisch-Neurologische Wochenschrift* 13, no. 4 (22 April 1911), 31–32.

Sommerfelt, Søren Christian. *Physisk-oeconomisk Beskrivelse over Saltdalen.* Trondheim: Kgl. Norske Vidsksabers Selskab, 1827.

Stern, Curt. *Principles of Human Genetics.* San Francisco: Freeman 1949.

Stewart, Hugh Grainger. "On Hereditary Insanity." *Journal of Mental Science* 10 (1864), 50–67.

Stolz, Alban. *Ueber die Vererbung sittlicher Anlagen.* Freiburg: Universitäts-Buchdruckerei von Hermann Meinhard Poppen & Sohn, 1859.

Strohmeyer, Wilhelm. "Die Bedeutung des Mendelismus für die klinische Vererbungslehre." *Die deutsche Klinik am Eingänge des zwanzigsten Jahrhunderts in akademischen Vorlesungen* 14 (1913), 331–350.

Strohmeyer, Wilhelm. "Ueber die Bedeutung der Individualstatistik bei der Erblichkeitsfrage in der Neuro- und Psychopathologie." *Münchener medicinische Wochenschrift* 48, no. 45 (5 Nov. 1901), 1786–1789 and no. 46 (12 Nov. 1901), 1824–1844.

Strohmeyer, Wilhelm. "Ueber den Wert genealogischer Betrachtungsweise in der psychiatrischen Erblichkeitslehre." *Monatsschrift für Psychiatrie und Neurologie* 22 (1907), 115–131.

Strohmeyer, Wilhelm. "Ziele und Wege der Erblichkeitsforschung in der Neuro- und Psychopathologie." *Allgemeine Zeitschrift für Psychiatrie* 61 (1904), 355–370.

Stumpfl, Friedrich. *Erbanlagen und Verbrechen: Charakterologische und psychiatrische Sippenuntersuchungen.* Berlin: Julius Springer, 1935.

Sundt, Eilert. *Om Ædrueligheds-Tilstanden i Norge.* Christiania: J. Chr. Adelsted, 1859.

Taylor, Alfred S. *Medical Jurisprudence.* 2nd ed. (from 3rd London ed.). Philadelphia: Lea and Blanchard, 1850.

Thurnam, John. *Observations and Essays on Statistics of Insanity and on Establishments for the Insane to which are added the Statistics of the Retreat near York.* London: Simpkin, Marshall; York: John L. Linney, 1845.

Thurnam, John. *Statistics of the Retreat: consisting of A Report and Tables exhibiting the Experience of that Institution for the Insane from its Establishment in 1796 to 1840.* York: John Lewis Linney, 1841.

Tigges, Wilhelm (Guilelmus). "Bericht über die Irren-Heilanstalt Sachsenberg vom Jahre 1871–1875 mit vergleichender Statistik." *Beiträge zur Statistik Mecklenburgs vom Grossherzoglichen statistischen Bureau zu Schwerin* 8 (1876): 68–94.

Tigges, Wilhelm (Guilelmus). "Die Lunier'schen Vorschläge für die Statistik der Geisteskrankheiten." *Allgemeine Zeitschrift für Psychiatrie* 26 (1869), 667–706.

Tigges, Wilhelm (Guilelmus). "Statistik der Erblichkeit, betreff. die Kinder und die Geschwister der in die Anstalt Aufgenommenen." *Allgemeine Zeitschrift für Psychiatrie* 35 (1879), 485–512.

Tocher, J. F. "The Necessity for a National Eugenic Survey." *Eugenics Review* 2 (July 1910), 125–141.

Tocher, J. F. "Pigmentation Survey of School Children in Scotland." *Biometrika* 6 (1908–9), 130–235.

Tredgold, A. F. *Mental Deficiency (Amentia).* London: Baillière, Tindall and Cox, 1908.

Trélat, Ulysse. "Des causes de la folie." *Annales médico-psychologiques*, 2d ser., 6 (1856), 7–23, 174–190.

Tuke, D. Hack. "Presidential Address, delivered at the Annual Meeting of the Medico-Psychological Association . . . August 2nd, 1881." *Journal of Mental Science* 27 (1881), 305–342.

Tuke, Samuel. *Description of the Retreat, an Institution near York, for Insane Persons of the Society of Friends. Containing an Account of its Origin and Progress, the Modes of Treatment, and a Statement of Cases.* York: W. Alexander, 1813.

Urquhart, A. R. "The Morison Lectures—On Insanity, with Special Reference to Heredity and Prognosis." *Journal of Mental Science* 53 (April 1907), 233–321.

Van Gieson, Ira. "The Pathological Institute of the New York State Hospitals." In *Ninth Annual Report of the State Commission in Lunacy for the Year 1896–97*, 73–240. Albany: James B. Lyon, 1897. Republished with revisions in *Archive of Neurology and Psychopathology* 1 (1898), 25–262.

Vorster, Johannes. "Ueber die Vererbung endogener Psychosen in Beziehung zur Classification." *Monatsschrift für Psychiatrie und Neurologie* 9 (1901), 161–176, 301–315, 367–392.

Wagner-Jauregg, Julius. "Einiges über erbliche Belastung." *Wiener klinische Wochenschrift* 19, no. 1 (4 Jan. 1906), 1–6.

Wagner-Jauregg, Julius. "Ueber erbliche Belastung." *Wiener klinische Wochenschrift* 15 (30 Oct. 1902), 1153–1159.

Weinberg, Wilhelm. "Aufgabe und Methode der Familienstatistik bei medizinisch-biologischen Problemen." *Zeitschrift für soziale Medizin* 3–4 (1907), 4–26.

Weinberg, Wilhelm. "Bemerkungen zur Reform der deutschen Bevölkerungs- und Gesundheitsstatistik." *Oeffentliche Gesundheitspflege* 4 (1919), 520–524.

Weinberg, Wilhelm. "Die familiäre Belastung der Tuberkulösen und ihre Beziehung zu Infektionen und Vererbung." In Ludolph Brauer, ed., *Beiträge zur Klinik der Tuberkulose.* Vol. 7, 257–289 (Würzburg: A. Stuber's Verlag), 1907).

Weinberg, Wilhelm. "Die neuere psychiatrische Vererbungsstatistik." *Archiv für Rassen- und Gesellschaftsbiologie* 10 (1913), 303–312.

Weinberg, Wilhelm. "Die rassenhygienischen Bedeutung der Fruchtbarkeit." *Archiv für Rassen- und Gesellschaftsbiologie* 7 (1910), 684–696, and 8 (1911), 25–32

Weinberg, Wilhelm, comp. "Die Tuberkulöse in Stuttgart, 1873–1901: Ergebnisse der Untersuchungen einer vom Stuttgarter ärztlichen Vereine eingesetzten Kommission." *Medizinisches Correspondenz-Blatt des Württembergischen ärztlichen Landesvereins* 76 (1906), offprint.

Weinberg, Wilhelm. "Einige Tatsachen der experimenteller Vererbungslehre." In Robert Sommer, ed., *Bericht über den II. Kurs mit Kongreß für Familienforschung, Vererbungs- und Regenerationslehre in Gießen vom 9. bis 13. April 1912*, 71–80. Halle: Carl Marhold Verlagsbuchhandlung, 1912.

Weinberg, Wilhelm. "Methodologische Gesichtspunkte für die statistische Untersuchung der Vererbung bei Dementia praecox." *Zeitschrift für die gesamte Neurologie und Psychiatrie* 59 (1920), 39–50.

Weinberg, Wilhelm. "Pathologische Vererbung und genealogische Statistik." *Deutscher Archiv für klinische Medizin* 78 (1903), 521–540.

Weinberg, Wilhelm. "Sterblichkeit, Lebensdauer, und Todesursachen der württembergischen Aerzte von 1810–1895 und der Ärzte überhaupt." *Württembergische Jahrbücher für Statistik und Landeskunde* (1896), 104–170.

Weinberg, Wilhelm. "Über den Nachweis der Vererbung beim Menschen." *Jahreshefte des Vereins für vaterländische Naturkunde in Württemberg*, 64 (1908), 369–382.

Weinberg, Wilhelm. "Über Methoden der Vererbungsforschung beim Menschen." *Berliner klinische Wochenschrift* 49 (1912), 646–649 and 697–701.

Weinberg, Wilhelm. "Über Vererbungsgesetze beim Menschen." *Zeitschrift für inductive Abstammungs- und Vererbungslehre* 1 (1908), 377–397, 440–460, and 2 (1909), 277–330.

Weinberg, Wilhelm. "Unfruchtbarkeit vom Standpunkte der Statistik." In Siegfried Placzek, ed., *Künstliche Fehlgeburt und künstliche Unfruchtbarkeit, ihre Indikationen, Technik und Rechtslage*, 437–448. Leipzig: Georg Thieme, 1918.

Weinberg, Wilhelm. "Vererbungsforschung und Genealogie: Eine nachträgliche Kritik des Lorenzschens Lehrbuches." *Archiv für Rassen- und Gesellschaftsbiologie* 8 (1911), 753–760.

Weinberg, Wilhelm. "Vererbungsstatistik und Dementia praecox (Vorläufige Mitteilung)." *Münchener medizinischen Wochenschrift* 67, no. 3 (4 June 1920), 667–668.

Weinberg, Wilhelm. "Vererbung und Soziologie." *Berliner klinische Wochenschrift* 49 (27 May 1912), 1080–1084.

Weinberg, Wilhelm. "Verwandtenehe und Geisteskrankheit." *Archiv für Rassen- und Gesellschaftsbiologie* 4 (1907), 471–475.

Weinberg, Wilhelm. "Weitere Beiträge zur Theorie der Vererbung." *Archiv für Rassen- und Gesellschaftsbiologie* 7 (1910), 35–49.

Weinberg, Wilhelm. "Zur Bedeutung der Mehrlingsgeburten für die Frage der Bestimmung des Geschlechts." *Archiv für Rassen- und Gesellschaftsbiologie* 6 (1909), 28–32, 322–339, 470–482, 609–630.

Weinberg, Wilhelm. "Zur Vererbung bei manisch-depressive Irresein." *Zeitschrift für angewandte Anatomie und Konstitutionslehre* 6 (1920), 380–388.

Weldon, W.F.R. "On the Ambiguities of Mendel's Categories." *Biometrika* 2 (1902), 44–55.

Wille, Ludwig. "Die Aufgaben und Leistungen der Statistik der Geisteskranken." *Jahrbücher für Nationalökonomie und Statistik* 35 (1879), 307–331.

Wille, Ludwig. "Ueber Einführung einer gleichmässigen Statistik der schweizerischen Irrenanstalten." *Zeitschrift für Schweizerische Statistik* 1872, 249–254.

Wilson, James. "Presidential Address: Ninth Annual Meeting." *American Breeders Magazine* 4 (1913), 53–58.

Wittermann, Ernst. "Psychiatrische Familienforschungen." *Zeitschrift für die gesamte Neurologie und Psychiatrie* 10 (1913), 153–278.

Zählkarten und Tabellen für die Statistik der Irrenanstalten aufgestellt von dem Verein der deutschen Irrenärzte. Berlin, 1874. Beilage to *Allgemeine Zeitschrift für Psychiatrie* 30, no. 6 (1874).

Zeller, Ernst Albert von. "Bericht über die Wirksamkeit der Heilanstalt Winnenthal vom 1. März 1840 bis 28 Febr. 1843." *Allgemeine Zeitschrift für Psychiatrie* 1 (1844), 1–79.

Scholarly Sources

Ackerknecht, Erwin. "Diathesis: The Word and the Concept in Medical History." *Bulletin of the History of Medicine* 56 (1982), 317–325.

Alborn, Timothy. "Quill-Driving: British Life-Insurance Clerks and Occupational Mobility, 1800–1914." *Business History Review* 82 (2008), 31–58.

Alborn, Timothy. *Regulated Lives: Life Insurance and British Society, 1800–1914.* Toronto: University of Toronto Press, 2009.

Allen, Garland E. "The Eugenics Record Office at Cold Spring Harbor, 1910–1940: An Essay in Institutional History." *Osiris* 2 (1986), 225–264.

Andrews, Jonathan. "Bedlam Revisited: A History of Bethlem Hospital, c1634–c1770." PhD diss., London University, 1991.

Andrews, Jonathan. *They're in the Trade . . . of Lunacy. They "cannot interfere"—they say: The Scottish Lunacy Commissioners and Lunacy Reform in Nineteenth-Century Scotland.* Wellcome Institute for the History of Medicine, Occasional Publications No. 8. London: Wellcome Trust, 1998.

Andrews, Jonathan, Asa Briggs, Roy Porter, Penny Tucker, and Keir Waddington. *The History of Bethlem* London: Routledge, 1997.

Ayers, Gwendoline. *England's First State Hospital and the Metropolitan Asylums Board, 1867–1930.* London: Wellcome Institute of the History of Medicine, 1971.

Bangham, Jenny. "Blood Groups and Human Groups: Collecting and Calibrating Genetic Data after World War II." *Studies in the History and Philosophy of Biological and Biomedical Science* 47A (2014), 74–86.

Banzhaf, Katharina. "Vorläufer der psychiatrischen Genetik: die psychiatrischen Erblichkeitsforschung in der deutschsprachigen Psychiatrie im Spiegel der Allgemeinen Zeitschrift für Psychiatrie 1844 bis 1911." Inauguraldissertation zur Erlangung des Grades eines Doctors des Medizin des Fachbereichs Medizin der Justus-Liebig Universität Gießen, 2014.

Barker, David. "The Biology of Stupidity: Genetics, Eugenics and Mental Deficiency in the Inter-War Years," *British Journal for the History of Science* 22 (1989), 347–375.

Becker, Peter, and William Clark, eds. *Little Tools of Knowledge: Historical Essays on Academic and Bureaucratic Practices.* Ann Arbor: University of Michigan Press, 2001.

Beltrán, Carlos Lopez. "In the Cradle of Heredity: French Physicians and L'Hérédité Naturelle in the Early 19th Century." *Journal of the History of Biology* 37 (2004), 39–72.

Benbassat, Carlos A. "Kallmann Syndrome: Eugenics and the Man behind the Eponym." *Rambam Maimonides Medical Journal* 7, no. 2, (2016), doi:10.5041/RMMJ.10242.

Berghof, Winfried. "Heinrich Damerow (1798–1866)—Ein bedeutender Vertreter der deutschen Psychiatrie des 19. Jahrhunderts." Dissertation zur Erlangung des akademischen Grades Dr. med., Karl-Marx-Universität Leipzig, 1990.

Bix, Amy Sue. "Experiences and Voices of Eugenics Fieldworkers: 'Women's Work' in Biology." *Social Studies of Science* 27 (1997), 625–668.

Bonneuil, Christophe. "Pure Lines as Industrial Simulacra: A Cultural History of Genetics from Darwin to Johannsen." In Müller-Wille and Brandt, *Heredity Explored*, 213–242.

Bowler, Peter. *The Mendelian Revolution.* Baltimore: Johns Hopkins University Press, 1989.

Braun, Salina. *Heilung mit Defekt: Psychiatrische Praxis an den Anstalten Hofheim und Siegburg, 1820–1878.* Göttingen: Vandenhoeck & Ruprecht, 2009.

Burkhardt, Marga Maria. "Krank im Kopf: Patienten-Geschichten der Heil- und Pflegeanstalt Illenau, 1842–1889." PhD diss., Albert-Ludwigs Universität, Freiburg i. Br., 2003.

Burleigh, Michael. *Death and Deliverance: Euthanasia in Germany, 1900–1945*. Cambridge: Cambridge University Press, 1994.

Carbonel, Frédéric. "L'Asile pour aliénés de Rouen: Un laboratoire de statistiques morales de la Restauration à 1848." *Histoire et mesure* 20 (2005), 97–136.

Caron, Jean-Claude. *Les feux de la discorde: Conflits et incendies dans la France du XIXe siècle*. Paris: Hachette, 2006.

Carson, John. *The Measure of Merit: Talents, Intelligence, and Inequality in the French and American Republics, 1750–1940* (Princeton, NJ: Princeton University Press, 2007).

Cartron, Laure. "Degeneration and 'Alienism' in Early Nineteenth-Century France." In Müller-Wille and Rheinberger, *Heredity Produced*, 155–174.

Cassedy, James. *American Medicine and Statistical Thinking, 1800–1860*. Cambridge, MA: Harvard University Press, 1984.

Chadarevian, Soraya de. "Chromosome Surveys of Human Populations: Between Epidemiology and Anthropology." *Studies in the History and Philosophy of Biological and Biomedical Sciences* 47A (2014), 87–96.

Chadarevian, Soraya de. "Putting Human Genetics on a Solid Basis: Human Chromosome Research, 1950s–1970s." In Gausemeier, Müller-Wille, and Ramsden, *Human Heredity*, 141–152.

Chadarevian, Soraya de and Harmke Kamminga. *Representations of the Double Helix*. Cambridge: Whipple Museum of the History of Science, 2001.

Cock, Alan G., and Donald R. Forsdyke. *Treasure Your Exceptions: The Science and Life of William Bateson*. New York: Springer, 2008.

Cohen, Patricia Cline. *A Calculating People: The Spread of Numeracy in Early America*. Chicago: University of Chicago Press, 1982.

Cole, Simon A. *Suspect Identities: A History of Fingerprinting and Criminal Identification*. Cambridge, MA: Harvard University Press, 2001.

Comfort, Nathaniel. *The Science of Human Perfection*. New Haven, CT: Yale University Press, 2013.

Cottebrune, Anne. *Der planbare Mensch: Die Deutsche Forschungsgemeinschaft und die menschliche Vererbungswissenschaft, 1920–1970*. Stuttgart: Franz Steiner Verlag, 2008.

Cottebrune, Anne. "Franz-Josef Kallmann (1897–1965) und der Transfer psychiatrisch-genetischer Wissenschaftskonzepte vom NS-Deutschland in die USA." *Medizinhistorisches Journal* 44 (2009), 269–324.

Cottebrune, Anne. "Zwischen Theorie und Deutung der Vererbung psychischer Störungen: Zur Übertragung des Mendelismus auf die Psychiatrie in Deutschland und in den USA, 1911–1930." *NTM* 17 (2009), 35–54.

Crook, Thomas, and Glenn O'Hara, eds. *Statistics and the Public Sphere: Numbers and the People in Modern Britain c. 1800 to 2000*. London: Routledge, 2011.

Daston, Lorraine ed. *Science in the Archives: Pasts, Presents, Futures*. Chicago: University of Chicago Press, 2017.

Daston, Lorraine, and Elizabeth Lunbeck, eds. *Histories of Scientific Observation*. Chicago: University of Chicago Press, 2011.

Deringer, William. *Calculated Values: Finance, Politics, and the Quantitative Age*. Cambridge, MA: Harvard University Press, 2018.

Digby, Ann. *Madness, Morality and Medicine. A Study of the York Retreat, 1796–1914*. Cambridge: Cambridge University Press, 1985.

Dowbiggin, Ian Robert. *Keeping America Sane: Psychiatry and Eugenics in the United States and Canada, 1880–1940*. Ithaca, NY: Cornell University Press, 1997.

Dowbiggin, Ian. *Inheriting Madness: Professionalization and Psychiatric Knowledge in Nineteenth-Century France*. Berkeley: University of California Press, 1991.

Dwyer, Elllen. *Homes for the Mad: Life inside Two Nineteenth-Century Asylums*. New Brunswick, NJ: Rutgers University Press, 1987.

Edwards, A.W.F. "G. H. Hardy (1908) and Hardy-Weinberg Equilibrium." *Genetics* 179 (1908), 1143–1150.

Endelman, Todd M. "Anglo-Jewish Scientists and the Science of Race." *Jewish Social Studies* 11 (2004), 52–92.

Engstrom, Eric J. *Clinical Psychiatry in Imperial Germany*. Ithaca, NY: Cornell University Press, 2003.

Engstrom, Eric J. "Die Ökonomie klinischer Inskription. Zu diagnostischen und nosologischen Schreibpraktiken in der Psychiatrie." In Cornelius Borck and Armin Schäfer, eds., 219–240. *Psychographien*. Zürich/Berlin: Diaphanes, 2005.

Engstrom, Eric J., "'Organizing Psychiatric Research in Munich (1903–1925): A Psychiatric Zoon Politicon between State Bureaucracy and American Philanthropy." In Volker Roelcke, Paul Weindling, and L. Westwood, eds., *International Relations in Psychiatry: Britain, Germany, and the United States to World War II*, 48–66. Rochester, NY: University of Rochester Press, 2010.

Espeland, Wendy Nelson, and Michael Sauder. *Engines of Anxiety: Academic Rankings, Reputation, and Accountability*. New York: Russell Sage Foundation, 2016.

Falk, Raphael. "Three Zionist Men of Science: Between Nature and Nurture." In Ulrich Charpa and Ute Deichmann, eds., *Jews and Sciences in German Contexts*, 129–154. Tübingen: Mohr Siebeck, 2007.

Fleckner, Uta. "Emanuel Mendel (1839–1907). Leben und Werk eines Psychiaters im Deutschland der Jahrhundertwende." Medical diss., Freie Universität Berlin, 1994.

Forsbach, Ralf. *Die Medizinische Fakultät der Universität Bonn in "Dritten Reich."* Munich: Oldenbourg, 2006.

Fortun, Mike. *Promising Genomics: Iceland and deCODE Genetics in a World of Speculation*. Berkeley: University of California Press, 2008.

Foucault, Michel. *Psychiatric Power: Lectures at the Collège de France, 1973–74*. Edited by Arnold Davidson, translated by Graham Burchell. London: Palgrave Macmillan, 2008.

Fox, Richard W. *So Far Disordered in Mind: Insanity in California, 1870–1930*. Berkeley: University of California Press, 1978.

Friedman, Judith E. "The Disappearance of the Concept of Anticipation in the Post-War World." In Gausemeier, Müller-Wille, and Ramsden, *Human Heredity*, 153–164.

Gateaux-Mennecier, Jacqueline. *Bourneville et l'enfance aliénée*. Paris: Centurion, 1989.

Gateaux-Mennecier, Jacqueline. *La débilité légère: Une construction idéologique*. Paris: Éditions du CNRS, 1990.

Gaudillière, Jean-Paul, and Ilana Löwy. "The Hereditary Transmission of Human Pathologies between 1900 and 1940: The Good Reasons Not to Become, 'Mendelian.'" In Müller-Wille and Brandt, *Heredity Explored*, 311–335.

Gausemeier, Bernd. "Auf der 'Brücke zwischen Natur- und Geschichtswissenschaft': Ottokar Lorenz und die Neuerfindung der Genealogie um 1900." In Florence Vienne and Christina Brandt, eds., *Wissensobjekt Mensch: Humanwissenschaftliche Praktiken im 20. Jahrhundert*, 137–164. Berlin: Kulturverlag Kadmos, 2008.

Gausemeier, Bernd. "Borderlands of Heredity: The Debate about Hereditary Susceptibility to Tuberculosis, 1882–1945." In Gausemeier, Müller-Wille, and Ramsden, *Human Heredity*, 13–26.

Gausemeier, Bernd. "In Search of the Population: The Study of Human Heredity before and after the Mendelian Break." In Müller-Wille and Brandt, *Heredity Explored*, 337–363.

Gausemeier, Bernd. "Pedigrees of Madness: The Study of Heredity in Nineteenth- and Early Twentieth-Century Psychiatry." *History and Philosophy of the Life Sciences* 36 (2015), 467–483.

Gausemeier, Bernd, Staffan Müller-Wille, and Edmund Ramsden, eds., *Human Heredity in the Twentieth Century*. London: Pickering & Chatto, 2013.

Gere, Cathy. "Evolutionary Genetics and the Politics of the Human Archive." In Lorraine Daston, ed., *Science in the Archives: Pasts, Presents, Futures*, 203–222. Chicago: University of Chicago Press, 2017.

Gillham, Nicholas Wright. *A Life of Sir Francis Galton: From African Exploration to the Birth of Eugenics*. Oxford: Oxford University Press, 2001.

Goffman, Erving. *Asylums: Essays on the Social Situation of Mental Patients and other Inmates*. Garden City, NY: Anchor Books, 1961.

Goldberg, Ann. *Sex, Religion, and the Making of Modern Madness: The Eberbach Asylum and German Society, 1815–1849*. New York: Oxford University Press, 1999.

Goldstein, Jan. *Console and Classify: The French Psychiatric Profession in the Nineteenth Century*. Cambridge: Cambridge University Press, 1987.

Gollaher, David. *Voice for the Mad: The Life of Dorothea Dix*. New York: Free Press, 1995.

Goodheart, Lawrence B. *Mad Yankees: The Hartford Retreat for the Insane and Nineteenth-Century Psychiatry*. Amherst: University of Massachusetts Press, 2003.

Gottesman, Irving, and Peter McGuffin. "Eliot Slater and the Birth of Psychiatric Genetics in Great Britain." In Hugh Freeman and German Barrios, eds., *150 Years of Psychiatry, Volume II: The Aftermath*, 537–548. London: Athlone, 1996.

Grier, David Alan. *When Computers Were Women*. Princeton, NJ: Princeton University Press, 2005.

Grob, Gerald N. "Introduction" to Edward Jarvis, *Insanity and Idiocy in Massachusetts: Report of the Commission on Lunacy, 1855*, 1–71. Cambridge, MA: Harvard University Press, 1971.

Grob, Gerald N. *Edward Jarvis and the Medical World of Nineteenth-Century America*. Knoxville: University of Tennessee Press, 1978.

Grob, Gerald N. *Mental Illness and American Society*. Princeton, NJ: Princeton University Press, 1983.

Grob, Gerald N. *The State and the Mentally Ill: A History of Worcester State Hospital in Massachusetts, 1830–1920*. Chapel Hill: University of North Carolina Press, 1966.

Haberer, Anja. *Zeitbilder: Krankheit und Gesellschaft in Theodor Fontanes Romanen* Cécile *(1886) und* Effi Briest *(1895)*. Würzburg: Königshausen & Neumann, 2012.

Hagner, Michael. *Homo Cerebralis: Der Wandel vom Seelenorgan zum Gehirn*. Berlin: Berlin Verlag, 1997.

Harper, Peter S. "Julia Bell and the Treasury of Human Inheritance." *Human Genetics* 116 (2005), 422–432.

Harper, Peter S. *A Short History of Medical Genetics*. Oxford: Oxford University Press, 2008.

Hartmann, Heinrich. *Der Volkskörper bei der Musterung. Militärstatistik und Demographie in Europa vor dem Ersten Weltkrieg*. Göttingen: Wallstein Verlag, 2011.

Harwood, Jonathan. *Styles of Scientific Thought: The German Genetics Community, 1900–1933*. Chicago: University of Chicago Press, 1993.

Hearnshaw, Leslie. *Cyril Burt, Psychologist*. Ithaca, NY: Cornell University Press, 1979.

Hess, Volker. "Die Buchhaltung des Wahnsinns: Archiv und Aktenführung zwischen Justiz und Irrenreform." In Cornelius Borck and Armin Schäfer, eds., *Das psychiatrische Aufschreibesystem*, 55–76. Paderborn: Fink, 2015.

Higgs, Edward. *The Information State in England: The Central Collection of Information on Citizens since 1500*. London: Palgrave Macmillan, 2004.

Irgens, Lorentz M. "The Fight against Leprosy in Norway in the 19th Century." *Michael Quarterly* 7 (2010), 296–306.

Judson, Horace Freeland. *The Eighth Day of Creation: The Makers of the Revolution in Biology*. New York: Simon and Schuster, 1979.

Karlsen, Marianne Berg. "Den første norske telling av sinnsvake." *Nytt Norsk Tidsskrift* 17, no. 3 (2000), 276–293.

Keller, Evelyn Fox. *The Century of the Gene*. Cambridge, MA: Harvard University Press, 2000.

Kevles, Daniel J. *In the Name of Eugenics: Genetics and the Uses of Human Heredity*. New York: Knopf, 1985.

Kimmelman, Barbara A. "The American Breeders' Association: Genetics and Eugenics in an Agricultural Context, 1903–13." *Social Studies of Science* 13 (1983), 163–204.

Krajewski, Markus. *Paper Machines: About Cards & Catalogs 1548–1929*. Cambridge, MA: MIT Press, 2011.

Kuper, Adam. "Incest, Cousin Marriage, and the Origin of the Human Sciences in Nineteenth-Century England." *Past and Present* 174 (2002), 158–183.

Lakoff, Andrew. *Pharmaceutical Reason: Knowledge and Value in Global Psychiatry*. Cambridge: Cambridge University Press, 2005.

Lamb, S. D. *Pathologist of the Mind: Adolf Meyer and the Origins of American Psychiatry*. Baltimore: Johns Hopkins University Press, 2014.

Ledebur, Sophie. "Schreiben und Beschreiben: Zur epistemischen Funktion von psychiatrischen Krankenakten, ihrer Archivierung und deren Übersetzungen in Fallgeschichten." *Berichte zur Wissenschaftsgeschichte* 34 (2011), 102–124.

Levitan, Kathrin. *A Cultural History of the British Census: Envisioning the Multitude in the Nineteenth Century*. New York: Palgrave Macmillan, 2011.

Lie, Einar, and Hege Roll-Hansen. *Faktisk Talt: Statistikkens historie i Norge*. Oslo: Universitetsforlaget, 2001.

Lindee, Susan. *Moments of Truth in Genetic Medicine*. Baltimore: Johns Hopkins University Press, 2005.

Lindee, Susan. *Suffering Made Real: American Science and the Survivors at Hiroshima*. Chicago: University of Chicago Press, 1994.

Lötsch, Gerhard. *Von der Menschenwürde zum Lebensunwert: Die Geschichte der Illenau von 1842–1940*. Kappelrodeck: Achertäler Verlag, 2000.

Macalpine, Ida, and Richard Hunter. *George III and the Mad Business*. New York: Pantheon Books, 1969.

MacKenzie, Donald. *Statistics in Britain, 1865–1930: The Social Construction of Scientific Knowledge*. Edinburgh: Edinburgh University Press, 1981.

Matthews, J. Rosser. *Quantification and the Quest for Medical Certainty*. Princeton, NJ: Princeton University Press, 1995.

Mayr, Ernst. *The Growth of Biological Thought: Diversity, Evolution, and Inheritance*. Cambridge, MA: Harvard University Press, 1982.

Mazumdar, Pauline M. H. *Eugenics, Human Genetics and Human Failings: The Eugenics Society, Its Sources and Its Critics in Britain*. London: Routledge, 1992.

Mazumdar, Pauline M. H. "Two Models for Human Genetics: Blood Grouping and Psychiatry between the Wars." *Bulletin of the History of Medicine* 70 (1996), 609–657.

Meier, Marietta, Brigitta Bernet, Roswitha Dubach, and Urs Germann. *Zwang zur Ordnung: Psychiatrie im Kanton Zürich, 1870–1970*. Zürich: Chronos, 2007.

Mellett, D. J. "Bureaucracy and Mental Illness: The Commissioners in Lunacy 1845–1890." *Medical History* 23 (1981), 221–250.

Mildenberger, Florian. "Auf der Spur des 'scientific pursuit,' Franz Joseph Kallmann (1897–1965) und die rassenhygienische Forschung." *Medizinhistorische Journal* 37 (2002), 183–200.

Mills, James H. *Madness, Cannabis, and Colonialism: The "Native-Only" Lunatic Asylums of British India, 1857–1900*. New York: St. Martin's Press, 2000.

Müller, Thomas, Bobo Rüdenburg, and Martin Rexer, eds. *Wissenstransfer in der Psychiatrie: Albert Zeller und die Psychiatrie Württembergs im 19. Jahrhundert*. Zwiefalten: Verlag Psychiatrie und Geschichte des Zentrums für Psychiatrie, 2009.

Müller-Hill, Benno. *Murderous Science: Elimination by Scientific Selection of Jews, Gypsies and Others, 1933–1945*. Translated by George Fraser. Oxford: Oxford University Press, 1988. Originally published as *Tödliche Wissenschaft*. Reinbek bei Hamburg: Rowohlt Verlag, 1984.

Müller-Wille, Staffan, and Christina Brandt, eds., *Heredity Explored: Between Public Domain and Experimental Science, 1850–1930*. Cambridge, MA: MIT Press, 2016.

Müller-Wille, Staffan, and Hans-Jörg Rheinberger. *A Cultural History of Heredity*. Chicago: University of Chicago Press, 2012.

Müller-Wille, Staffan, and Hans-Jörg Rheinberger, eds. *Heredity Produced: At the Crossroads of Biology, Politics, and Culture, 1500–1870*. Cambridge, MA: MIT Press, 2007.

Murat, Laure. *The Man Who Thought He Was Napoleon: Toward a Political History of Madness*. Translated by Deke Dusinberre. Chicago: University of Chicago Press, 2014. Originally published as *L'homme qui se prenait pour Napoléon: pour une histoire politique de la folie*. Paris: Gallimard, 2011.

Noll, Richard. *American Madness: The Rise and Fall of Dementia Praecox*. Cambridge, MA: Harvard University Press, 2011.

Norton, Bernard, and E. S. Pearson. "A Note on the Background to, and Refereeing of, R. A. Fisher's 1918 Paper 'On the Correlation between Relatives on the Supposition of Mendelian Inheritance.'" *Notes and Records of the Royal Society of London* 31 (1976), 151–162.

Nye, Robert. *Masculinity and Male Codes of Honor in Modern France*. New York: Oxford University Press, 1993.

Oertzen, Christine von. "Machineries of Data Power: Manual versus Mechanical Census Compilation in Nineteenth-Century Europe." *Osiris* 32 (2017), 129–150.

Olby, Robert C. "Constitutional and Hereditary Disorders." In W. F. Bynum and Roy Porter, eds., *Companion Encyclopedia of the History of Medicine*. Vol. 1, 412–437. New York: Routledge, 2001. First published in 1993 by Routledge, New York.

Olby, Robert. *Origins of Mendelism*, 2nd ed. Chicago: University of Chicago Press, 1985.

Panofsky, Aaron. "From Behavior Genetics to Postgenomics." In Sarah Richardson and Hallam Stevens, eds., *Postgenomics: Perspectives on Biology after the Genome*, 150–173. Durham, NC: Duke University Press, 2015.

Panofsky, Aaron. *Misbehaving Science: Controversy and the Development of Behavior Genetics*. Chicago: University of Chicago Press, 2014.

Patriarca, Silvana. *Numbers and Nationhood: Writing Statistics in Nineteenth-Century Italy*. Cambridge: Cambridge University Press, 1996.

Paul, Diane B. "Eugenics and the Left." In Paul, *Politics of Heredity*, 12–35.

Paul, Diane B. *The Politics of Heredity: Essays on Eugenics, Biomedicine, and the Nature-Nurture Debate*. Albany: State University of New York Press, 1998.

Paul, Diane B. "The Rockefeller Foundation and the Origins of Behavior Genetics." In Paul, *Politics of Heredity*, 53–89.

Paul, Diane B., and Barbara A. Kimmelman. "Mendel in America: Theory and Practice, 1900–1919." In Ronald Rainger, Keith R. Benson, and Jane Maienschein, eds., *The American Development of Biology*, 281–310. New Brunswick, NJ: Rutgers University Press, 1988.

Paul, Diane B., and Hamish G. Spencer. "Did Eugenics Rest on an Elementary Mistake?" In Paul, *Politics of Heredity*, 117–132.

Paul, Diane B., and Hamish G. Spencer. "Eugenics without Eugenists? Anglo-American Critiques of Cousin Marriage in the Nineteenth and Early Twentieth Centuries." In Müller-Wille and Brandt, *Heredity Explored*, 40–79.

Pemberton, Stephen. "'The Most Hereditary of all Diseases': Haemophilia and the Utility of Genetics for Haematology, 1930–70." In Gausemeier, Müller-Wille, and Ramsden, *Human Heredity*, 165–178.

Pick, Daniel. *Faces of Degeneration: A European Disorder c. 1848–c. 1918*. Cambridge: Cambridge University Press, 1989.

Pomata, Gianna. "Observation Rising: Birth of an Epistemic Genre." In Lorraine Daston and Elizabeth Lunbeck, eds., *Histories of Scientific Observation*, 45–80. Chicago: University of Chicago Press, 2011.

Porter, Roy. *Mind-Forg'd Manacles: A History of Madness in England from the Restoration to the Regency*. London: Athlone Press, 1987.

Porter, Theodore M. "Asylums of Hereditary Research in the Efficient Modern State." In Müller-Wille and Brandt, *Heredity Explored*, 81–109.

Porter, Theodore M. "The Curious Case of Blending Inheritance." *Studies in History and Philosophy of Biological and Biomedical Sciences* 46 (2014), 125–132.

Porter, Theodore M. "Genres and Objects of Social Inquiry from the Enlightenment to 1890." In *Cambridge History of Science*. Vol. 7, *Modern Social Sciences*, Theodore M. Porter and Dorothy Ross, eds., 13–39. Cambridge: Cambridge University Press, 2003.

Porter, Theodore M. "*Irrenärzte aller Länder!* Tabular Unity and the Nineteenth-Century Struggle to Comprehend Insanity." *Soziale System* 18, no. 2 (2012), 204–224.

Porter, Theodore M. *Karl Pearson: The Scientific Life in a Statistical Age*. Princeton, NJ: Princeton University Press, 2004.

Porter, Theodore M. *The Rise of Statistical Thinking, 1820–1900*. Princeton, NJ: Princeton University Press, 1986.

Porter, Theodore M. "The Social Sciences." In David L. Cahan, ed., *From Natural Philosophy to the Sciences: Writing the History of Nineteenth-Century Science*, 254–290. Chicago: University of Chicago Press, 2003.

Porter, Theodore M. "Statistics and the Career of Public Reason: Engagement and Detachment in a Quantified World." In Tom Crook and Glen O'Hara, eds., *Statistics and the Public Sphere: Numbers and the People in Modern Britain, c. 1800–2000*, 32–47. New York: Routledge, 2011.

Porter, Theodore M. "Thin Description: Surface and Depth in Science and Science Studies." *Osiris* 27 (2012), 209–226.

Proctor, Robert N. *Racial Hygiene: Medicine under the Nazis*. Cambridge, MA: Harvard University Press, 1988.

Provine, Will. *The Origins of Theoretical Population Genetics*. Chicago: University of Chicago Press, 1971.

Ramsden, Edmund. "Remodelling the Boundaries of Normality: Lionel S. Penrose and Population Surveys of Mental Ability." In Gausemeier, Müller-Wille, and Ramsden, *Human Heredity*, 39–54.

Richebächer, Sabine. *Sabina Spielrein: "Eine fast grausame Liebe zur Wissenschaft."* Zurich: Dörlemann, 2005.

Ritter, Hans Jakob. *Psychiatrie und Eugenik: Zur Ausprägung eugenischer Denk- und Handlungsmuster in der schweizerischen Psychiatrie, 1850–1950*. Zürich: Chronos Verlag, 2009.

Ritter, Hans Jakob. "Von den Irrenstatistiken zur "erblichen Belastung" der Bevölkerung: Die Entwicklung der schweizerischen Irrenstatistiken zwischen 1850 und 1914." *Traverse: Zeitschrift für Geschichte* 10 (2003), 59–70.

Ritter, Hans Jakob, and Volker Roelcke. "Psychiatric Genetics in Munich and Basel between 1925 and 1945: Programs—Practices—Cooperative Arrangements." *Osiris* 20 (2005), 263–288.

Roelcke, Volker. "Die Etablierung der psychiatrischen Genetik in Deutschland, Großbritannien und den USA, ca. 1910–1960: Zur untrennbaren Geschichten von Eugenik und Humangenetik." *Acta historica Leopoldina* 48 (2007), 173–190.

Roelcke, Volker. "Politische Zwänge und individuelle Handlungsspielräume: Karl Bonhoeffer und Maximinian de Crinis im Kontext der Psychiatrie im Nationalsozialismus." In Sabine Schleiermacher and Udo Schagen, eds., *Die Charité im Dritten Reich*, 68–84. Paderborn: Ferdinand Schöningh, 2008.

Roelcke, Volker. "Programm und Praxis der psychiatrischen Genetik an der Deutschen Forschungsanstalt für Psychiatrie unter Ernst Rüdin." In Hans-Walter Schmuhl, ed., *Rassenforschung an Kaiser-Wilhelm-Instituten vor und nach 1933*, 38–67. Göttingen: Wallstein Verlag, 2003.

Roelcke, Volker. "Psychiatry in Munich and Yale, ca. 1920–1935: Mutual Perceptions and Relations, and the Case of Eugen Kahn (1887–1973)." In Volker Roelcke, Paul J. Weindling, and Louise Westwood, eds., *International Relations in Psychiatry: Britain, Germany, and the United States to World War II*, 156–178. Rochester, NY: University of Rochester Press, 2010.

Roelcke, Volker. "Quantifizierung, Klassifikation, Epidemiologie: Normierungsversuch des Psychischen bei Emil Kraepelin." In Werner Sohn and Herbert Mehrtens, eds., *Normalität und Abweichung: Studien zur Theorie und Geschichte der Normalisierungsgesellschaft*, 183–200. Opladen: Westdeutscher Verlag, 1999.

Roelcke, Volker. "Unterwegs zur Psychiatrie als Wissenschaft: Das Projekt einer 'Irrenstatistik' und Emil Kraepelin's Neuformulierung des psychiatrischen Klassifikation." In Eric J. Engstrom and Volker Roelcke, eds., *Psychiatrie im 19. Jahrhundert. Forschungen zur Geschichte von psychiatrischen Institutionen, Debatten und Praktiken im deutschen Sprachraum*, 169–188. Mainz: Akademie der Wissenschaften und der Literatur, 2003.

Rogers, Alan. *Murder and the Death Penalty in Massachusetts*. Amherst: University of Massachusetts Press, 2008.

Rosenberg, Charles E. "The Bitter Fruit: Heredity, Disease, and Social Thought." In Rosenberg, *No Other Gods: On Science and American Social Thought*, 25–53. Baltimore: Johns Hopkins University Press, 1976.

Rosenberg, Charles E. *The Trial of the Assassin Guiteau: Psychiatry and the Law in the Gilded Age*. Chicago: University of Chicago Press, 1968.

Rosental, Paul-André. *Destins de l'Eugenisme*. Paris: Seuil, 2016.

Rosner, David J. *The Discovery of the Asylum: Social Order and Disorder in the New Republic*. Rev. ed. New York: Alding de Gruuyter, 1990. First published in 1971 by Little, Brown, Boston.

Rothmaler, Christiane. "Von 'haltlosen Psychopathinnen' und 'konstitutionellen Sittlichkeitsverbrechen': Die kriminalbiologische Untersuchungs- und Sammelstelle der Hamburgischen Gefangenenanstalten 1926 bis 1945." In Heidrun Kauper-Haas and Christian Saller, eds., *Wissenschaftlicher Rassismus: Analysen einer Kontinuität in den Human- und Naturwissenschaften*, 257–303. Frankfurt: Campus Verlag, 1999.

Rushton, Alan R. *Genetics and Medicine in Great Britain, 1600–1939*. Victoria, BC: Trafford Publishing, 2009.

Rushton, Alan R. *Genetics and Medicine in the United States, 1800 to 1922*. Baltimore: Johns Hopkins University Press, 1994.

Rushton, Alan R. "Nettleship, Pearson, and Bateson: The Biometric-Mendelian Debate in a Medical Context." *Journal of the History of Medicine and Allied Sciences* 55 (2000), 134–157.

Rusnock, Andrea A. *Vital Accounts: Quantifying Health and Population in Eighteenth-Century England and France.* New York: Cambridge University Press, 2002.

Schaffer, Wolfgang. "Erbbiologische Bestandsaufnahme im Rheinland—Das Institut fur Psychiatrisch-Neurologische Erbforschung in Bonn." In Erik Gieseking and Herbert Hörmig, eds., *Zum Ideologieproblem in der Geschichte*, 419–444. Lauf a. d. Pegnitz: Europaforum-Verlag, 2006.

Schlich, Thomas. "Die Konstruktion der notwendigen Krankheitsursache: Wie die Medizin Krankheit beherrschen will." In Cornelius Borck, ed., *Anatomien medizinischen Wissens. Medizin Macht Moleküle*, 201–229. Frankfurt am Main: Fischer Taschenbuch Verlag, 1996.

Schmiedebach, Heinz-Peter. *Psychiatrie und Psychologie im Widerstreit: Die Auseinandersetzungen in der Berliner medicinisch-psychologischen Gesellschaft (1867–1899).* Husum: Matthiesen Verlag, 1986.

Schmuhl, Hans-Walter. "Die Charité und die Forschungspolitik der Kaiser-Wilhelm-Gesellschaft und der Deutsche Forschungsgemeinschaft in der Zeit des Nationalsozialismus." In Sabine Schleiermacher and Udo Schagen, eds., *Die Charité im Dritten Reich*, 226–245. Paderborn: Ferdinand Schöningh, 2008.

Schmuhl, Hans-Walter. *The Kaiser Wilhelm Institute for Anthropology, Human Heredity, and Eugenics, 1927–1945.* Translated by Sorcha O'Hagan. Dordrecht: Springer, 2008.

Schneider, Michael C. "Medizinalstatistik im Spannungsfeld divergierender Interessen: Kooperationsformen zwischen statistischen Ämtern und dem Kaiserlichen Gesundheistamt/Reichsgesundheitsamt." In Axel C. Hüntelmann, Johannes Vossen, und Herwig Czech, eds., *Gesundheit und Staat: Studien zur Geschichte der Gesundheitsämter in Deutschland, 1870–1930*, 49–62. Husum: Matthiesen Verlag, 2006.

Schneider, Michael C. *Wissensproduktion im Staat: Das königlich preußische statistische Bureau 1860–1914.* Frankfurt: Campus Verlag, 2013.

Schor, Paul. *Compter et classer: Histoire des recensements américains.* Paris: Éditions de l'École des Hautes Études en Sciences Sociales, 2009.

Scull, Andrew. *Madness in Civilization: A Cultural History of Insanity.* Princeton, NJ: Princeton University Press, 2015.

Scull, Andrew. *The Most Solitary of Afflictions: Madness and Society in Britain, 1700–1900.* New Haven, CT: Yale University Press, 1993.

Scull, Andrew. *Social Order/Mental Disorder: Anglo-American Society in Historical Perspective.* Berkeley: University of California Press, 1989.

Scull, Andrew, Charlotte MacKenzie, and Nicholas Hervey. *Masters of Bedlam: The Transformation of the Mad-Doctoring Trade.* Princeton, NJ: Princeton University Press, 1996.

Simon, Jürgen. "Kriminalbiologie und Strafrecht von 1920 bis 1945." In Heidrun Kauper-Haas and Christian Saller, eds., *Wissenschaftlicher Rassismus: Analysen einer Kontinuität in den Human- und Naturwissenschaften*, 226–256. Frankfurt: Campus Verlag, 1999.

Soloway, Richard A. *Demography and Degeneration: Eugenics and the Declining Birthrate in Twentieth-Century Britain.* Chapel Hill: University of North Carolina Press, 1990.

Spencer, Hamish G., and Diane B. Paul. "The Failure of a Scientific Critique: David Heron, Karl Pearson, and Mendelian Eugenics." *British Journal for the History of Science* 31 (1998), 441–452.

Stark, Alan, and Eugene Seneta. "Wilhelm Weinberg's Early Contribution to Segregation Analysis." *Genetics* 195 (September 2013), 1–6.

Staum, Martin. *Nature and Nurture in French Social Sciences, 1859–1914, and Beyond.* Montreal: McGill-Queens University Press, 2011.

Stern, Alexandra Minna. *Eugenic Nation: Faults and Frontiers of Better Breeding in Modern America.* Berkeley: University of California Press, 2005.

Stern, Alexandra Minna. *Telling Genes: The Story of Genetic Counseling in America.* Baltimore: Johns Hopkins University Press, 2012.

Stigler, Stephen. *The History of Statistics: The Measurement of Uncertainty before 1900.* Cambridge, MA: Harvard University Press, 1986.

Suzuki, Akihito. "Framing Psychiatric Subjectivity: Doctor, Patient, and Record-Keeping at Bethlem in the Nineteenth Century." In Joseph Melling and Bill Forsythe, eds., *Insanity, Institutions, and Society, 1800–1914: A Social History of Madness in Comparative Perspective*, 115–136. New York: Routledge, 1999.

Szreter, Simon. *Fertility, Class, and Gender in Britain, 1860–1940.* Cambridge: Cambridge University Press, 1996.

Tewarson, Heidi Thomann. *Die ersten Zürcher Ärztinnen: Humanitäres Engagement und wissenschaftliche Arbeit zur Zeit der Eugenik.* Basel: Schwabe Verlag, 2018.

Theunissen, Bert. "Breeding without Mendelism: Theory and Practice of Dairy Cattle Breeding in the Netherlands, 1900–1950." *Journal of the History of Biology* 41 (2008), 637–676.

Thomson, Mathew. "Disability, Psychiatry, and Eugenics." In Alison Bashford and Philippa Levine, eds., *The Oxford Handbook to the History of Eugenics*, 116–133. New York: Oxford University Press, 2010.

Thomson, Mathew. *The Problem of Mental Deficiency: Eugenics, Democracy, and Social Policy in Britain c. 1870–1959.* Oxford: Clarendon Press, 1998.

Thurtle, Philip. *The Emergence of Genetic Rationality: Space, Time, and Information in American Biological Science, 1870–1920.* Seattle: University of Washington Press, 2007.

Tomes, Nancy. *A Generous Confidence: Thomas Story Kirkbride and the Art of Asylum-Keeping, 1840–1883.* Cambridge: Cambridge University Press, 1984.

Trent, James W., Jr., *Inventing the Feeble Mind: A History of Mental Retardation in the United States.* Berkeley: University of California Press, 1994.

Vial, Monique. *Les enfants anormaux à l'école: Aux origines de l'éducation spécialisée, 1882–1909.* Paris: Armand Colin, 1990.

Waller, John. "Ideas of Heredity, Reproduction, and Eugenics in Britain, 1800–1875." *Studies in History and Philosophy of Biological and Biomedical Sciences* 32, 2001, 457–489.

Weber, Matthias M. "'Ein Forschungsinstitut für Psychiatrie . . . ' Die Entwicklung der Deutschen Forschungsanstalt für Psychiatrie in München zwischen 1917 und 1945." *Sudhoffs Archiv* 75 (1991), 73–89.

Weber, Matthias M. *Ernst Rüdin: Eine kritische Biographie.* Berlin: Springer Verlag, 1993.

Weber, Matthias M., and Eric J. Engstrom. "Kraepelin's 'Diagnostic Cards': The Confluence of Clinical Research and Preconceived Categories." *History of Psychiatry* 8 (1997), 375–385.

Weindling, Paul. *Health, Race and German Politics between National Unification and Nazism, 1870–1945.* Cambridge: Cambridge University Press, 1989.

Weiner, Dora B. *Comprendre et soigner: Philippe Pinel (1745–1826). La médecine de l'esprit.* Paris: Fayard, 1999.

Weingart, Peter, Jürgen Kroll, and Kurt Bayertz, *Rasse, Blut, und Gene: Geschichte der Eugenik und Rassenhygiene in Deutschland.* Frankfurt am Main: Suhrkamp, 1988.

Weiss, Sheila Faith. *The Nazi Symbiosis: Human Genetics and Politics in the Third Reich.* Chicago: University of Chicago Press, 2010.

Wetzell, Richard F. *Inventing the Criminal: A History of German Criminology, 1880–1945.* Chapel Hill: University of North Carolina Press, 2000.

Wexler, Alice. *The Woman Who Walked into the Sea: Huntington's and the Making of a Genetic Disease.* New Haven, CT: Yale University Press, 2008.

Wise, M. Norton, ed. *The Values of Precision.* Princeton, NJ: Princeton University Press, 1995.

Whitrow, Magda. *Julius Wagner-Jauregg (1857–1940)*. London: Smith-Gordon, 1993.
Wright, David. *Mental Disability in Victorian England: The Earlswood Asylum, 1849–1901*. Oxford: Clarendon Press: 2001.
Yates, JoAnne. *Control through Communication: The Rise of System in American Management*. Baltimore: Johns Hopkins University Press, 1989.
Zenderland, Leila. *Measuring Minds: Henry Herbert Goddard and the Origins of American Intelligence Testing*. Cambridge: Cambridge University Press, 1998.

INDEX

Page references in italics refer to illustrations and their captions.